The Ecology of the Fish Pond Ecosystem

Developments in Hydrobiology 72

Series editor
H. J. Dumont

The Ecology of the Fish Pond Ecosystem

with special reference to Africa

by

Guy Delincé

Institute of Animal Ecology,
University of Ghent, Belgium

SPRINGER-SCIENCE+BUSINESS MEDIA, B.V.

Library of Congress Cataloging-in-Publication Data

Delincé, Guy.
The ecology of the fish pond ecosystem : with special reference to Africa / Guy Delincé.
p. cm. -- (Developments in hydrobiology ; 72)
Includes bibliographical references and index.
ISBN 978-90-481-4132-6 ISBN 978-94-017-3292-5 (eBook)
DOI 10.1007/978-94-017-3292-5
1. Fish-culture--Africa. 2. Fish ponds--Africa--Fertilization. 3. Fresh water--Fertilization. 4. Pond ecology--Africa. I. Title. II. Series.
SH121.D45 1992
639.3'11--dc20 92-54

ISBN 978-90-481-4132-6

Printed on acid-free paper

Contents

Preface

This exhaustive review on the potential of fertilization as a means to increase fish production in ponds under African tropical conditions is much overdue. Pond fish farming has been introduced to Africa over the past forty years. Although it is generally agreed that this continent has a considerable potential for freshwater aquaculture, past development efforts have met with variable but generally limited success. Several constraints have recently been identified while suggesting future research topics and ranking their respective priority. From the technical point of view, pond fertilization has been recognized as the main limiting factor for the development of tilapia farming.

In fact, even outside Africa, most of the present aquaculture production depends at least partly on the use of fertilizers. Over 70 % of the world production of farmed finfish and shrimp is realized with extensive to semi-intensive pond cultural systems, where natural productivity is enhanced through the controlled additions of fertilizers.

Most of the basic scientific research on pond fertilizations was carried out in temperate latitudes, mainly in Central and Eastern Europe, and more recently in the Southern United States and Israel. Not only should much of this past research be open to criticism on the design of experiments, but its results should only be applied in tropical Africa with the greatest caution, if at all. It is a fact that our understanding - and therefore our possible control - of the intermediate stages through which fertilizers are converted into fish food and fish flesh is still far from complete, particularly in African ponds.

The principles of pond fertilization are similar to those applied to agriculture. Fish production is increasing by the addition of certain essential substances present in limited concentrations and by the stimulating effect of these substances on the production of natural live fish food. Such effects relies on the improved operation of several interlinked food chains based on highly complex chemical and biological interactions taking place simultaneously in the bottom soil and the overlying water of the pond. This review is unique in bringing together and synthesizing the available knowledge on all factors involved and their multiple interactions. The importance of the mud surface as the "chemical laboratory" of the pond is duly stressed. Water quality, nutrient cycling and natural productivity are fully discussed. Such a complete illustration of the complexity of pond fertilization pathways towards fish production should hopefully promote further research on these particular subjects, in Africa as well as in other tropical regions.

The application of fertilizers to tropical waters, although showing great promise, is still in its infancy. For example no standard procedure has yet been devised for the on-farm determination of fertilizer requirements. Therefore the process is still largely empirical and depends on the fish farmer's judgement. In Far Eastern Asia, the latter is often based on the evolution, by trial and error over several centuries, of productive farming methods well suited to local conditions. In Africa, however, where the people have no background of practical experience and where both the farming conditions and the fish are different, such judgement can only be based on locally-acquired knowledge. Undoubtedly, this is one of the main reasons, which has prompted the author to prepare this multidisciplinary review. It should be a standard reference text to enable African scientists to become increasingly interested, it is hoped, in removing the numerous questions marks still attached to pond fertilization.

Dr. André G. Coche
Former Senior Fishery Resources Officer (Aquaculture)
Food and Aquaculture Organization of the U.N.
Rome, Italy

Acknowledgements

Words cannot express my profound gratitude to Prof. Dr. H. J. Dumont for the many ways in which he assisted me in bringing this manuscript to the press. I am grateful to him for his critical reading and commenting of the manuscript, and for the many suggestions for improvements, which helped me to shape the manuscript into its final form. I also thank him for his constant availability, whenever his assistance was required, and his patience with me.

My thanks go to Dr. A. G. Coche, former Senior Fishery Resource Officer (Aquaculture), FAO, for writing the foreword and for his useful comments and advice on the manuscript.

I also wish to gratefully acknowledge the material provided by Prof. Dr. Tilzer on primary productivity.

This handbook could only be completed thanks to the logistical assistance of a number of people, especially Mr. D. Claeys, my brother, Jean-Marie, Mr. K. Roche, Mr. E. Bruyneel, Dr. F. Fiers, Ms. N. Moortgat. Mr. R. S. Coppola, from the Fishery Information, Data and Statistics Service, FAO, computed the statistical analysis of the pond production data. I gratefully acknowledge his help.

Most of the information was gathered at various public libraries in Brussels, in particular the Library of the Royal Institute for Natural Sciences, the Library of the Institute for Hygiene and Epidemiology, the Centre for Research and Information on Tropics (CERDAT), and the library of the Laboratory for Ecology, University of Ghent. I wish to thank them all for making publications available to me.

I also thank Kluwer Academic Publishers for agreeing to publish this volume in the Developments in Hydrobiology series.

Lastly I wish to thank my wife, Martine, for her cheerful support throughout the work.

This volume is dedicated to my parents.

Introduction

Production in African fish farming relies largely on the natural fertility of ponds. Use of artificial or composite feed in fish culture in Africa is, for a number of economic and social reasons, limited. This lack of artificial fish feed has long been known, and research programmes on the use of fertilizers in fish farming were begun by Fisheries Administrations during the fifties and sixties in almost all African countries. Fish production relying on natural food produced within ponds, enhanced by manuring with livestock dung and organic compost, may achieve high yields. Therefore, there is a need to develop a technology based on management of the natural fertility of ponds. The specific conditions leading to such high productions, as well as the means to achieve them, have to be properly defined.

At the FAO Conference on Warmwater Aquaculture in Rome in 1966, one of the objectives had already been to review the status of knowledge on pond fertilization and the role of soil in pond productivity. The Conference concluded that more research was needed on the destiny of fertilizers within ponds, the processes leading to nitrogen-fixation in ponds, the influence of fertilization on the abundance of desirable algae, the need for lime, the effects of purity of phosphorus fertilizers on phosphorus mobilization, and the aquatic flora and fauna within ponds. Unfortunately, during the Symposium on Aquaculture in Africa in Accra, Ghana, in 1975, the emphasis on pond fertilization was weak, and, in the specific recommendations of the meeting, no statement on problems related to pond fertility was issued.

Previous approaches to fertilization needs in ponds considered differences between ponds to be so wide as to render general recommendations for the application of fertilizers impossible. The need for fertilizers had to be defined experimentally by comparing yields obtained with increasing amounts of such fertilizers as lime, and nitrogen and phosphorus compounds in various amounts and combinations, and assessing the cost on production. These trials did not take soil properties into account, nor did they consider the physical, chemical and biological variables of the water. The studies were conducted under various stocking conditions, and results achieved were not conclusive on fertilizer needs, and did not lead to a general procedure for assessing the need for fertilizers in pond culture.

To achieve a good fish yield in pond production based only on natural food, one needs a good understanding of the processes of plankton production and the relevant limnetic conditions. Water quality is still an abstract concept for fish farmers, as the criteria behind it have never been adequately described. Some attempts had been made to define a biogenic capacity to assess fish biomass, such as the Léger-Huet index proposed in 1949 and 1964, but never really refined or used. Plankton production is dependent on the availability of nutrients, which is determined by the cycling of nutrients within the environment. Nutrient cycling is the result of all processes consuming and producing nutrients. Nutrient production is achieved by recycling and regeneration processes, which are either physical, chemical or biological. The concentration of nutrients in water is the result of all processes mobilizing and immobilizing nutrients in soil and water, including the activities of aquatic organisms.

Maintaining optimum water quality means keeping the water at an optimum for the physiological requirements of fish and for plankton production.

The next step is to stimulate natural productivity, increasing the production of plankton, so as to ensure higher yields of fish. Intensification of farming requires a good understanding of the ecological processes of natural productivity, so as to ensure control of production factors.

Nutrients are the basic elements for the elaboration of organic matter by autotrophic production; understanding the nutrient cycling in ponds is therefore fundamental. The establishment of nutrient budgets in ponds will lead to the modelling of nutrient cycling and eventually to the setting up of indicators which will enable the calculation of fertilizer needs.

At the end of the seventies, Prof. C. E. Boyd, of the University of Auburn, Alabama, USA, set up a procedure for the establishment of the lime requirements of fish ponds. At that time, there was some hope that procedures for the requirements of nitrogen and phosphorus in ponds would follow. But, unfortunately, until now, no further step had been made. This contribution aims at bringing together the latest information on the subject of soil-water interactions, water chemistry and natural productivity, hoping to stimulate new ideas and foster anew research on the fertilizer needs of ponds.

*
* *

The ecology of aquatic environments has made much progress during the last decades. But, information gathered in limnology does not seem to be used in aquaculture, the two disciplines seemingly appearing isolated. There is, however, much to gain in bringing them together: the information gathered in limnology can help to improve production in ponds, while ponds can be used as research sites for limnological research. The pond is an environment which is very convenient for study, because it is of small size and many of its factors are easily controlled. An exchange of information between the hydrobiological domain and aquacultural research would therefore be mutually beneficial.

The objective here is to bring together, in one book, information which concerns all aspects of natural productivity in ponds. First, there is a need to summarize the latest information on nutrient cycling and to outline how this is related to production factors. The aquatic environment is partitioned into several compartments, which are often separately investigated without setting the information together in the overall context of the aquatic environment. Between those working on the dynamics of the sediments, the specialists of water chemistry, the plankton specialists, and the fisheries biologists, exchange of information is poor. It seems interesting to combine in one book the information and ideas of all these specialities, even when taking the risk of being considered too superficial by each of these specialists.

*
* *

The book is divided into three parts. The first is devoted to the sediments, with emphasis on the exchange of nutrients and substances across the sediment-water interface, and the transformations occurring within the sediments.

The second part is devoted to water chemistry, to the cycling of nutrients within the water column, and to effects of fertilizers on water quality.

The third part is devoted to the analysis of productivity, to the factors involved in plankton production, and to the interrelation of nutrients with aquatic organisms, which compose the aquatic community. The specific factors involved in fish production are described and discussed in the light of existing information. The maximum fish yields obtained in the ponds of some African farms are shown and compared with maximum yields gained outside the region.

Chapter 1.

Pond soil

1. Soil characteristics

Importance of soil for water productivity

The soil lining pond bottoms is the border zone between the terrestrial and the aquatic environment. It is an area of exchange between both, which links them together and makes them interdependent. Soil is more dense owing to its solid state and is rich in nutrients, unlike water which is liquid and with a more diluted nutrient content. The upper part of the sediment is microbially active compared to the water column and to lower level sediments. According to Doremus & Clesceri (1982), one centimeter of surface sediment in Lake George, New York, U.S.A., is approximately two times more productive than 18 m of water column and four orders of magnitude more productive than sediments at 5 or more metres depth.

The interpenetration of the two environments (Fig. 1) results in a transformation of the soil into a semi-liquid mud, soaked with water, which is chemically very active. Some nutrients are released from it after transformation, but mud also acts as a sink for chemicals, which are immobilized and retained by it. The water supplies the soil with energy in the form of organic matter. Solar radiation, the primary source of energy for living organisms, usually does not reach down to the pond bottom. The biological activity of the mud is therefore dependent on supply of energy-yielding organic matter from the overlying water, in combination with the physical, chemical and biological properties of the soil itself.

Life in the pond soil depends on specific conditions, related essentially to the capability of the environment to transform the organic compounds and recover some of their energy content. This capability relies on the presence of electron acceptors, needed for the redox reactions of decomposition. Hence, the specific nature of the soil is of crucial importance.

Pond bottom constitution

The pond bottom is a composite environment made up of solid matter, interstices, water and living organisms. It becomes more abiotic with depth. The water surrounding the soil particles plays an important role in the aquatic ecosystem and is discussed below (p. 7).

The solid matter is a conglomerate of mineral and organic particles. The mineral particles originate from the underlying soil and rocks or are brought in by water entering the pond (washload). They sediment onto the pond bottom in the course of time. These are the sediments *sensu stricto*. Mineral constituents of the soil appear as particles of sand, loam, silt or clay and range over a variety of mineralogical compounds, such as silica, clay, aluminium and iron oxides, carbonates, sulphides, etc. Their importance is dependant on their chemical reactivity. Some of these constituents, such as quartz sand (crystalline silica), are chemically inert, while others are reactive, such as colloidal clay. A soil is characterized by its major constituents.

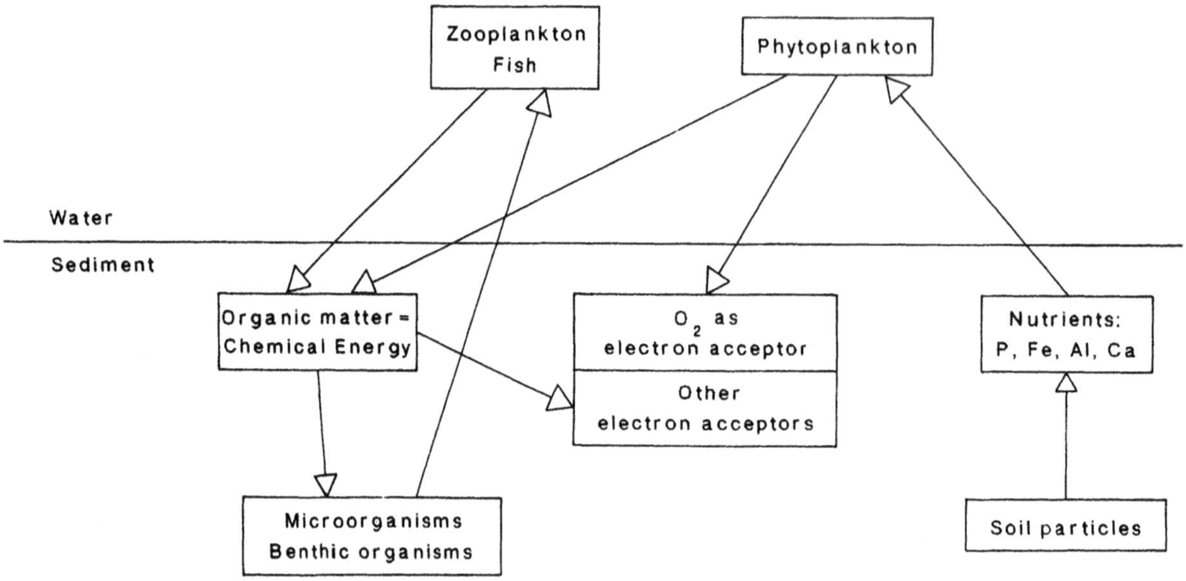

Fig. 1. Exchanges between water and soil. The aquatic environment provides the soil with (1) organic matter, which supplies energy and electron acceptors for life in the soil and (2) oxygen. The soil supplies nutrients of geochemical origin and some food to benthic organisms and microorganisms consumed by benthophagous organisms.

Morphological features of a soil

The morphological feature taking into account the size of soil particles is the texture of the soil or its granulometric composition, i. e. the relative frequency of particles of different sizes. Each soil has a specific and characteristic size distribution. The sizes of the mineral particles determine whether they are sand, silt, loam or clay. Different classifications exist. They vary in the limits of the size ranges of each class. According to the USDA (U. S. Department of Agriculture) classification, soil types are as follows (Coche, 1986):

type of soil	particles sizes
stones or boulders	: above 50 mm
gravel	: from 2 to 50 mm
sand	: from 0.05 to 2 mm
silt	: from 0.002 to 0.05 mm
clay	: below 0.002 mm

Particles smaller than 0.1 μm are colloids.

Other properties, such as structure, colour and mottling, specific weight or density, porosity, permeability, compressibility or plasticity, moist consistence, and shearing stress, are also used to characterize a soil. These properties are extensively discussed in Coche (1986).

The clay and organic matter content of a soil play key roles in the interaction between soil and water.

Some clay properties

Clay particles are small, of plate- and flake-like, tabular or even spheroidal shape, and many of their properties stem from their small size. The latter entails an elevated surface to volume ratio, defined by the specific surface or surface area, i.e. the external surface area of one unit weight of particles. A colloidal clay has a specific surface varying between 10 to 1000 $m^2 g^{-1}$, compared to that of silt or sand, which is 1 and 0.1 $m^2 g^{-1}$, respectively (Brady, 1984). Some other properties result from these particle size characteristics: for instance, water retention by capillarity, and adsorption of nutrients. Dried clay forms soil lumps very hard to break. Moist clay soil can easily be squeezed into a ball or a roll. The water retention capacity of clay soils is high as indicated by the high shrink-swell potential.

Clay minerals are alumino-silicates (feldspars) or iron or aluminium silicates (micas), made up of thin sheets of tetrahedral silica or octahedral alumina, arranged like the pages of a book. The minerals are classified according to the number of silica and aluminium/magnesium sheets contained in the crystal structure. Common clay minerals are kaolinite, smectite, illite, montmorillonite. Clay minerals are characterized by the variety of their chemical composition, wherein silicon and aluminium atoms may be substituted by other atoms of similar size. Montmorillonite is a mixed silicate of aluminium, sodium, iron and magnesium. Illite is a mixed silicate of potassium and aluminium. Clays found in tropical areas are rich in iron and aluminium oxides (Brady, 1984).

Fine clays are colloids and their chemical properties are related to surface chemistry, in connection with surface phenomena and exchange sites. Their importance and role in soil are preeminent.

Organic matter in soil

Organic particles are either dead or living organic matter. Sediments are often considered as abiotic material, with no account being taken of the microorganisms or benthic macroinvertebrates which live there. Yet, large populations of bacteria are associated with detritus in soils (Hart, 1982). They are not easy to isolate from their substrate, or to identify and enumerate. These microorganisms are a dynamic element of sediments, as they transform dead organic matter into living material. The dead organic particles are the remains of organisms living in the water column of the pond and sedimented on the pond floor after death and of organisms which live in the mud itself. Their dimensions vary over a large range of sizes, from coarse material almost the original size of the organism from which they originated, to transformed and aggregated matter. They are often characterized by a diffuse outline; in this case they are often called aggregates. The variety of compounds is great owing to the diversity of compounds available in living organisms and of patterns of decomposition. The ultimate form of organic material is humus, a group of organic compounds, in a dissipated (colloidal) state, coating particles of silica and clay and difficult to isolate. Its presence is revealed by the dark-brown colouration of a soil. Organic matter in colloidal form plays an important role in the exchange capacity of soils. It can be likened to a polyelectrolyte which adsorbs and releases solutes.

A soil is characterized by its proportion of organic matter versus mineral content (percentage of organic matter). It must contain more than 80 % of organic matter to be qualified as organic. Organic soils are common in marshes, wetlands, floodplains, bogs, swamp areas, and estuary zones, where they are subjected to continuous or long periods of inundations, which retard the oxidation of organic matter. Organic soil has a dark brown or black colour.

Stratification of flooded soil

The composition of pond bottom sediment also varies with depth. On the surface and in the upper layers, the influence of the overlying water column is strong, but decreases strongly at greater depths. The organic matter content of a soil is mainly generated by the water column. This content increases with depth, indicating accumulation over time (Twinch & Breen, 1982).

Three zones can be distinguished in flooded soil. Their characteristics are summerized in Table 1:

- a surficial oxidized microzone, with recently deposited, loose, uncompacted material, unstructured flocculent and fluffy flakes, approximately 1 cm thick (between 1 to 20 mm), with a patchy and intermittent distribution, easily disturbed by turbulent mixing (Twinch & Breen, 1982; Doremus & Cleresci, 1982; Sweerts *et al.*, 1986);
- a superficial layer (-1.5 to -5 cm stratum), more consolidated, still aerobic or oxidized, of variable thickness. (Levitan (1987) observed an oxidized capping extending at least 10 cm deep in Lake Mitchell, U.S.A.);
- a deeper layer (beyond -5 cm), anaerobic or anoxic, with reducing conditions, less microbially active.

Table 1. Stratification of soil - Characteristics of the strata.

	Physical state	Chemical state		Biological status		
				Microbiots	Benthic algae	Zoobenthos
Superficial microzone	loose	oxidized	aerobic	+++	+++	+++
Superficial layer	slightly compacted	oxidized	aerobic	+	+	++
Deeper layer	compacted	reduced	anaerobic	-	-	+

	Type of transformation within sediments	EXCHANGES WITH OTHER LAYERS AND WATER:	
		supplies	exports
Superficial microzone	Organic	- sedimentation of organic and mineral particles - adsorption - absorption	- resuspension - predation - transfer to superficial layer - diffusion
Superficial layer	Organic/chemical	- transfer from the microzone - upward migration	- resuspension - leaching
Deeper layer	Organic/chemical/ physical	- accumulation of leached nutrients	- gaseous ebullition

The three strata play an analogous role to those of a typical soil: the upper layer contributes most to biological activity. It accumulates organic material sedimenting from the water column and colonies of bacteria, benthic algae, zoobenthos, and a vast spectrum of decomposers, with a high heterotrophic potential (also shown by its high uptake potential). The layer below is a sediment layer, made up of partially transformed material, of poor nutritional quality, while the deepest layer is an original soil layer, where leached nutrients, such as calcium, magnesium, and phosphate accumulate. These nutrients may be lost from the environment if sequestered but otherwise may diffuse upwards by means of the interstitial water and be returned to the aquatic cycle.

The penetration of oxygen into the soil reaches a certain depth where it forms the aerobic/anaerobic borderline. This coincides with a change in chemical conditions, switching from an oxidizing to a reducing environment. This border is called the redox-cline or the oxic/anoxic boundary (or oxycline) and may not be sharply delineated, if the redox environment changes gradually. These two frontiers are of importance for life in the soil.

The borderline between the pond bottom and the water column is called the sediment-water interface and is more a concept than a distinct physical barrier. It is the exchange area between soil and water, allowing the transport of solutes and nutrients in and out of the sediments.

Pore water or interstitial water

Due to impoundment, the soil of pond bottoms becomes waterlogged. The pores between soil particles fill up with water. The total volume of pore water contained in sediments can be significant when compared with the amount of water in a pond, but depends largely on the underlying soil properties (its permeability) and the proximity of the water table. In Lake Södra Bergundasjön, Sweden, the water contained in the upper layers of soil represents up to 5 % of the total water volume (Enell & Löfgren, 1988). The major part of this water is a mobile liquid medium, which surrounds the soil particles and connects deeper layers of sediment with the aquatic environment through a complex network of tiny channels.

This water is named interstitial or pore water and contains solutes, which are transferable to the overlying water (Fig. 2). The composition of interstitial water is controlled by complex interactions between the ground-water recharge system (diluting the concentration of some of the more concentrated solutes released inside the sediments and carrying some solutes along with it), mineralogical dissolution and precipitation (due to the occurrence of inter-reactive solutes leading to authigenic mineral formation), biological activity (sequestering and release of some specific solutes) and physical interactions between mud and water (adsorption and desorption of solutes) (Enell & Löfgren, 1988).

Interstitial water as a transition compartment

Water in soil is predominantly interstitial or pore water. Only a small portion of the water is bound to solid material. The greatest part is in the chemically active form of a solvent.

Interstitial water is a transition compartment for solutes between soil and water. As pore water surrounds the soil particles, nutrients are released from the soil, or are removed from water. The composition and concentration of solutes in interstitial water are the result of concentration gradients and solubility kinetics of the solutes they encounter, together with the movement of the pore water. Changes in composition appear with depth (Nembrini *et al.*, 1982). The concentration gradients and mobility of water in soil ensure the transport of nutrients across the sediment-water interface. Exchange of nutrients across the interface is of considerable importance for the trophic status and productivity of the pond.

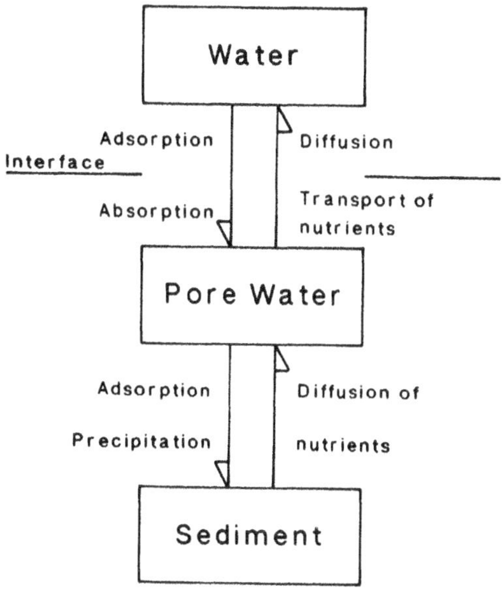

Fig. 2. Exchanges between water, pore water and sediment.

Movement in the groundwater influences the retention capacities of a soil. In laboratory experiments, Van Liere & Mur (1982) observed an increase of nutrient concentration (in their specific case, essentially phosphorus release) in the pore water, whenever there was an upward or downward movement of groundwater. During downward movement of groundwater, release of phosphorus occurred in an opposite direction to water movement, whereas diffusion and upward migration were working in the same direction.

The high water content (over 90 %) of sediment prompted Davison (1982) to conclude that sediments sometimes behave as liquids with a high concentration of particles. In the upper layers of the sediment, biological, chemical and physical activities are high, concentration gradients appear and the solute and nutrient content of the pore water changes in the course of time, whereas in the lower layers concentrations stabilize.

Mechanisms for transport across the interface

Transport processes are grouped according to the direction of the fluxes (Table 2). Sedimentation processes are downward transport mechanisms and encompass the sedimentation of mineral particulate matter suspended in water entering the pond as well as the sedimentation of living organisms after death. Sedimentation of plankton is a phenomenon of quite some magnitude in rich productive ponds. The sedimentation of a high load of poor nutritional value may contribute significantly to the silting-up of a pond. All these processes result in the growth of a sediment column in the pond and are macroscopic.

With regard to upward fluxes, the most important process is the transport of particulate matter and solutes owing to turbulence, bioturbation (turbulence caused by organisms) and other reworkings of sediment, while other upward migrations are essentially molecular diffusion phenomena, caused by concentration gradients or by changes in the flow of pore water. Upward flow as a macroscopic phenomenon is an erratic and random process, while the diffusion of solutes and nutrients occurs steadily.

When attempting to assess the importance of exchange between sediment and water, one should take into consideration the nature of the material exchanged. Sediments contain various types of material and their individual reactions to transport vary. For example, inert material, such as soil particles of sand, loam or silt, and fibrous organic material, such as remains of rice bran, will behave differently, according to the size of the particles and the cohesion forces (soil consistency and compaction) which hold them together.

Movements of particles with an electrical charge such as clay and organic particulate matter will depend on electrical forces (electrical potential) and fields and their electrophoretic mobility. In the absence of an electrical potential, they carry a cloud of ions of opposite charge. In the absence of an electrical field, the transport of ions, much smaller than colloidal electrolytes, depends on the intrinsic movement of the pore water (movements in the groundwater, modification of the level of the water table), and also on diffusion, the latter being dependent for each ion considered on the concentration gradient (chemical potential) generated and the ionic mobility of the ion species. Ions transported in the pore water, moreover, will react with other ion species they meet to form new compounds.

Table 2. Transport processes between sediment and water

TRANSPORT ACROSS THE INTERFACE (U = upward; D = downward)		
Physical transport	Chemical transport	Biological transport
Molecular diffusion (U, D)	Electrophoretic migration (U, D)	Bioturbation (U)
Gaseous ebullition (U)		Predation (U)
Turbulences: effect of wind and waves (U)		
Sedimentation (D)		

Concentration of some ions will lead to formation of stable or metastable mineral compounds.

Finally, the redox boundary separates the oxic from the anoxic region, and forms a chemical barrier, implying changes to the ions and compounds crossing it. Although this boundary is not commonly present at the sediment-water interface, its presence in deeper layers affects global transport across the interface. A typical example of ion species converted while crossing this redox boundary are iron and manganese. Oxidized forms do not have the same solubility as reduced ions. The ferrous (Fe^{2+}) and manganese ions (Mn^{2+}) are soluble and mobile in their reduced form, but are insoluble and precipitate when oxidized (Fe^{3+}, Mn^{3+}) (see p. 45). Once precipitated they form particles and are transported as particulate matter. The speeds at which manganese and iron are reduced are quite different. Manganese is reduced 50 times more slowly than iron (Davison, 1982).

Importance of soil for the productivity of water

1. Soil is rich in solutes and nutrients and exchanges them with the water across the soil-water interface.
2. Nutrients and solutes released by the soil and the rocks accumulate in the soil and can be supplied to the water column.
3. Organic matter sedimenting from the water column is recycled and returned to the water column in the form of living organisms or nutrients.
4. The recycling role of a soil depends on the chemical reactivity of the soil, essentially surface chemistry and low-energy bonds of ions of mineral, mainly clay, and organic colloids,and on the activity of decomposers.
5. Chemical transformations inside the soil are dependent on the prevailing oxidizing or reducing conditions, which change with depth.
6. Exchange of nutrients between water and sediment occurs in the interstitial or pore water, which transports dissolved solutes. The sediment-water interface plays a key role in the exchange.

2. Soil properties

Soil properties related to nutrient cycling are of direct interest to the fish farmers, as they influence the productivity of a pond. Soil activities are related to the quantity and quality of transformable material and to conditions conducive to reaction.

Clay and organic matter are amongst the most active components of soils. Nutrients and solutes are adsorbed and retained on the surface of colloidal particles and are released when conditions change. Besides exchange properties, a variety of chemical and biological transformations occur in a soil. These transformations are subject to the chemical conditions of the soil, which are either reducing or oxidizing, acidic or alkaline.

Organic or peaty soils have properties different from mineral soils; a notable difference is their retention capacity for water (and nutrients). A waterlogged mineral soil retains about 2 to 4 times its weight in water, while an organic soil has the capacity to hold 12 to 20 times its weight (Brady, 1984).

2.1. Colloids and their adsorption properties

In the past, the study of soil colloids was approached analytically. The main soil components suspected to be responsible for the colloidal properties were extensively studied *in vitro* and their adsorption capacities deduced from textural analysis. Soil fractions which exhibit colloidal properties are mainly clay particles and organic compounds, and so colloids were classified as mineral or organic. However, electron microscopic observations revealed that organic compounds and metallic oxides coat clay or silica particles and clay is likely to act only as a substrate (Hart, 1982).

Adsorption capacities of soil are further complicated by interference from biological absorption by soil microorganisms. Biological absorption and excretion of nutrients by microorganisms takes place rapidly and can cause confusion (Björk-Ramberg, 1985).

Thus, the analytical point of view has been substituted by a more functional approach, wherein total sorption capacities (all types of physical or chemical adsorption combined) are described using sorption isotherms and where the specific sorption of each species, for example phosphate or a heavy metal (in the case of pollution), is considered independently.

2.1.1. Mineral colloids

Even though clay does not play the role in nutrient sorption formerly ascribed to it, it still merits consideration.

Colloidal clay has particle dimensions ranging from 0.5 to 0.2 μm. The particles are negatively charged and the charges are localized on the surfaces, which, in the case of clays, are external and internal, between the crystal units or the silica and alumina layers. Free ions in solution in the pore water are adsorbed onto the external and internal surfaces of the colloid. Ions adsorbed onto internal surfaces are not easily removable. Water molecules, because of their polar character, penetrate the crystalline layers of some clays, such as smectite or montmorillonite, making the clay swell. The fixation sites on the external surface of the particles are surrounded by a swarm of electronic charges and those sites are essential for the exchange capacity of clays.

Electric charges on clays

From a chemical point of view, clays are alumino-silicates. Negative charges originate from atomic substitution inside the silicate tetrahedrons or the alumina octahedrons (Fig. 3). Atomic isomorphous substitution of silicon or aluminium atoms by heterovalent atoms leads to the appearance of negative non-neutralized charges. In silicate tetrahedrons, the silicon atom can be replaced by an aluminium or iron atom and in the alumina octahedrons, the aluminium atom by an iron, magnesium or zinc atom. These charges are permanent and do not depend on the pH of the environment. When such replacements occur, oxygen and hydroxyl ions, previously neutralized by the silicon or the aluminium, are so no longer, hence the appearance of negative charges.

Another cause of the appearance of negative charges on clay is the ionization of the hydroxyl groups at the broken edges of the clay lattice. These charges are pH-dependent and appear within certain pH ranges:

	Clay—Al—OH_2^+	Clay—Al—OH	Clay—Al—O^-
pH:	below 3	around 3	above 3

The pH at which clay has no net charge, called the point of zero charge, differs from clay to clay. Soils which are not highly acidic are mainly cation exchangers. In slightly to highly acidic soil, hydrogen ions (H^+) are tightly held and can be exchanged only with acidic cations, such as Al^{3+} or Fe^{3+}. At pH 6 and more, the hydrogen cation is exchanged against basic cations, such as Ca^{2+}, Mg^{2+}, K^+ and Na^+, or the hydroxyl group, which is more mobile than most other anions, is replaced by SO_4^{2-}, $H_2PO_4^{2-}$ or other dissolved anions (ligand exchange). The number of exchange sites depends on the type of clay.

Other types of charges

Metallic oxides, especially alumina, Al_2O_3, FeOOH (amorphous), and MnO_2, originating from iron and aluminium oxide clays, common in tropical soils, play an important role in adsorption mechanisms. In particulate material, iron and manganese hydroxides or oxides occur as surficial armouring on material and finely dispersed particles. They are present in amorphous, microcrystalline and more 'aged' crystalline forms (Förstner, 1982). In water, the metal oxides are covered with surface hydroxyl groups and adsorption of ions, cations or anions,

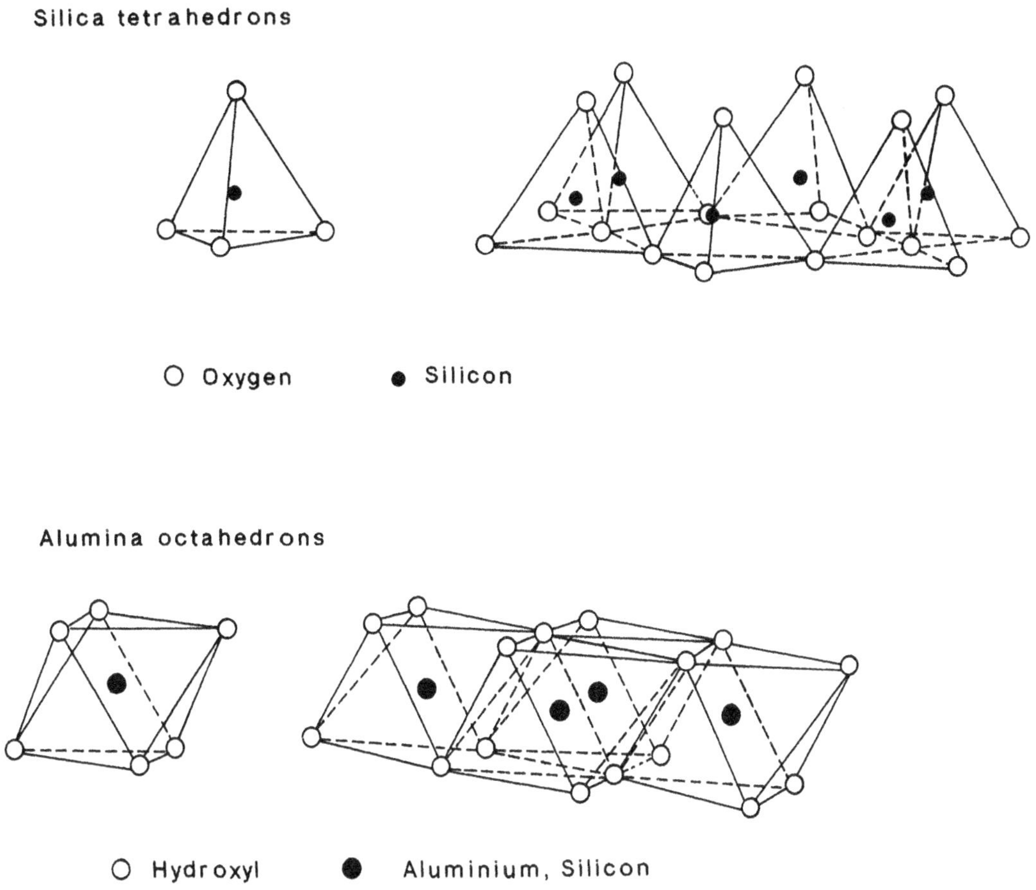

Fig. 3. Silica tetrahedrons and alumina octahedrons from the clay mineral lattice.

occurs at the water-oxide interface (Hart, 1982). In lake Léman, Switzerland, iron (III) forms polynuclear complexes which are highly reactive and have high sorption capacities (Nembrini *et al.*, 1982). The behaviour of aluminium or iron hydrous oxide is as follows (Holtan *et al.*, 1982):

$$\begin{array}{c} OH_2^+ \\ \backslash\ / \\ Al \\ /\ \backslash \\ OH \end{array} \underset{+H^+}{\overset{-H^+}{\rightleftharpoons}} \begin{array}{c} OH \\ \backslash\ / \\ Al \\ /\ \backslash \\ OH \end{array} \underset{+H^+}{\overset{-H^+}{\rightleftharpoons}} \begin{array}{c} OH^- \\ \backslash\ / \\ Al \\ /\ \backslash \\ OH \end{array}$$

The charge (positive, negative or neutral) of the oxide depends on pH. The surface hydroxyl groups are the reactive sites and are binding ions. Exchange of cations or ligands occurs depending upon pH and redox potential, upon the formation constants of the complexes, and upon the concentration of the reacting species, the ionic strength, and competing ions (Hart, 1982). The strength of binding sites depends also on the sorption densities, sites not all being equal in their binding capacities. At higher sorption densities, the number of strongest binding sites decreases. Freshly precipitated hydrous iron and manganese oxides adsorb more than aged hydroxides (Nembrini *et al.*, 1982; Förstner, 1982).

Metallic oxides are also active in adsorbing organic matter. Adsorption on metal oxides is explained as surface complexing or ion exchange with other cations (Förstner, 1982); in Lake Neusiedler, Austria, organically complexed Fe^{3+} and Al^{3+} are the most likely sites for adsorption on organic matter surfaces (Gunatilaka, 1982). Various compounds, such as amino acids as well as natural fulvic and humic material, appear to be adsorbed to oxides. In Lough Neagh, Northern Irland, a eutrophic lake, adsorption of hydrous iron oxide on organic matter suggests involvement of ligand exchange with surface $FeOH^+$ or FeOH groups (Tipping *et al.*, 1982). In the presence of calcium and magnesium ions, competition between cations and the oxide for negative fixation sites on the humics increases with pH. The oxide adsorbs more humic material, in reducing the fixation sites of the humics (Hart, 1982).

Soil carbonates also show some adsorption properties. In the Neusiedlersee, Austria, freshly precipitated calcium carbonate has a high affinity for binding phosphate by coprecipitation (Gunatilaka, 1982). Phosphate anions are sorbed in a monolayer at the surface of the calcite crystals, replacing sorbed water molecules, bicarbonates or hydroxyl ions (Holtan *et al.*, 1988). The sorption mechanisms are summarized in Table 3.

2.1.2. Organic colloids

Solid organic matter in soil is present as various compounds, depending on its origin and transformation. There is no easy way to classify this organic matter, because of the variability of the compounds. Some are well defined, such as amino acids and organic acids, while most are uncharacterized and belong to the so-called natural fulvic and humic material. Fish farming is often sited in marshland areas where the acid water is rich in humic material, revealed by the dark colour of the water. Sometimes organic material is grouped according to molecular weight.

Organic material coats silica and clay particles and forms a biofilm. Organic surfaces originate in various ways:

- remains of substrate-dwelling microorganisms such as bacteria, algae and fungi
- condensation of organic compounds of low molecular weight forming organic aggregates
- sorption of organic compounds onto clay or metallic oxides.

According to Hart (1982), surfaces of solid matter in aquatic systems are more homogenous than might be expected from the finding that the particles all have a negative charge and that their electrophoretic mobility varies only slightly, indicating similar surfaces; the negative charge is the result of the presence of an organic film covering most of the particles.

The binding groups of organic material involve several chemical forms, but are predominantly carboxylic and phenolic groups, and in particular salicylic acid (Hart, 1982). Humic matter is normally negatively charged and binds preferably cations. However, in association with certain iron and aluminium cation species, it is able to form complexes which also bind anions. This mechanism of sorption is quite important as it supports the

Table 3. Different sorption mechanisms occurring in soil

SORPTION MECHANISMS		
Physical sorption	Adsorption (reversible)	- Clay - Metallic (iron and aluminium) oxides and hydroxides - Carbonate (calcite) - Organic matter (alone or with metals): humic and fulvic material
Chemical sorption	Chemisorption (partly reversible) Precipitation (non reversible)	- Clay - Carbonate and calcium salts
Biological sorption	Absorption, followed by excretion	- Microorganisms

fixation of phosphate in sediments. Organic ligands or compounds can block and reduce the number of sorption sites on mineral surfaces. The organic matter acts in two ways: in sorbing ions and in blocking sites for fixation.

Interaction between metallic oxides and organic compounds is complex, as iron and aluminium oxides adsorb organic matter and vice-versa. The mechanism is not well understood.

Binding between organic matter and soluble organic compounds or ligands also occurs. This leads to condensation of organic compounds. The nature of these reactions often makes it difficult to draw the line between physical and chemical sorption.

2.1.3. Ions sorbed by colloids

A colloidal particle is a polyelectrolyte micelle, carrying electric charges at its surface, and surrounded by adsorbed ions. The micelle as a whole is negatively charged, though the surface displays fixation sites for positive and negative ions. Charges associated with soil particles are simple cations and anions as well as complex organic and inorganic charge complexes.

The strength of the retention force of the cations to a micelle depends on the hydrated ionic radii and on the valency of the cation species, cations with short radii and those with higher valency being more tightly sorbed, and on the ionic strength of the pore water solution surrounding the particles. The composition and the ionic strength of the pore water change considerably locally and with depth. This explains why calcium and magnesium concentrations in Lake Léman, Switzerland, increase with depth, while nitrates and sulphates decrease (Nembrini *et al.*, 1982).

Interaction between different species of cations is of importance for the sorption properties of soils, though the mode and range of interactions are not well understood. For instance, magnesium and calcium ions significantly affect the amounts of organic matter sorbed to a solid surface, increasing the degree of sorption of fulvic material to goethite (Hart, 1982); iron, aluminium and manganese are essential for the availability of nutrients (Nembrini *et al.*, 1982). Competitive reactions between solid surfaces, colloids and pore water solutions are not well understood.

Colloids in suspension are surrounded by water molecules, and monovalent hydrated ions, which are not strongly held, but increase the electronegativity of the colloidal particles. This permits individual particles to repel each other and stabilizes them. This state is termed dispersion as opposed to flocculation, which is a reduction of electronegativity accompanied by an increase of the size of the colloidal particles and settling rate. Flocculation is enhanced by divalent and trivalent ions. These ions are more tightly adsorbed by the colloids and enhance the attractive forces between particles. Aluminium, hydrogen and calcium ions favour flocculation (Brady, 1984). Flocculation of mineral particles and organic solids modifies the real or effective particle size. In Wilton Creek, Ontario, U.S.A., this phenomenon is seasonal and related to biological activity in sediments (Ongley *et al.*, 1982).

A distinction should be made between acidic, and alkaline ions: hydrogen, aluminium and iron are the acidic, and magnesium, calcium, potassium and ammonium alkaline cations. Acidification of soil is the result of two different processes: a release of acidity caused by decomposition of organic matter by microorganisms (nitrification and sulphate reduction); and leaching of basic cations from the upper layers of the soil towards deeper horizons. The acidic or alkaline character of a soil will be discussed later (p. 15).

2.1.4. Ionic exchange capacity

Cation exchange capacity

The capacity of a soil to sorb and release ions is of paramount importance for the fertility of the overlying water. As soil particles are negatively charged, ions withheld by the colloids are primarily cations. The amount of cations sorbed to a soil can be determined and is termed its cation exchange capacity (CEC). It represents the quantity of cations adsorbed to the soil and is expressed as milliequivalents of cations per 100 g of dry soil or as centimoles of positive charge per kg of soil (cmol kg^{-1}), according to the International system unit. The conversion is: 1 meq $(100\ g)^{-1}$ is equal to

1 cmol kg^{-1}. Exchangeable cations are Al^{3+}, H^+, Fe^{3+}, Ca^{2+}, Mg^{2+}, K^+, Na^+, NH_4^+.

The CEC of soil is related to its clay content. Some types of clays have higher CEC than other: illite has higher exchange capacities than kaolinite, owing to mean particle size or to surface area per gram of particles. The CEC of a few clays is listed below (expressed as meq (100 g of dry soil^{-1})):

Type of clay mineral	meq (100 g of dry soil^{-1})
Kaolinite	3-10
Illite	15-40
Montmorillonite	80-125
Vermiculite	85
Smectite	118

The CEC also depends on the organic content of a soil; the greater its proportion of organic matter, the higher its CEC. In soils rich in humus, CEC's in excess of 180 meq (100 g of dry soil^{-1}) are not uncommon. Samples of soil taken from deeper parts of fish ponds at Auburn, Alabama, U.S.A., have higher CEC's than from shallower parts (Boyd, 1976b). This is explained by a higher organic content in deeper parts of the pond compared to the shallower zones, where the mineralization processes are faster. In sediments below cages installed in two Polish lakes (Szczytno Małe and Łętowo Lakes) used for trout culture, CEC's were some 20 to 30 meq (100 g of dry soil)$^{-1}$) higher than in the adjacent sediments. Similarly, the organic carbon content of the soil underlying the cages was some 0.5 to 9.5 % higher than in littoral sediments, reaching a concentration of 12 to 17 % (Korzeniewski & Moczulska, 1985).

Changes in acidity or alkalinity of sediments affect the CEC. In acidic conditions, hydrogen ions are more tightly sorbed and not exchanged against other cations as easily as in alkaline conditions. Furthermore, many pH-dependant charges are no longer available and the amount of exchange sites is reduced. Acidification of a sediment can considerably reduce its amounts of exchangeable cations, principally alkaline ones such as calcium and magnesium; along with these, sulphate or chloride anions are also washed out. This process is called leaching. The relative proportion of acid cations such as aluminium increases and the sediment develops an acid reaction. Restoration of alkaline conditions is achieved through liming.

The fraction of CEC (expressed as percentage) occupied by basic cations is called base saturation and the fraction of acidic cations is the base unsaturation. Base saturation, pH of the soil and total hardness of pond water are closely related (Boyd, 1979).

Anion exchange

Most nutrients essential for the productivity of a pond, such as carbonates, nitrates, and phosphates, are anions. Their availability in sediments as exchangeable anions is important. Other common anions present in soil are chloride, sulphate, sulphide and hydroxide ions. In some tropical sediments, colloids carrying positive charges are present. They are mostly clays rich in hydrous aluminium or iron oxides, of reddish or yellowish colour. Minerals found in such soils are gibbsite, $Al_2O_3.3H_2O$, goethite, α-FeOOH, and hematite, α-Fe_2O_3. In most soils, anion exchange is more complex, because sorption of anions is associated with cations such as Fe^{2+}, Al^{3+} and Ca^{2+} which are themselves exchangeable (Holtan *et al.*, 1988). Measurement of the anion exchange capacity is not commonly carried out, the availability of specific anions being estimated with the help of sorption isotherms.

Sorption isotherms

Sorption isotherms express the relation between the quantity of ion sorbed and the concentration of this ion in solution. This relation was first developed to account for the sorption of gases on solids. The Langmuir isotherm assumes that the adsorbed ions cover the surface until full saturation of the exchange sites occurs, after which there is no further adsorption. This assumes that there are no interactions between adsorbed ions, that interchange of ions can proceed as readily as on a bare site, and that there is no variation in sorption energy with the degree of surface saturation. The Freundlich isotherm introduces a correction factor into the relation; when the exchange sites are almost saturated, the exchange constant is reduced and when the exchange sites are free the exchange constant is increased.

The experimental data fit these isotherms relatively well, provided the concentration ranges are not too wide.

Importance of the colloidal properties of a soil

1. Soil colloids bind solutes and nutrients by low-energy bounds and form a store of exchangeable nutrients.

2. Soil colloids are usually classified as mineral or organic. Mineral colloids consist mainly of clays and metallic oxides, while organic colloids are composed of a group of largely unidentified compounds, of which fulvic and humic material are the most important.

3. Some observations by electron microscopy give evidence that most soil particles are coated by an organic film, which carries negative charges able to bind cations. This gives a certain homogeneity to the surface of solid matter.

4. The sorption of cations occurs at the negative fixation sites, while the sorption of anions occurs in association with iron, aluminium and calcium ions. Cations are separated into acid and alkaline species.

5. The amount of cations sorbed to a soil is measured by the CEC, the cation exchange capacity. The proportion of alkaline cations sorbed is termed the base saturation and the proportion of acid cations sorbed is the base unsaturation.

6. The amount of anions sorbed is determined by isotherm sorption curves.

7. Investigations on the sorption properties of the soil are hampered by interference from biological absorption by microorganisms.

2.2. The acid or alkaline reaction of soil

Many soil properties depend on its acidic or alkaline character, as this determines its oxidizing or reducing properties. Furthermore, in an acidic soil, the fixation sites on colloidal particles are saturated with ions not suitable for productivity, while the ions essential for productivity are unavailable. A soil is said to be acidic if its pH is below 7, and alkaline if above 7.

A compound is defined as an acid if it yields dissociated hydrogen ions in solution. A base yields hydroxyl ions in solution and per extension a solution is defined as acidic when the concentration of hydrogen ions is higher than that of hydroxyl ions. pH is thus a measure of the concentration of hydrogen ions, expressed as the negative logarithm of the hydrogen ion concentration.

The pH of a soil is determined not only by the presence of hydrogen ions but by all other acidic ions, mostly aluminium and iron cations, or by the hydroxyl and alkaline ions, such as calcium, magnesium, potassium, and sodium.

Soil pH is a measure of its acidity at a given moment, but as a soil is a dynamic system, changes occur continuously as reactions proceed. It is therefore important to determine the factors responsible for modifying soil pH. In some cases, pH is able to resist changes and is said to be buffered. The buffer systems of soils are of much interest.

Many factors affect soil pH: texture, with the contribution of soil minerals generated by weathering; organic matter content, which has is own acidic or alkaline reaction and which, through decay processes, acidifies the soil; biological activities of microorganisms, releasing carbonic acid, and whose densities are correlated to organic matter content; pore water concentration and the supply of alkaline and acidic ions to the soil; supply of oxygen to the soil which favours oxidizing or, in its absence, reducing reactions; soil drainage, which washes away alkaline ions, such as calcium, magnesium, sulphate and nitrate, which are not tightly sorbed. Organic soils have pH with high buffering capacities and shifting the acidic pH of these peaty soils towards values well suited to production is not easy.

In short, all processes generating protons acidify a soil, while all proton-consuming processes have an alkalizing effect.

Acidifying role of aluminium and iron

Aluminium and iron are said to be acidic cations. In soil with pH below 4.5, aluminium is in the soluble form of Al^{3+} ions; in pH between 4.5 to 9, aluminium is precipitated as $Al(OH)_3$; above 9, it appears as an aluminate ion, $Al(OH)_4^-$. In water, all aluminium species are subjected to hydration. The Al^{3+} ion, for example, is surrounded by water molecules and forms:

$$Al^{3+} + 6H_2O \longrightarrow [Al(H_2O)_6]^{3+}$$

The aluminium ion in soil of pH above 4.5 reacts with water to form two soluble species of hydrous ions. They can be written as follows, leaving out the hydrating water molecules and considering only the anhydrous ions:

$$Al^{3+} + H_2O \rightleftharpoons AlOH^{2+} + H^+$$

$$AlOH^{2+} + H_2O \rightleftharpoons Al(OH)_2^+ + H^+$$

$$Al(OH)_2^+ + H_2O \rightleftharpoons Al(OH)_3 + H^+$$

At each step, hydrogen ions are released into solution adding, to the acidification. These ions formed are partly adsorbed and partly free in solution, and their relative concentrations depend on the equilibrium constants and reaction rates. Moreover, these ions are able to polymerize and form chelates.

In the presence of silica, which is common in soils, the aluminium hydrous oxides react to form:

$$H_4SiO_4 + AlOH^{2+} \longrightarrow H_3SiO_4\text{-}AlOH^+ + H^+$$

which is an alumino-silicate ion, and

$$H_4SiO_4 + Al(OH)_2^+ \longrightarrow H_3SiO_4\text{-}Al(OH)_2 + H^+$$

which is the base molecule of kaolinite.

In acidic soil, hydrogen ions are not the only exchangeable ions responsible for acidity, as aluminium Al^{3+} ions are also involved. When these exchangeable aluminium ions are removed from the soil, they are replaced after a short time by some other aluminium ions which spontaneously appear at the same locations. They are probably extracted from the clay lattice to occupy positions at the exchange sites (Ségalen, 1973).

Iron reacts in a similar way, forming hydrous oxides. It is present as Fe (II) and Fe (III) ions, depending on the redox potential of the soil. Ferrous ions are soluble and relatively free from complexation, while ferric ions are insoluble (Davison, 1982) and in Lake Erie, U.S.A. and L. Windermere, U.K., tend to form colloidal complexes with organics (Förstner, 1982). Ferrous ions begin precipitating as ferrous hydroxide, $Fe(OH)_2$ at pH 6.0, and the reaction is completed at pH 7.8, while precipitation of ferric ions as $Fe(OH)_3$ occurs at a pH between 2.3 and 3.4. Free ferric ions are not likely to be found in soil, as acidity rarely drops so low. The transformation of Fe^{3+} and Fe^{2+} ions into hydrous oxides gives rise to intermediary ions similar to the aluminium-oxide reaction, which increase acidity, but to a lesser extent.

Moreover, iron easily forms chelates and other complexes with organic compounds. Complex aggregates between clay minerals, ferric ions and humic compounds can be formed. Included in these aggregates, iron loses its property as an acidic ion.

With increasing pH, the influence of aluminium and iron diminishes and they are transformed into precipitated hydrous oxides. The exchange sites they occupied are freed and made accessible to other ions, especially alkaline ones.

Active and potential acidity

Active acidity measures the amount of acidic ions in soil, present and active at the time of record. It is an intensity measure of the acidity at a given time. This acidity depends on a potential, an exchange or a reserve acidity, which is the total amount of acidic ions in reserve or adsorbed to soil particles and which will eventually be released. The potential acidity is related to CEC, base unsaturation and buffering capacity of the soil. Correction of the acidity of a pond mud must take into consideration the total, i.e. active and potential acidity.

Acidification and alkalinization factors

Acidification is generally attributed to three factors: high concentrations of free acidic ions, leaching of alkaline ions, and release of acidic and hydrogen ions by microorganisms, through respiration, nitrification (see p. 24) and sulphide oxidation (Table 4). Decaying organic matter releases organic and inorganic acids, carbonic acid and nitrate ions. Acidic ions displace alkaline ions from the adsorbed sites, which are subjected to leaching.

Alkalinization is caused by an increase of alkaline and hydroxyl ions in the mud. It results from a neutralization of acidic ions or a proton-consuming process. Ageing of sediments causes a neutralization of the soil. Soils subjected to high evaporation in arid regions display an alkaline pH. Proton-consuming processes are: solubilization of limestone by carbonic acid freeing alkaline ions, denitrification, and sulphate reduction.

Nutrient availability and pH

The increase of aluminium, iron, and manganese ion concentrations with decreasing pH entails a decrease in the availability of phosphate, which reacts with the former to form insoluble compounds such as $FePO_4$ (Golterman, 1982). Furthermore, metallic ions are toxic at high concentrations. In alkaline pH, the concentration of calcium increases and with phosphate forms calcium apatite, $Ca_5(PO_2)_6(OH)$, which is insoluble. Iron and manganese are precipitated at high pH, thus becoming unavailable. The requirements of the microorganisms for those oligo-elements are no longer met, and deficiencies arise.

An optimum pH is found within a median zone where the solubility, and thus availability, of the different nutrients is at a maximum. At a suitable pH, microbial decomposition is stimulated and nutrient regeneration optimized. pH acts on the composition of the benthic fauna as well (see p. 50).

2.3. Redox potential

Redox potentials indicate the type of oxidations likely to take place in sediments. The redox or oxidation-reduction potential of a solution is the proportion of oxidized to reduced substances. It is called E_h and measured by means of a hydrogen electrode (Boyd, 1979). Reducing conditions are indicated by negative potentials and oxidizing conditions by positive values. The redox potential of natural water shows values around 0.5 V (Boyd, 1979).

Values of redox potential are standardized to pH 7, using the following formula:

$$E_7 = E_m + 244 - 57 (7 - pH)$$

where E_m is the value of the redox potential

Table 4. Acidification and alkalinization factors occurring in soils.

ACIDIFICATION FACTORS	ALKALINIZATION FACTORS
Leaching - soil drainage	Evaporation
Organic decomposition (release of carbon dioxide)	Ageing of sediment
Production or release of acidic ions: Al^{3+}, Fe^{2+}, SO_4^{2-}, NO_3^-, HCO_3-	Production of alkaline or hydroxyl ions (solubilization of limestone): Ca^{2+}, Mg^{2+}
Nitrification	Denitrification
Sulphide oxidation	Sulphate reduction

recorded, all values of potentials being expressed in mV (Smayda, 1990). Each change of one pH unit is accompanied by a change of E_h of 57 mV, 58 mV or 59 mV according to Smayda (1990), Pearsall & Mortimer (1939) and Boyd (1979), respectively. For pH values above and below 7, the redox potential is respectively increased or decreased by 57 mV per unit of pH.

As indicated earlier, most transformations in sediments are related to the content of organic matter which settles from the overlying water. Non-photosynthetic organisms meet their energy requirements by transforming reduced organic matter, recovering part of the energy of the high energy bonds. The most common oxidizing agent or electron acceptor is the oxygen molecule, but a number of other electron acceptors is available. The organic loading of sediment and mineralization appear to be responsible for changes in redox potential depth profile (Rippey & Jewson, 1982; Lijklema & Hieltjes, 1982). As the load increases, the change in redox potential at different depths indicates a shift from an oxic to an early anoxic environment below the sediment surface, which later again becomes oxic (Rippey & Jewson, 1982). The free energy released by each of these reactions decreases as reducing conditions increase or redox potential decreases.

The sequence of microbially mediated redox processes is as follows:

- O_2 reduction to O^{2-}
- denitrification, NO_3^- to NO_2^- and NH_4^+
- MnO_2 reduction to Mn(II)
- FeOOH reduction to Fe(II), $FeCO_3$ for example
- SO_4^{2-} reduction to S^{2-}
- methane fermentation or methanogenesis, CO_2 or acetate to CH_4.

The redox potentials at which these reactions take place are listed below (Seitzinger, 1988).

Table 5. Redox potentials associated with the principal electron acceptors.

Electron acceptors	Redox potential (mV)
O_2	720-740
NO_3^-	710
MnO_2	470
FeO(OH)	60
SO_4^{2-}	-200
CO_2	-250

Oxygen concentrations required for nitrate-nitrite and nitrite-ammonium reductions to proceed are 4 mg l^{-1} and 0.4 mg l^{-1}, respectively. For the ferric-ferrous reduction, the oxygen concentration is 0.1 mg l^{-1}, and nil for the sulphate-sulphide reduction.

Efficient - in terms of energy-yield - transformation of the organic matter will depend on the prevailing redox conditions or on the supply of oxygen to the sediments. Redox conditions are also important for the mobilization-immobilization of phosphate in sediments. Ferric iron precipitates as hydrous ferric oxide, which adsorbs phosphate, whereas ferrous iron sets phosphate free.

Importance of pH and redox potential properties

1. The acid or alkaline character of a soil indicates the type of prevailing ions (alkaline or acidic) in solution and measured by the pH.
2. Two types of acidity occur: active and potential. Active acidity is the actual value recorded, whereas potential acidity is an indication of the total amount of acidic ions adsorbed to the soil and likely to be released in the course of time. It corresponds to the acid buffer capacity of the soil.
3. Aluminium and iron play an important role as acidic ions in soils.
4. The pH of the sediment acts essentially on the availability of nutrients, especially phosphate.

5. The pH results from a series of processes proceeding simultaneously in the soil. Microbiological activity in the soil has acidifying and alkalinizing effects on soils.

6. The redox potential indicates the oxidizing or reducing character of the soil.

7. The bulk of reducing reactions which occur in sediments results from activity of heterotrophic organisms degrading organic matter.

8. The amount of energy yielded by transformation depends on the availability of electron acceptors.

9. Redox potential plays an important role in the immobilization or mobilization of phosphate in sediments.

3. Mineral and organic components of a soil

A soil is a huge store-house for nutrients, provided by the parental soil or rock, by water, by decomposition of organic matter. The cycling and availability of nutrients in mud makes the latter of interest to fish farmers.

Soils are a mixture of chemical substances, some in large quantities, some in minute amounts. A soil can be characterized by its carrier substances, which are the major constituents onto which are attached or into which are incorporated trace elements (Jacquet *et al.* 1982), and major nutrients. The most important carriers are clays, metallic oxides, silicates, limestone and dolomite.

Some of the major elements of soil, such as phosphorus, calcium and magnesium are closely associated with living organic matter, while others, such as aluminium, iron and potassium, are linked to inorganic matter, especially clays (Ongley *et al.*, 1982).

Models, particularly mass-balance models, are often used to study interactions between water, sediments, and cycling of a particular element, such as phosphorus. They help to test assumptions underlying our knowledge, bring about an awareness of the complexity of the system, and identify areas in which more knowledge is required. Assessment of parameters needs to be made on a set of observations, while validation necessitates an independent data set. Only then can predictions and extrapolations be made (Lijklema & Hieltjes, 1982).

These models divide the environment into a series of boxes, simulating the major compartments of a water body. The fluxes between the different boxes are formulated mathematically. Such models vary in complexity and in resolution and represent a more or less comprehensive description of the real environment. A discussion of the use of a hierarchy of models is found in Lam *et al.* (1982). The ability of models to cope with reality is better when considering only physical and chemical phenomena than when including complex biological situations. Examples of failures in the use of models as a result of incomplete biological information are given by Fryer (1987).

3.1. Mineral component of soil

With regard to the mineral component of soil, nutrients are of most relevance for pond productivity. However, one must consider not only elements such as calcium, phosphorus and nitrogen, but also others such as sulphur and iron, which play a more significant role than their importance in living organisms might suggest.

3.1.1. Calcium

Calcium is present in soils as calcareous sediments, e.g. calcium carbonate or calcite, the mineral form. This element is found in a great variety of common soil minerals such as apatite or gypsum, and also in feldspars, zeolites, amphiboles, garnets, epidotes, pyroxenes, and clay minerals such as illites, smectite, vermiculites, etc., and in a solid form.

The adsorption of dissolved calcium ions onto soil colloids strengthens the base saturation or base buffer capacity of the soil. As free ions, they contribute to the concentration of alkaline ions in the pore water. Dissolution of calcium carbonate releases calcium and carbonate ions. These latter ions participate in the carbonate-bicarbonate system buffering the pH against sudden changes. The bulk of carbonate ions in this alkalinity system does not have a geochemical origin, but originates mainly from organic matter and decaying processes. In these ways, calcium participates in the regulation of soil pH.

The availability of phosphate is reduced in the presence of high concentrations of calcium, as apatite is formed, which is insoluble. Calcium acts on the flocculation of dissolved organic matter.

Calcium and magnesium influence sorption of other cations, especially trace metals, such as manganese, zinc, copper, on metallic oxide (Al_2O_3 and FeOOH) surfaces. This is so because the former are normally present at concentrations many orders of magnitude higher than the trace metals and thus occupy most surface binding sites, even though they form less stable surface complexes. Calcium and magnesium have a significant effect on the amount of natural organics sorbed onto the solid surface of the oxide. Twice as much natural fulvic material was sorbed to goethite when calcium and magnesium were present than when absent (Hart, 1982). Furthermore, calcium promotes flocculation of colloids.

Lastly, calcium is also biologically important as it is required by organisms for their physiological needs, such as in enzyme regulation.

Calcium is subject to leaching and therefore its concentration in soil increases with depth. Magnesium follows a similar trend (Nembrini *et al.*, 1982).

Control of soil pH is effected through liming, a common practice in fish ponds. The objective is to secure a lasting increase of the pH of the pond water; this subject will be discussed on p. 78.

Importance of calcium in soil

1. Calcium is present in numerous minerals in soils, mostly in crystalline and occluded forms, limestone being the most common form.

2. When dissolved, calcium carbonate generates calcium ions, which, being alkaline, strengthen the base saturation of the soil, and carbonate ions, which participate in the buffer capacity. Calcium thus has an important role in soil pH regulation.

3. Calcium acts on the flocculation of organic matter.

4. Many organisms require calcium for their physiological needs.

5. Calcium ions are subject to leaching.

3.1.2. Nitrogen

Nitrogen compounds present in pond muds

Usually, both inorganic and organic nitrogen compounds are present in soils. Organic nitrogen is associated with decaying organic matter, while inorganic nitrogen is essentially present as ammonium, nitrate and nitrite ions. These ions are soluble and occur free in pore water or adsorbed on colloids. The quantities of ions adsorbed depend on availability of fixation sites for sorption on the colloids and on the equilibrium between ions adsorbed and in solution. Ammonium at high concentrations and at alkaline pH forms free ammonia; however, a high pH is not common in soils. Ammonium and nitrate ions in solution never account for more than 1 to 2 % of total nitrogen, with the main proportion of nitrogen in an immobilized state.

Adsorption of ammonium to soil is not homogenous and is related to the organic matter content of the sediment: areas rich in organic matter tend to accumulate nitrogen. The amounts of ammonia adsorbed onto soil particles may be significant: sediments in fish ponds in three different fish breeding regions in Israel (Beith Shean valley,

Jordan valley and the coastal plain) contained from 300 to 900 μg N g^{-1} dry sediment (Fig. 4) (Diab & Shilo, 1986).

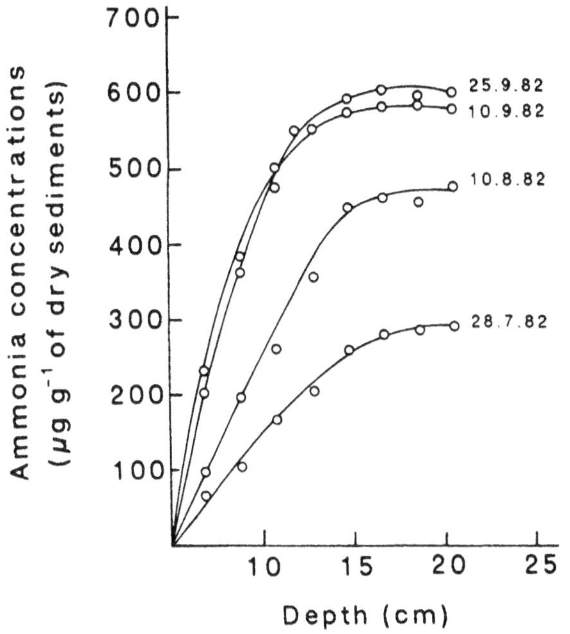

Fig. 4. Changes in ammonia concentration in the sediment profile of a fish pond in the course of a production period (adapted from Diab & Shilo, 1986).

Origin of nitrogen in sediments

Almost all nitrogen in pond mud is of biotic origin, stemming from the decay of organic matter produced in the pond. A small portion of organic matter may be of external origin, falling into the pondwater or imported by water fed into the pond. Depending on the state of decomposition, this nitrogen is in an organic or inorganic form. Only a small portion of the organic matter sedimenting on the pond bottom is mineralized every year.

While humic material is considered as organic matter, it is refractory, of poor nutritional value for decomposers, and supplies little nitrogen to the sediments.

Transfer of nitrogen across the interface

Most of the nitrogen ion species are well soluble and easily transferred from sediment to water. The means of transport are either diffusion across concentration gradients, or sediment perturbations, due to stirring of sediments and bioturbation. Blue-green algae make use of this upward transport of nitrogen compounds at the interface. When concentrations of nitrogen ions were low in the water, these algae were observed in shallow brackish water fish ponds in Israel to make daily downward migrations to the interface, where they took advantage of the diffusion of nitrogen compounds from the sediments (Van Rijn *et al.*, 1987).

Conditions at the interface

Pond sediments are predominantly anaerobic when a pond is impounded and in production, and aerobic when the pond is harvested and dried out.

In ponds in the Beith Shean and Jordan valleys, Israel, used for tilapia and carp rearing, Diab & Shilo (1986) showed an accumulation of ammonium in the sediments to a depth of 15 cm (Fig. 4). Klapwijk & Snodgrass (1982) observed that ammonium concentrations in pore water of Hamilton Harbour, Lake Ontario, Canada, subject to large ammonia and carbon loadings from a waste-water treatment plant and from steel industries, were some 2 to 100 times higher than in the overlying water (0.5-5.0 mg l^{-1} and 10-50 mg l^{-1} ammonia for water and sediment, respectively). This is in concordance with observations of Schindler *et al.* (1987) in Lake 227, of the Experimental Lake Area, Ontario, Canada (Fig. 5). Ammonium concentrations of sediments are dependent on sedimentation of organic matter from the water and on the ammonification potential of the sediments. In Lake Gunillajaure, northern Sweden, Björk-Ramberg (1984) showed a significant correlation between organic matter content and nitrogen content. At the surface (0 to -3cm), ammonium concentrations were lower owing to diffusion across the interface, and to microbial nitrification and denitrification.

Denitrification in sediments

Denitrification is the reduction of nitrate to gaseous molecular nitrogen or even nitrogen oxide, N_2O (Fig. 6). The concentration of the latter is always

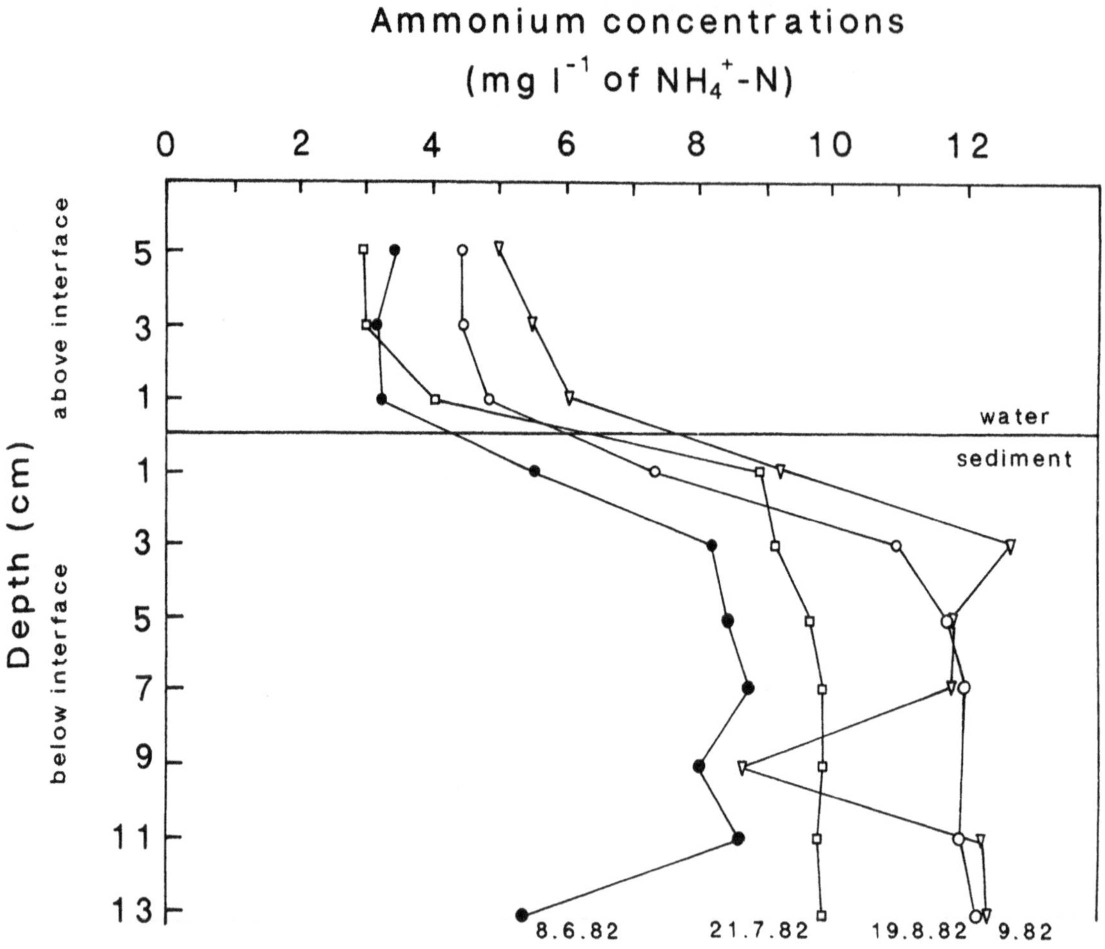

Fig. 5. Seasonal changes in ammonia concentration in the pore water of a sediment profile and in the overlying water of a Canadian lake. The samples were taken in the sediment under 9 m of water (modified from Schindler *et al.*, 1987).

low (Seitzinger, 1988). In this reduction, which is microbially mediated, nitrate is used as a final electron acceptor in the absence of oxygen. Nitrate ions migrate through aerobic sediments to the reaction site where the reduction takes place. Diffusion alone cannot account for the flux observed: bioturbation is partly responsible for the transport of nitrate, ammonium and molecular nitrogen (Klapwijk & Snodgrass, 1982). Nitrate concentration also decreases with depth (Nembrini *et al.*, 1982).

Nitrate for denitrification is essentially provided by sediments and not by water, and the sediments do not act as a sink for nitrate from the water (Klapwijk & Snodgrass, 1982; Gardner *et al.*, 1987; Seitzinger, 1988). During mineralization of organic matter in sediments, a major portion of the nitrogen is denitrified, but considering the external supply of nitrogen to aquatic ecosystems, the amount denitrified does not represent more than one quarter to one third of the supply (Klapwijk & Snodgrass, 1982; Seitzinger, 1988).

The activities of denitrifiers in a sediment vary according to the nature of the sediment and with depth: in loamy sediments, activities are five times higher and extend deeper than in sandy ones. Redox potentials required must not be over +300 mV and maximum activity of the nitrate reductase is around +210 mV; dissolved oxygen concentrations must not be in excess of 0.06 mmol dm^{-3} (1.92 mg l^{-1}) (Cooke & White, 1987a). Nitrate in soil is provided by nitrification of ammonium resulting from ammonifi-

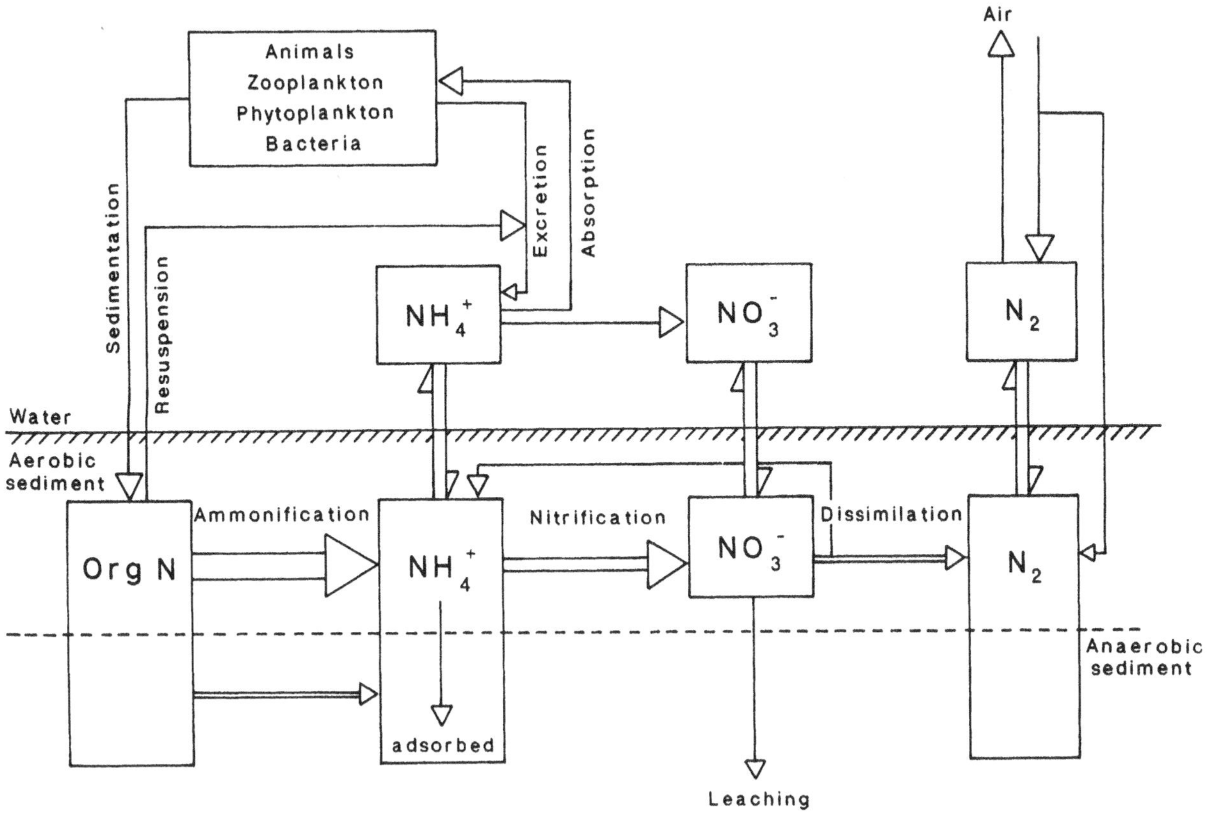

Fig. 6. Transformation of the nitrogen compounds within the sediment of an impounded pond. The nitrogen-fixation process is not shown: the molecular nitrogen is fixed into organic nitrogen.

cation of organic matter. Because of denitrification, ammonium concentration in deeper layers is depleted (Cooke & White, 1987b).

Nitrogen fixation

Nitrogen fixation in sediments has been intensively investigated by Blauw (1988). Molecular nitrogen is used by various microorganisms and blue-green algae to cover their needs of organic nitrogen (Fig. 6). Cellular enzymes controlling the process are nitrogenases and the activity of such enzymes in sediments is correlated to the availability of organic substrate, namely carbohydrates, organic alcohols and acids, and to the maximum rate of glucose absorption by the heterotrophic population. Nitrogen fixation in sediments is essentially an activity of heterotrophic organisms. Such fixation in anaerobic conditions is less efficient than in aerobic conditions and therefore more substrate is needed: *Clostridium pasteurianum* needs about 10 mg nitrogen per g of sugar, while *Azotobacter chroococcum* in microaerobic conditions is able to incorporate 46.5 mg of nitrogen per g of sugar consumed. The abundance of microorganisms in the rhizosphere of macrophytes where excretions from the roots are available is explained in the same way, these excretions providing the energy supply for nitrogen fixation. Such organic excretions are the limiting factor for nitrogen fixation, which is thus linked to the carbon cycle.

Nitrogenase synthesis is activated when interstitial ammonium concentrations drop below a minimum level. Ammonium represses the synthesis and inhibits the activity of the aerobic fixation enzyme, while the synthesis of the anaerobic fixation enzyme is repressed, but its activity is not inhibited. The intracellular concentration of ammonium has a greater effect on the regulation of the synthesis of nitrogenase than the concentration of ammonium in pore water. However, sediments with high ammonium contents still show some nitrogen fixation. This apparent contradiction could result from the huge sorption capacities for ammonium of sediments, but this does not seem a sound explanation. Probably the areas where nitrogen fixation takes place are protected from interstitial ammonium, and are located on microsites in a microenvironment not in direct contact with interstitial water. Protection or barriers consist of quaternary structures of macromolecules, attached to particular fractions or could be bacterial mucus or excretion products. Both bacteria and benthic organisms secrete mucus, within sediments and along burrow walls, which bind sediment particles (Ongley *et al.*, 1982). The barriers constituted by soil particles contribute to the maintenance and renewal of microgradients. Epipelic diatoms are also known to produce extended matrices covering sediments. These consist of mucopolysaccharides, which consolidate the sediments (Paterson, 1989).

Nitrogenase activity is highest at the mud-water interface, where organic matter is in abundant supply and of good nutritional quality; the organic matter content of sediments is reduced by half in the first cm of depth (Blauw, 1988).

Nitrification

Nitrification is the microbial transformation of ammonium into nitrate. It takes place in the upper 2 to 3 mm of the sediment, i.e. the aerobic layer (Fig. 6). This microzone or biofilm is thin, sometimes even less than a fraction of a millimetre. The thickness of the layer is dependent on the biological activity of the sediment, the dissolved oxygen concentration and the depth of overlying water. The shallower the water, the greater the quantity of oxygen reaching the sediment.

Denitrification depends on oxygen transport and biological and chemical oxygen demand. A good supply of oxygen is needed to allow nitrification to proceed. If oxygen consumption exceeds supply, anaerobic conditions appear; the oxygen demand of the sediment can no longer be met and denitrification commences (Bouldin *et al.*, 1974). Denitrification activities occur just below nitrification sites, which indicates a coupling of the two processes (Cooke & White, 1987a).

Nitrification stops when dissolved oxygen concentrations drop below 2.24 to 0.96 mg l^{-1} (Cooke & White, 1987a). The penetration of oxygen into the sediment results in inactivation of the denitrifying enzymes and decreases the amount of mineralizable organic carbon. In deeper layers, denitrifiers, which are facultative anaerobes, are the main consumers of organic carbon and the main mineralization agents (Cooke & White, 1987a). The activity of nitrifying bacteria is maximum above pH 6.5, decreasing with pH.

Ammonification

Organic matter settling from the water column, such as phytoplankton, remains of fish feeds, and fish excreta, decays (Fig. 6). The transformation of organic nitrogen into ammonium is called ammonification. As a consequence, the sediment of a fish pond steadily accumulates ammonium in the upper 10 to 20 cm layer, where sorption onto soil particles takes place till full saturation of sorption sites. Very little ammonium is in solution in the pore water as long as the sediments are not saturated with ammonium (the adsorption properties of sediment for ammonium are high). At the same time, the concentration of $N\text{-}NH_3$ in the water generally remains below 0.1 mg l^{-1}. Sediments act as a sink for the ammonium in the water. Concentrations of nitrate or nitrite in the pond mud are virtually nil.

Modifications occurring during the drying of pond bottoms

In a harvested pond, left to dry, sediments are exposed to the air and become oxidized. Organic matter accumulated during the production period is mineralized (Fig. 7). Liming of the pond bottom speeds up the process (Legendre *et al.*, 1987). Ammonium trapped during the production period is transformed into gaseous (volatile) ammonia if soil pH increases above 9.5; otherwise, it is nitrified and transformed into nitrate. Bacterial activity is dependent on water content of the sediment and on oxygenation. Nitrate in sediments is then used by denitrifiers or leaches to deeper horizons. Concentrations of nitrate in surface layers decrease rapidly

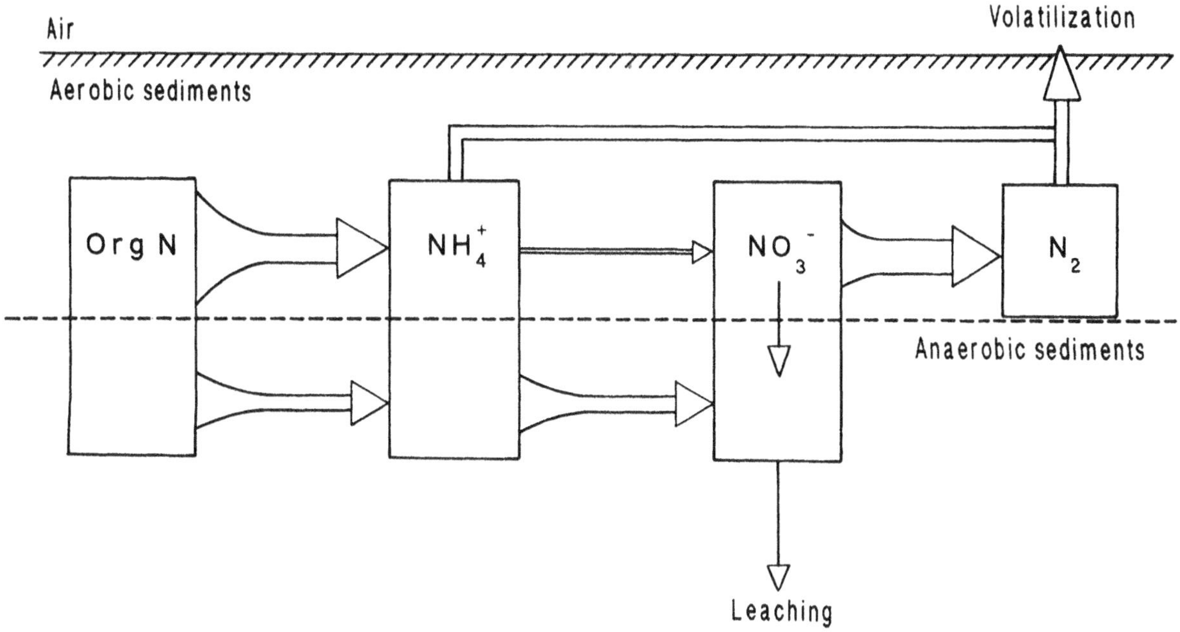

Fig. 7. Transformation of nitrogen compounds within the air-exposed sediment of a pond, during the period of drying.

and are soon virtually insignificant (Fig. 8).

Ammonium, nitrate and nitrite production, as well as nitrification, denitrification and ammonification activities are essentially concentrated in the upper layers of a sediment and decrease with depth (Diab & Shilo, 1986: 75, Fig. 5a). Nitrate increases with depth due to leaching (Diab & Shilo, 1986: 76, fig. 5b). Denitrification rates in deeper layers are low. Rewetting of sediments during the drying phase increases microbial activity and accelerates mineralization process after impoundment.

In soil exposed to air, some nitrogen fixation by bacteria and blue-green algae may occur, but this is not necessarily so if the drying period is short.

Changes occurring after refilling

In the refilled pond, nitrates and nitrites in the soil disappear quickly, owing to diffusion to the water column or to denitrification. Ammonification activities resume and ammonium accumulates in sediment and water. Immediately upon refilling, the ammonium present in the sediments diffuses to the water and high concentrations (up to 3.5 mg l^{-1} of N-NH_4^+, in the case reported) of ammonium and nitrite temporarily appear (Diab & Shilo, 1986). Nitrate is assimilated and its concentration in the water decreases. Ammonification, nitrification and denitrification are microbial processes, sited in the upper layers of the sediments, and occurring in such a sequence that ammonium concentration always remains low, being transformed into nitrate that is assimilated or transformed into molecular nitrogen. Volatilization of gaseous ammonia from sediments occurs only at alkaline pH or after transformation of ammonia into gaseous nitrogen by denitrification. The first process is merely a chemical phenomenon, while the second is microbial.

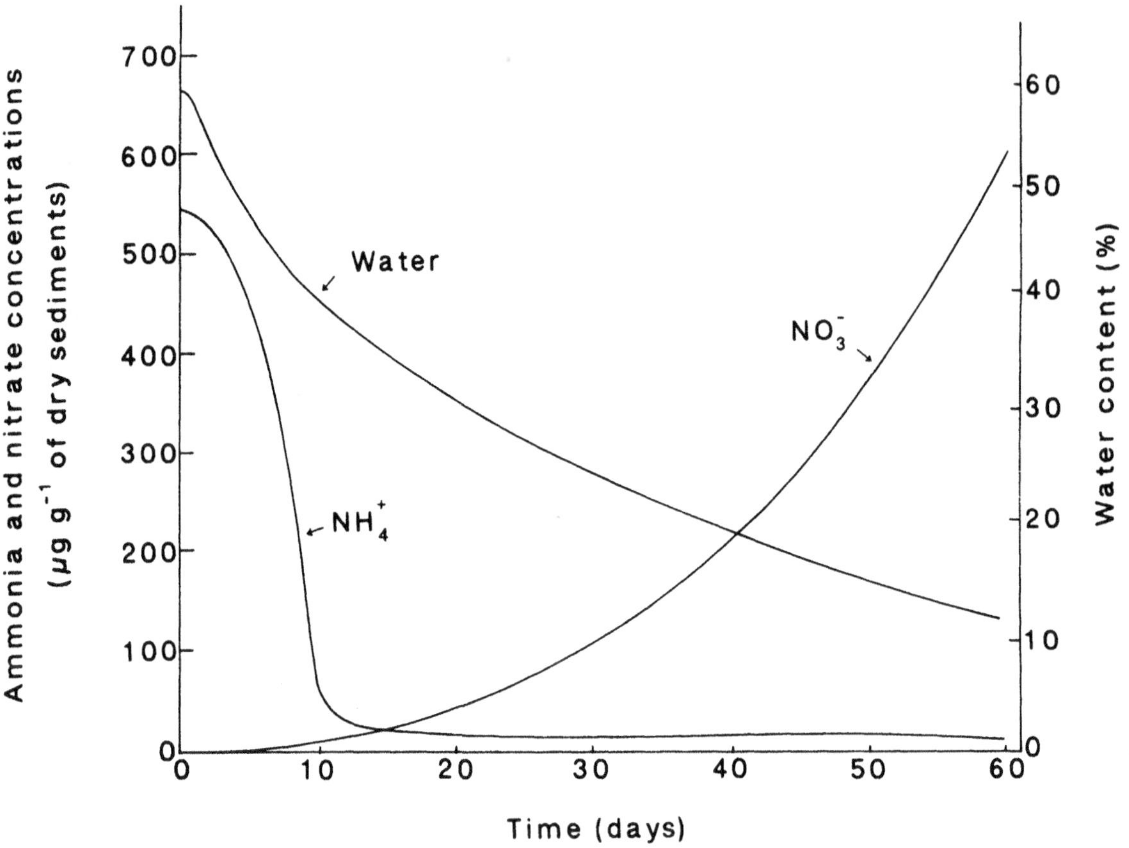

Fig. 8. Changes in ammonia and nitrate concentration and water content of the sediment of a fish pond with time. The values are the mean of four samples along a 10 cm core, and water content is expressed as percent of wet sediment weight (adapted from Diab & Shilo, 1986).

Importance of nitrogen in the sediments

1. Nitrogen is present in soil as the inorganic ions ammonium, nitrate, and nitrite, and as organic compounds. The total amount of nitrogen sorbed onto soil may be important, as ammonium is well soluble and easily sorbed onto colloids.

2. Most nitrogen is of organic origin and supplied by the organic loading of the pond.

3. The cycling of nitrogen in sediment is essentially microbial, different sets of microorganisms being responsible for the transformation of various nitrogen species.

4. Ammonification is the transformation of organic nitrogen into ammonium; nitrification is the transformation of ammonium into nitrate; denitrification is the transformation of nitrate into molecular nitrogen; nitrogen fixation is the transformation of molecular nitrogen into organic nitrogen.

5. During pond production, the most important process is ammonification, followed by nitrification in the aerobic strata of the sediment. The denitrification process is energy-yielding,

using nitrate as an electron acceptor in place of oxygen, and takes place in anaerobic portions of the sediment, where leached nitrate ions occur.

6. In drying pond soil, ammonification leads to nitrification.

7. Nitrogen ion species are easily transferred across the sediment-water interface.

3.1.3. Phosphorus

3.1.3.1. Introduction

Concentration of phosphorus in sediments

Great disparities between the phosphorus content of water and sediments exist. Phosphorus concentration in water is always low, while in sediments it can be found in large amounts. Hepher (1966) found up to 12 g m^{-2} of total phosphorus (P_{tot}) in the mud of a freshwater fish pond at Dor, in Israel. Schindler *et al.* (1987), studying lake sediments (Lake 227, Experimental Lake Area, Ontario, Canada), observed that amounts of phosphorus in solution in the upper layers of the sediments were much higher than in the water near the sediment-water interface (Fig. 9). Sediments are said to act as a sink for phosphorus.

Complexity of phosphorus dynamics in sediments

Phosphorus accumulates in sediments, but can also

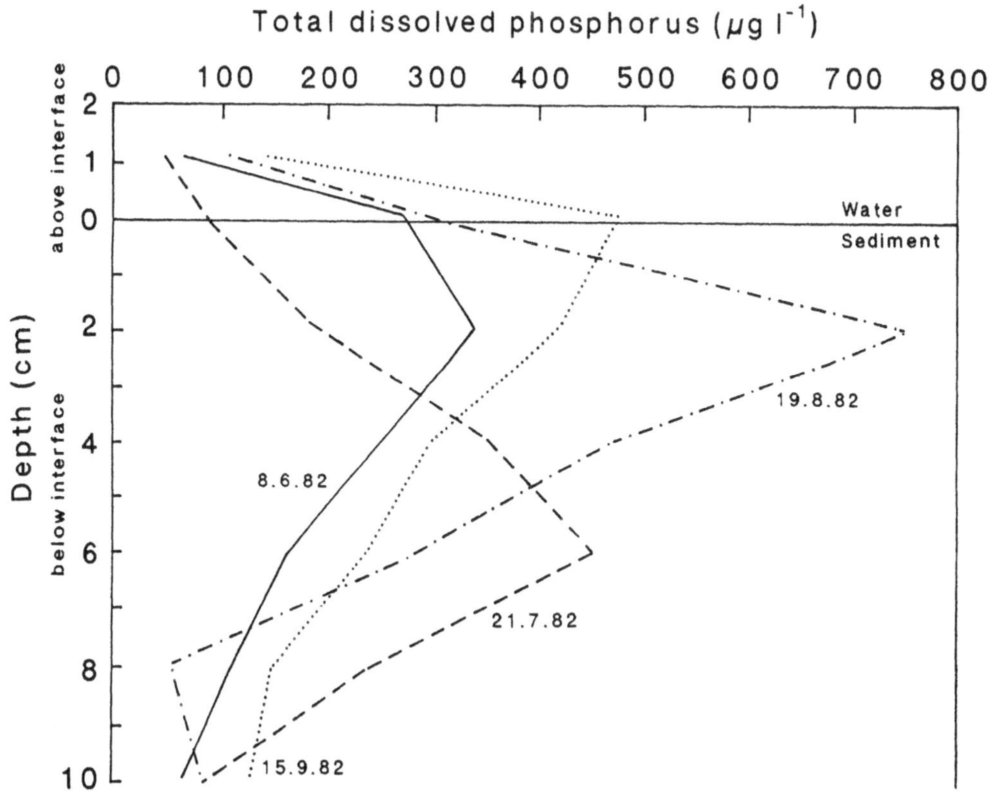

Fig. 9. Seasonal changes in total dissolved phosphorus concentration in the sediment profile and overlying water of a Canadian lake. The samples were taken in the sediment under 9 m of water (modified from Schindler *et al.*, 1987).

be released to the water. The mechanisms of sediment phosphorus release form a complex and dynamic system of which only some aspects are understood (Marsden, 1989). The processes responsible for phosphorus immobilization and mobilization differ in nature:

(1) physical and chemical sediment properties determine its phosphorus sorption capacities, and consequently, by way of sorption equilibria, the concentration of interstitial phosphorus;
(2) the sedimentation rate of organic matter, its mineralization rate, microbial assimilation and excretion rates, and to a lesser extent authigenic mineral formation, determine phosphorus loading to sediment and turn-over rates;
(3) disturbances of sediment causing reworking and resuspension of sediment is responsible for transport of sediment and phosphorus fluxes.

It is therefore impossible to assess, based only on total phosphorus of a sediment, the amounts which will be released in any period of time. There is also no simple procedure to establish the portion of phosphorus requirements, for fish production needs, which can be met by the sediments or the amount of phosphorus fertilizer which will be immobilized in pond mud.

Although phosphorus in sediments has been extensively studied and many limnologists have devoted their life to research on this subject, many aspects still remain poorly understood. A summary of present knowledge is given hereafter.

3.1.3.2. Different forms of phosphorus in sediments

Sediment-bound and interstitial phosphorus

Phosphorus present in pore water is termed interstitial phosphorus; that sorbed to particles is particulate phosphorus. Distinction between liquid and particulate phases is arbitrary (Logan, 1982) and is made by filtration through a membrane filter of 0.45 μm pore size. The filtrable material is defined as dissolved and the retained material as particulate. The filtration procedure neglects colloidal material less than 0.45 μm in size (Stumm & Morgan, 1981), the arbitrary size limit for colloids being 0.3 μm (Broberg & Persson, 1988). Dissolved phosphorus is mainly present as orthophosphate and is called soluble reactive phosphorus (SRP). (For the analytical determination of orthophosphate, see p. 109). The concentration of dissolved phosphorus is often overestimated, due to occurrence of particulate phosphorus in the filtrate (Logan, 1982).

Importance of interstitial phosphorus

Phosphorus in pore water is essentially soluble, and the most reactive species of phosphorus of sediments (Fig. 10). The pore water is an intermediary compartment between soil particles and the water column and is the transient cell for the phosphate carried in by the pond water to be sorbed onto particles or to be carried away after desorption. It is a communication network for solutes, connecting soil particles together. Because of higher concentrations than in pond water, transformations and chemical reactions between ions and solutes take place in the pore water.

The portion of soluble phosphorus is small (estimated to be 1 % (Boström *et al.*, 1988a)) compared to particulate phosphorus and is about 5 to 20 times more concentrated than in pond water. Concentration of interstitial phosphorus reflects the trophic state of the water body and the level of accumulation of phosphorus in the sediments. Solubilization of phosphorus in pore water depends on the solubility coefficients of the phosphorus compounds immobilized onto soil particles and chemical concentration gradients between soil particles and pore water. When the sediment concentration of P_{tot} exceeds 2 mg l^{-1}, phosphorus is not considered limiting (Böstrom *et al.*, 1982).

The concentration of interstitial phosphorus changes during the year. Highest concentrations are recorded when sediments are oxygenated. Interstitial phosphorus concentrations are sensitive to changes in the environment (Enell & Löfgren, 1988). Most changes occurring in soil are the result of microbial activities, reacting to changes in the availability (load) of organic matter, in temperature, in availability of electron acceptors, etc., which all modify the dynamics of phosphorus release.

Orthophosphate is the dominant form of phosphorus in the upper layers of sediments, while in deeper layers, condensed forms occur.

Data on concentrations of interstitial phosphorus

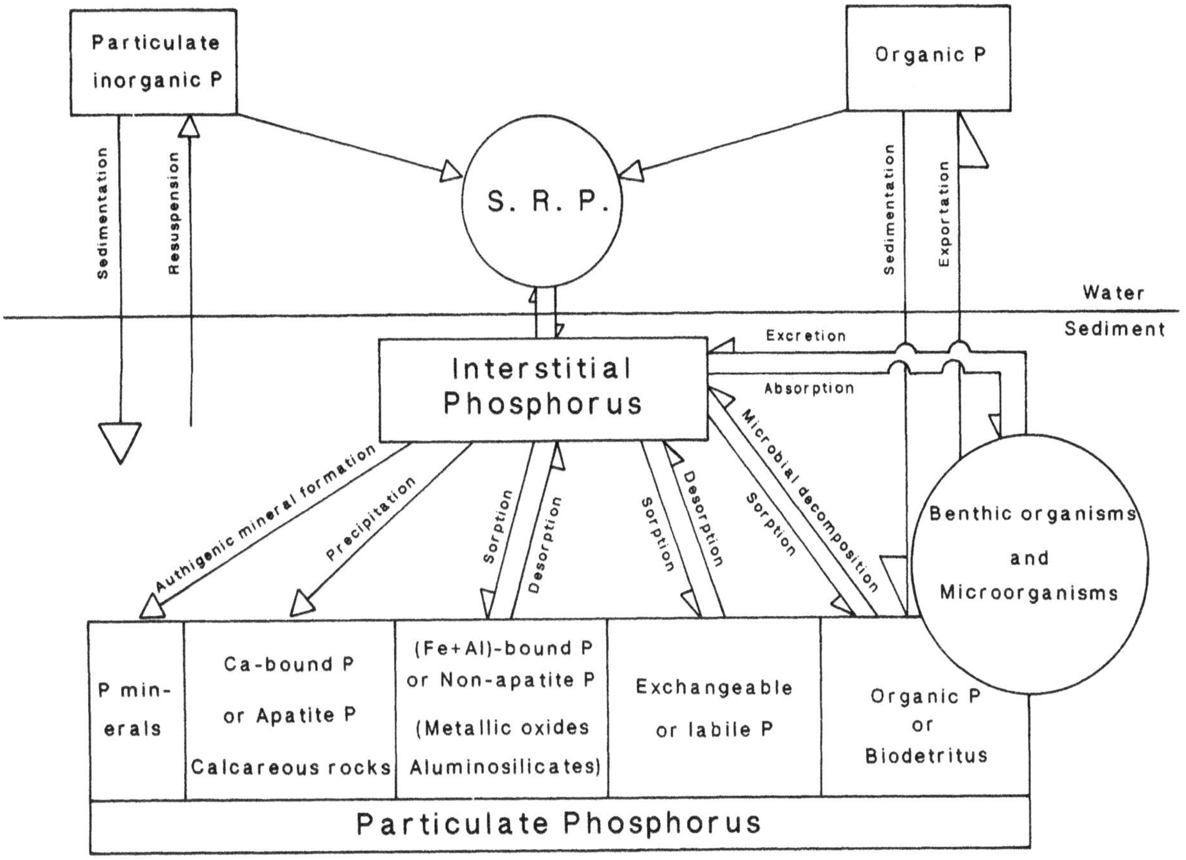

Fig. 10. Importance of interstitial phosphorus. Relations of interstitial phosphorus with soluble reactive phosphorus (S.R.P.) in water and with particulate sedimentary phosphorus. The overlap of organic phosphorus and benthic organism compartments indicates the difficulty in separating biotic organic phosphorus from non-living organic phosphorus.

are often more important than data on concentrations of total phosphorus in soil, but are more difficult to obtain. A review of methods for sampling interstitial water is given in Enell & Löfgren (1988).

Particulate phosphorus

Phosphorus associated with particles or sediment-bound phosphorus is a mixture of various chemical compounds, ranging from mineral, non-reactive (crystalline or amorphous) forms to labile or complex organic compounds (Fig. 10). Distinction is made between adsorbed and precipitated, and between mineral and organic phosphorus. Labile compounds include adsorbed phosphate, easily hydrolyzable or dissolved. Sequential phosphorus extraction schemes are numerous and their results difficult to compare (Pettersson *et al.*, 1988). Such procedures attempt to quantify the different forms of phosphorus present and, through fractionation, partition particulate phosphorus into major groups, which generally comprise (Logan, 1982; van Eck, 1982; Pettersson *et al.*, 1988; sequential extraction procedure according to Hieltjes & Lijklema, 1980):

- (Fe + Al) bound P (24 h extraction with 0.5 N HCl)
- calcium bound P (apatite) (17 h extraction with 0.1 N NaOH)

- reductant soluble P or exchangeable P (2 h extraction repeated twice with 1 M NH_4Cl)
- organic P, including labile and resistant organic P (digestion with $HNO_3/HClO_4$).

Comparisons between results concerning different phosphorus fractions obtained with different extraction methods are difficult and one should be cautious in interpreting such data.

Mineral phosphorus is usually separated into forms sorbed to apatite and non-apatite minerals. Apatite phosphorus is associated with calcium, while non-apatite phosphorus is essentially associated with iron or, to a lesser extent, aluminium. Acid-extracted phosphorus encompasses most of the phosphate bound to aluminium and iron; the most important forms of phosphorus in sediments are amorphous iron-hydroxy complexes with an atomic ratio of Fe to P between 5 and 10 (van Eck, 1982; Lijklema & Hieltjes, 1982: 230, Fig. 4), while Al-bound phosphorus plays only a minor role (van Eck, 1982). The atomic ratio of iron to phosphorus of the acid-extract is between 1.5 and 1.8. Besides the other compounds extracted (exchangeable-P, carbonate-P and labile organic-P followed by clay and other compounds), tinctite, $Fe_3(PO_4)_2(OH)_3.3H_2O$, or cacoxenite, $Fe_4(PO_4)3_2(OH)_3.12H_2O$, two stable Fe(III) phosphates under oxidizing conditions, are assumed to be present (van Eck, 1982). The abundance of iron in aerobic sediments is important in governing phosphorus sorption. The oxidation of ferrous iron should immobilize phosphorus only when the atomic Fe/P ratio in the interstitial water is above 1.8 (Marsden, 1989). A great number of phosphorus minerals - more than 200 have been described - are also likely to be found in sediments, as a result of diagenetic processes. (Diagenesis refers to abiotic and biogenic changes taking place within the sediment during and after burial).

In the humic lake Tjeukemeer, Netherlands, organic phosphorus compounds have high molecular weights, belonging to the fulvic or humic compound groups. Fulvic acids are aromatic polycondensates, with phenolic and carboxylate groups, which give them chelating properties. (The aromatic nature of the fulvic material could be an artefact). In contrast to iron, humic acids are not able to bind inorganic phosphorus directly to form a colloidal phase (De Haan & De Boer, 1986).

A distinction exists between the various fractions of phosphorus characterized in sediments and phosphorus available for production, designated as bioavailable. The fractions from which phosphorus is taken by microorganisms and autotrophs to satisfy their needs are not well defined and will be discussed on p. 42.

The concentrations of phosphorus in sediments are generally expressed in weight per unit weight of dry sediment, though it might be better to use weight per unit surface or per unit volume of sediment, specifying the depth of sediment considered (Holtan *et al.*, 1988).

Sorbed or immobilized sediment phosphorus

Immobilization of phosphorus in soil is the result of sorption (see Table 3, p. 12). Sorption is removal from solution to a solid phase and involves binding at fixation sites on the surface of colloids. Distinction is made between physical sorption and chemisorption. Physical sorption is due to forces between the solid surface and the adsorbed molecules or ions. These forces are similar to van der Waals forces between molecules and are undirected and relatively non-specific. The energies involved are of the order of a few hundred to a few thousand calories per mole. Physical adsorption is generally reversible. In contrast, chemisorption is the result of strong binding forces, comparable to those leading to the formation of chemical compounds; such an adsorption may be regarded as the formation of a surface compound and is seldom reversible. The energies of chemisorption range from 10^4 to 10^5 calories per mole (Morris, 1956; Holtan *et al.*, 1988).

Adsorption in sediments was first studied on clay minerals and on metallic oxides and hydroxides. Acid muds with high concentrations of hydrous iron (III) oxides, together with small quantities of alumina, silica and organic matter, show high sorption, with peaks at pH 2 to 5 (Hepher, 1958). Sorption capacities for phosphate of non-calcareous sediments are much higher than calcareous sediments. Crystalline iron and aluminium hydrous oxides sorb five to ten times more phosphorus than crystalline alumino-silicates or calcium carbonates. In contrast, short-range order components, such as iron oxide gel, sorb ten to one hundred times more phosphorus than their crystalline counterparts, i.e.

up to almost one thousand times more than crystalline alumino-silicates and calcium carbonate (Holtan *et al.*, 1988).

In alkaline soils with high concentrations of carbonates, phosphates are adsorbed to the surface of calcite crystals, where they displace water molecules, or carbonate or hydroxyl ions, or co-precipitate with calcium carbonate, or form calcium phosphate.

Organic matter can act in a twofold, antagonistic way. It sorbs phosphate directly onto fixation sites with positive charges or indirectly with the help of intermediary ions, but can also block sorption sites on oxides, preventing the sorption of phosphates. Organic anions prevent crystallization of aluminium oxide (Holtan *et al.*, 1988). Organic colloids, such as humic material, do not adsorb phosphates, as they carry negative surface charges, but in the presence of iron, aluminium and calcium, they are able to adsorb great amounts of phosphates. Humic acids have a high affinity for chelating iron, which is decreased in the presence of calcium. In a pH range of 2 to 5, phosphates react to form ferri-phosphate-humate complexes (Hepher, 1958 & 1966). These complexes are highly refractory, once formed.

Decomposition of organic matter releases carbon dioxide and phosphate, which raises the concentration of free phosphate and increases the concentration of phosphate sorbed. An increase in acidity causes a drop in pH, which modifies the composition of the pore water and decreases the sorption of phosphate on calcium carbonate, which in turn affects phosphate sorption.

Microorganisms involved in decomposition excrete enzymes responsible for hydrolysis of organic matter and these extracellular enzymes and chelating agents modify the mobilization of phosphorus.

Precipitation of phosphorus or authigenic mineral formation

Formation of stable or metastable minerals is a geochemical process taking place in soils, but still poorly understood since identification of mineral species in the solid phase is difficult (Nembrini *et al.*, 1982). Phosphates in soil react with cations to form insoluble salts. Ionic activity product calculations for different mineral phases in the pore fluid are used to estimate conditions under which precipitations may occur, though direct identification is the only proof of existence of a mineral. Supersaturation of the pore fluid with respect to the solubility products constant has been observed for a variety of minerals, including vivianite, apatite, and others (Nembrini *et al.*, 1982; Enell & Löfgren, 1988).

Examples of some minerals formed are as follows: with calcium, phosphate forms apatite; with aluminium, phosphate forms variscite and wavellite; with iron, vivianite, $Fe_3(PO_4)_2.8H_2O$, and strengite are formed. The number of phosphate minerals identified in soils reaches almost 200 (Psenner *et al.*, 1984), which amply shows the complexity of this topic. Chemical formulae of these minerals may be found in Pettersson *et al.* (1988).

3.1.3.3. Phosphorus loading to sediments

Phosphorus accumulation in sediments

As a result of sorption onto sediments, phosphorus accumulates (loading) and is trapped in the soil. Identification and quantification of the mechanisms responsible for immobilization and mobilization of phosphorus are important. Mobilization is also called desorption, but according to Golterman (1988), the latter term is unfortunate and is a *contradictio in terminis*, as it is in reality an adjustment of sorption equilibria.

Origin of phosphorus in sediments

Mechanisms of transfer of phosphorus from water to sediments are as follows (Boström *et al.*, 1988a):

- sedimentation of detrital phosphorus and particles in suspension, brought in by the water feeding the pond
- adsorption or precipitation of phosphorus with inorganic compounds, such as coprecipitation with iron, manganese or calcium, and adsorption on clays, carbonates, or oxyhydroxides, and sedimentation of these aggregates
- sedimentation with allochthonous or autochthonous organic material
- assimilation of phosphorus in the water column by plankton or other biota and sedimentation, as well as sedimentation of all dead organisms

- direct adsorption by sediments.

The largest portion of phosphate in soil originates from decomposition of organic matter (Fitzgerald, 1970; Sinke & Cappenberg, 1988) and from the fertilization of the pond. The contribution of phytoplankton to organic loading of sediment depends on natural productivity of the pond, and grazing upon phytoplankton. Decomposition and release of phosphates from organic matter is a function of temperature and activity of microorganisms (Sinke & Cappenberg, 1988). When considering immobilization of phosphorus in sediments, the first approach was to study the physical and chemical sorption processes occurring in soils, as if the latter were an abiotic environment. However, it soon became clear that biotic aspects of phosphorus cycling were essential for availability of phosphorus; these aspects involve microbial decomposition of organic matter sedimenting to the sediments. Compartmentalization of phosphorus between microbial, non-living organic matter and labile sorbed phosphorus is difficult, because of the rapid turnover of such compounds.

Loading models and kinetics of sorption

Continuous loading of phosphorus to the pond bottom leads to a saturation of its sediment. Sedimented phosphorus is transformed diagenetically and is sorbed. Sorption processes are saturation processes, which were first described with the help of models. But to be useful, these models should allow prediction of sorption as well as release of phosphorus from sediments.

Equilibrium models calibrated at various phosphate concentrations have been used, describing accumulationaccording to a saturation curve, levelling off at a maximum internal load (sigmoid curve). Phosphorus sorption onto soil is best described by sorption isotherms, of which those of Langmuir and Freundlich are most popular. They are based on equilibrium or non-equilibrium kinetics, and partition sorbed (or labile) and dissolved phosphorus. Most models do not, however, satisfactorily describe desorption (Logan, 1982). Quantitative comparisons of adsorption isotherms reported in the literature are often difficult, because experimental designs are not standardized (Enell & Löfgren, 1982).

When using adsorption isotherms to describe phosphate sorption, it often appears that using a single exponential function to describe phosphorus uptake kinetics is inadequate. Analysis of the kinetics leads to a separation into two, three or more distinct phases, each described by an exponential function, the observed kinetic being the combined effects of several reactions with different rates. Graphical resolution of uptake curves into distinct phases does not, however, constitute proof of the existence of an equivalent number of reactions (Twinch & Breen, 1982).

Diphasic Langmuir plots have often been reported to indicate dual phosphate adsorption mechanisms, separated into a fast and a slow uptake phase (Twinch & Breen, 1982; Logan, 1982; Gunatilaka, 1982; Froehlich, 1988). The significance of each phase remains obscure (Twinch & Breen, 1982). According to some authors, sorption proceeds at various binding energies at distinct sites of higher and lower energy (Gunatilaka, 1982).

A more realistic explanation suggests a biotic-abiotic partitioning of phosphorus uptake, which has been investigated with the help of sterilized versus unsterilized sediments (Doremus & Clesceri, 1982).

Sterilized sediments show much slower uptake kinetics. Abiotic sorption is determined by the relative presence of different carrier substances, i.e. metal oxides, calcium, carbonates, found in sediments. Sorption of phosphate to them will depend on their amounts and their binding forces.

The most mobile sediment phosphorus pool is linked to biodetritus and to lower bonding energies. Absorption of phosphate by microorganisms is rapid, followed after a short time by the excretion of phosphate, which accounts for the observed desorption rate. The rapid uptake matched by a corresponding efflux is indicative of a dynamic steady-state system. Turnover time of phosphate within the biodetritic compartment, i.e. the abiotic organic material and the microorganism population, is short, and the biodetritus acts as a transitory compartment between organic and inorganic phosphorus or in a buffer capacity for phosphate (Doremus & Clesceri, 1982; Twinch & Breen, 1982). This buffer capacity is expressed as the quantity of phosphate adsorbed per unit increase of phosphate in solution (Twinch & Breen, 1982). Inorganic phosphorus in excess of the phosphate sorption and buffering capacities of the biodetritus is adsorbed onto ferric oxides and other sorbent compounds (Doremus & Clesceri, 1982).

Phosphate sorption properties vary with depth and with changes from aerobic to anaerobic conditions. In oxidized sediments, microbial activity is at a maximum and available phosphorus is important, while in reduced sediments, microbial activity is less important and less efficient, an iron oxide sequestering mechanism is lacking and non-sorbed phosphate accumulates (Doremus & Clesceri, 1982).

Surface sediments have higher sorption properties than deeper layers; the actively exchanging layer in Midmar Dam, South Africa, consists almost exclusively of the upper 2 cm of sediment (Twinch & Breen, 1982). Langmuir adsorption isotherm curves show that phosphorus sorption and bonding energy constants are maximum in the uppermost sediment layers (the loose surface material) being some 20 % and 60 % respectively lower than those of sediments from the -3 to -5 cm strata (the original flooded terrestrial soil, which is more mineral).

Phosphorus sorption in sediments - the Einsele-Mortimer paradigm

Phosphorus is immobilized in sediments because surface sediments act as a barrier to exchange between water and sediment. Immobilization of phosphorus in sediments has been described by a now classical model, based on interaction between iron and phosphorus under aerobic and anaerobic conditions, or oxidized and reduced conditions (Einsele, 1936; Mortimer, 1941 & 1942) (Fig. 11). A few millimeters below the sediment-water interface, an oxidized microzone of amorphous or poorly crystallized ferric oxyhydroxides forms which has high sorption properties for phosphate and constitutes a protective oxidized surface (Hepher, 1958; Callender, 1982). These iron oxides sequester the phosphate and considerably reduce its concentration in pore water.

In deeper strata, iron is reduced owing to anoxic conditions to ferrous ions, which do not have sorbing properties, and release phosphate (Fig. 12). When anoxic conditions, i.e. redox conditions decreasing from 450 to 200 mV and dissolved oxygen concentration decreasing from 4 to 0.1 mg l^{-1}, extend to the sediment surface, phosphate is released into water. Optimum pH and redox conditions are shown in Fig. 13. The concentration of ferrous iron is negatively correlated with redox potential of sediments: oxic sediments sorb some 25 times more phosphorus than anoxic sediments (Callender, 1982).

Reduction of iron partly results from the use of iron as an electron acceptor in the absence of other electron acceptors (Boström *et al.*, 1988a). Redox conditions are mainly modified by the activity of microorganisms; the destruction of the protective oxidized microzone and shift to anoxic conditions is

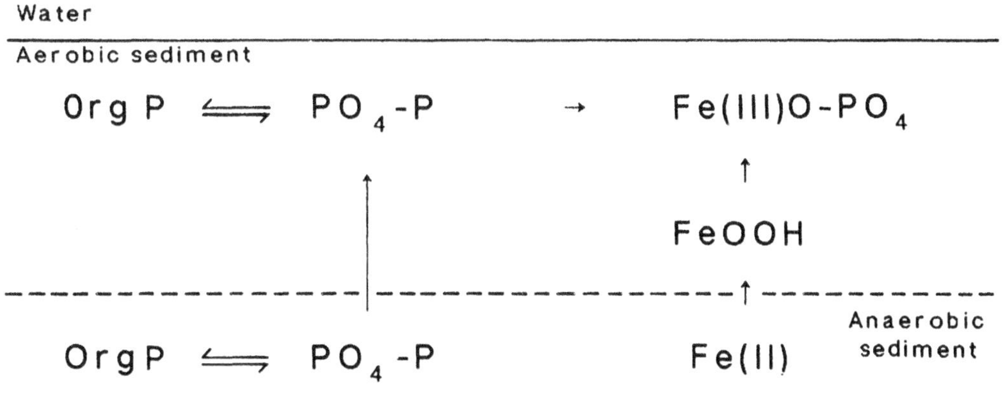

Fig. 11. Sorption of phosphate to ferric oxyhydroxide in aerobic sediment and oxidation of ferrous ions migrating from the anaerobic zone of the sediment.

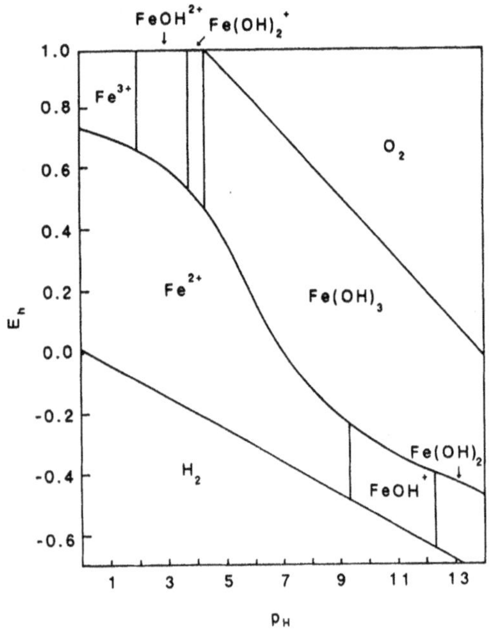

Fig. 12. Different iron ion species present at various pH and redox conditions (after Segalen, 1964).

the result of an organic loading in excess of the processing capabilities of microorganisms and dissolved oxygen supply to the sediments. It results in a change in the phosphate flux from sediment to water, the concentration gradient existing at the interface promoting the exchange.

The Einsele-Mortimer paradigm for phosphate sorption in the sediments has created the widespread opinion that phosphorus cycling is entirely linked to iron chemistry (Boström *et al.*, 1988). Though an important process, the contribution to phosphate sorption of other carrier substances should not be ignored.

Factors affecting phosphorus immobilization

Internal phosphorus loading or immobilization of phosphorus is dependent on sediment characteristics, such as pH, porosity, compaction, texture and sorption properties (ion exchange capacities), on

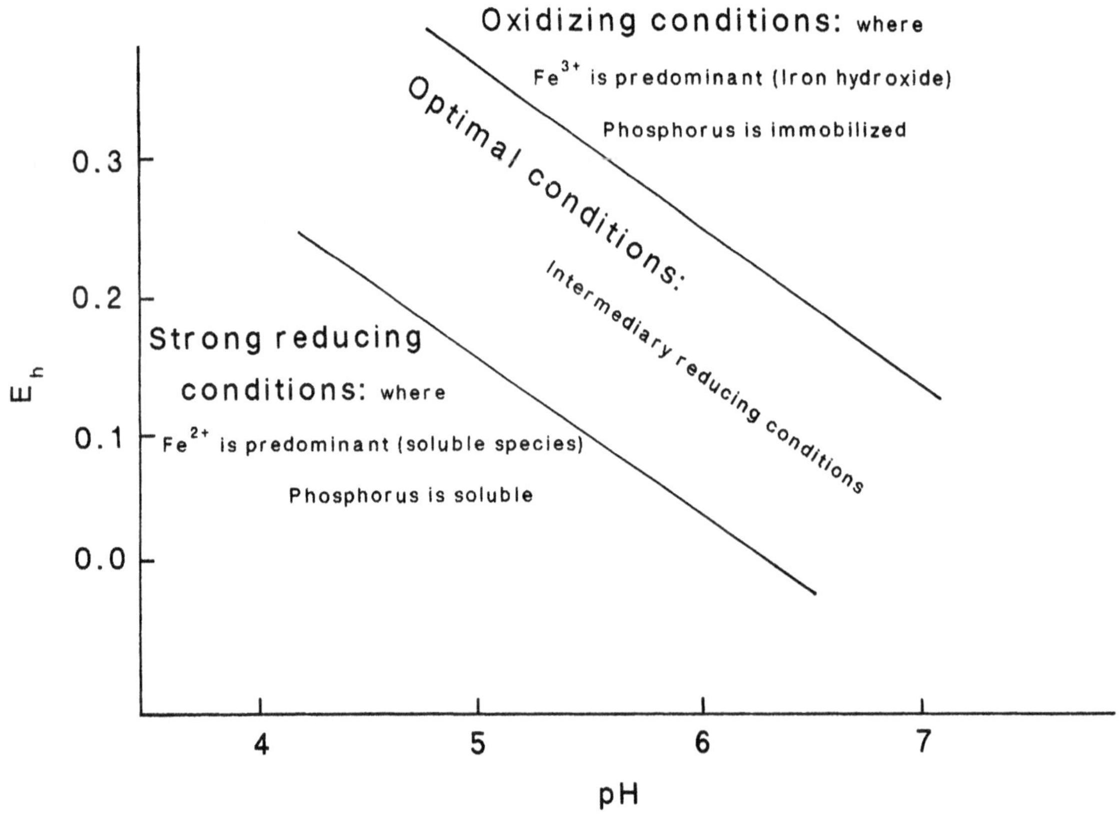

Fig. 13. Iron and phosphorus species at varying pH and redox conditions of the sediment (after Segalen, 1964).

concentrations of dissolved ions, and on the presence of carrier substances. Ostrofsky (1987) established, on data collected from 66 lakes (of different trophic levels) of eastern North America, significant correlations between iron and labile phosphorus (NH_4Cl-extractable), calcium-bound phosphorus (HCl-extractable) and iron plus aluminium-bound phosphorus (NaOH-extractable), total inorganic content, and total phosphorus in sediments. His results suggest that sediments with high sedimentary iron levels have higher concentrations of labile and iron and aluminium-bound phosphorus (Fig. 14).

Immobilization or trapping of phosphorus in sediments is essentially the result of microbiotic (Gächter *et al.*, 1988) and biogeochemical processes. Factors affecting phosphorus mobilization operate differently according to the type of process and must be discussed in connection with controlling environmental factors. Net phosphorus release is also typically seasonal, linked to microbial activities (Marsden, 1989).

A raise in temperature increases microbial activity and, as such, has an indirect effect on mobilization. Mineralization processes and metabolism of microorganisms, including all processes related with assimilation of phosphorus, excretion of phosphate from cells and transformation of organic detrital matter, are stimulated. Temperature also modifies sorption equilibria on sorbent substrates. Increase in microbial activity induces changes in the chemical environment of sediments: release of acidity and drop in pH, consumption of oxygen, release of reduced electron acceptors and drop in redox potential. The increase of microbial activity due to a raise in temperature may be locally restricted to patches where nutritionally rich organic matter is

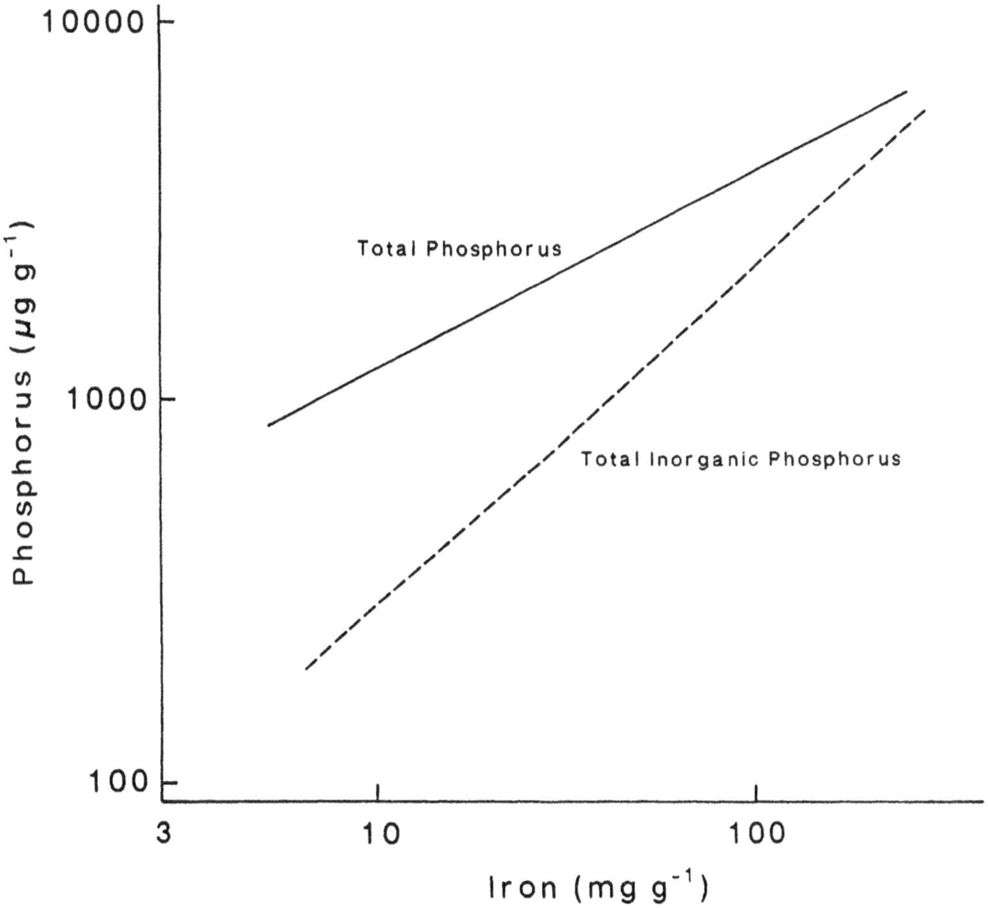

Fig. 14. Relations between the amount of total and total inorganic phosphorus and the amount of iron within the sediment (modified from Ostrofsky, 1987).

available. At those loci, anaerobic micropatches can appear which pierce the oxidized microzone and allow diffusion of phosphate.

An increase in pH reduces the sorption potential of iron and aluminium ions, primarily by competitive ligand substitution in which hydroxyl ions replace orthophosphate (Boström *et al.*, 1988a; Marsden, 1989). Changes in pH are a consequence of modifications in carbonic acid or dissolved oxygen concentrations. Microbial activity has specific pH requirements or pH ranges at which activity is maximum. The spectrum of a bacterial population changes according to prevailing pH conditions of mud. Moreover, pH changes affect sorption potential on pH-dependent fixation sites on colloids. Liming of the pond bottom increases pH, but also adds calcium carbonate to the sediment. Phosphate freed from its sorption to ferric hydrous oxides by a pH raise will coprecipitate with or be adsorbed onto calcium carbonate to form apatite.

A reduction in redox potential means an increase in reducing conditions and a reduction in dissolved oxygen concentration. Under reducing conditions, ferric iron is transformed into ferrous form, which is soluble. Sorption potentials for phosphate of the ferrous iron are greatly reduced in comparison to ferric compounds and interstitial phosphate concentrations increase (Holtan *et al.*, 1988; Søndergaard, 1989). Progressive changes from aerobic to anaerobic conditions induce modification of the spectrum of microorganisms capable of living in mud and making use of alternative electron acceptors. In ponds with high biomass and/or organic load, the water at night is not supplied with oxygen, and consumption of the latter leads to its depletion. The sediments, in which heterotrophic activity goes on, deplete their oxygen reserves and anaerobic conditions set in, which allows nightly release of phosphate. A diel diphasic phosphate release and immobilization may hence be present.

Efficiency of energy transfer in transformation of energy-yielding organic compounds is reduced and the diversity of microorganisms is lowered. Absorption of phosphate by microorganisms also differs according to aerobic and anaerobic conditions. Rapid uptake of phosphate, in excess of needs, by microorganisms is stored in polyphosphate granules; when conditions turn anaerobic, polyphosphates are used and phosphate, then in excess in the cell, diffuses out by increased osmotic pressure (Boström *et al.*, 1988a).

Among the geochemical factors responsible for immobilization of phosphorus, the abundance of iron in sediment is important. The oxidation of iron to the ferric form immobilizes phosphorus in aerobic sediments only when the atomic ratio of Fe/P in the interstitial water is greater than 1.8. Mechanisms that maintain low interstitial iron concentration (such as precipitation with sulphide ions) may be important for leakage of phosphorus from sediment to water (Marsden, 1989). Other mechanisms which reduce phosphorus sorption during its transport across aerobic sediments include: organic acids which act as chelating agents sorbing calcium, iron, manganese and aluminium; high concentrations of soluble silica, which compete with phosphate for adsorption onto iron (Marsden, 1989).

Authigenic mineral formation may change the availability of phosphate and other ions, removing them definitively from pore water. Geochemical transformations occurring in sediments have been studied intensively to assess whether there are any equilibrium reactions which can 'buffer' high pore water concentration by precipitation to a solid phase. Environments suitable for formation of these minerals are mostly found in highly reduced sediments. Moreover, kinetics of precipitation reactions are slow, while most sorption reactions involving active biological exchange, as occur in the uppermost layers, are rapid (Enell & Löfgren, 1988). The former are not likely to be a major importance in surficial strata.

Influence of mineralization on the availability of phosphate

The transformation of organic matter in sediments results in a transfer of detrital phosphorus, via living biomass, to the pool of interstitial or labile phosphorus. The cycling of phosphorus is not easy to quantify, owing to the rapidity at which it is transferred from one compartment of the soil to the other.

The effects of mobilization processes on phosphorus depend on (Boström *et al.*, 1988a):

- the state of decomposition of the organic matter sedimenting on the sediments and the prevailing type of transformation

- the initial phosphorus content of the organic matter and the C:N ratio of settling organic matter
- the concentration of interstitial phosphorus and the saturation percentage of sediments with regard to phosphorus
- the sedimentation rate of organic matter, and whether steady or irregular, as in the case of an algal crash (see p. 148)
- environmental conditions, such as temperature, redox potential, and dissolved oxygen concentration.

Microbial activities modify the concentrations of sedimentary phosphorus. Microorganisms provide for their needs by drawing phosphorus from organic matter. This involves a transfer of phosphorus from the biodetrital to the biotic compartment and is realized with the help of hydrolytic enzymes. During this extracellular hydrolysis, some phosphorus is released to the environment. The amount of phosphorus released depends on the nutritional value of organic matter. Nutrients are absorbed by micro-organisms in a given proportion; nutrient in excess of this ratio is wasted by the organisms and released to the environment.

Decomposition of substrata of poor nutritional value results in a net immobilization of nutrients (Doremus & Clesceri, 1982), while decomposition of substrata of better value results in nutrient regeneration (Boström *et al.*, 1988a). In lake George, New York, U.S.A., sedimented organic matter was often of poor nutritional quality, so that decomposers were nutrient-limited (Doremus & Clesceri, 1982).

The intensification of decomposition activities mobilizes various fractions of phosphorus between compartments and increases the amount of phosphorus available.

Effect of nitrate on mineralization

The presence of nitrate affects mineralization and mobilization of phosphorus by either promoting or inhibiting its release.

Release of phosphorus is inhibited when the concentration of nitrate is high: the presence of nitrate buffers the redox potential, preventing the use of iron as an electron acceptor.

On the other hand, high nitrate concentration increases the mineralization rate of phosphorus as a result of increased denitrification, which promotes more rapid transfer of phosphorus from degradable organic matter to the mobilized phosphorus pool (Boström *et al.*, 1988a).

Changes in phosphorus loading with depth

In an oligotrophic lake in northern Sweden, Björk-Ramberg (1984) noticed that the amounts of sediment phosphorus correlated positively with sediment depth and negatively with organic matter content. Phosphorus concentrations in surficial sediments were some 40 % lower than in deeper strata. Sedimentary phosphorus increases with depth, when expressed on a volumetric basis (weight per unit volume of sediment), but decreases when expressed as weight per dry weight of sediment. This is most likely explained by an increase in compaction of sediments, and a decrease in water content and of mineralization rates with depth. The concentration of phosphorus in surficial sediments increases as a result of:

- an increase in concentration of recently sedimented material, due to an increase in phosphorus loading
- a decrease in mineralization rate due to an increase in volume of settling matter (increased sedimentation rate), and/or a decrease in activities of microorganisms as a consequence of reduced oxygen supply
- an upward diffusion along a concentration gradient of sorbed phosphorus accumulated in deeper horizons, where oxygen depletion and redox potential reduce sorption capacity. The phosphate migrating to the surface layer is sorbed, due to changes in redox potential and pH (Holtan *et al.*, 1988).

On the other hand, reworking of the sediments by benthic macroorganisms opposes the establishment of this phosphorus gradient. Sediments from deeper strata are mixed with sediments from upper layers and thus brought in contact with the aerated water of the pond.

Buffering mechanisms of sediment phosphorus

Various buffering mechanisms of sediment phosphorus are as follows and are summarized in Table 6:

- the first or rapid step of the sorption process, in which a large pool of dissolved phosphate is retained by low-energy bonds and set aside as labile phosphorus, which is easily released into solution, should conditions change to promote desorption (Froehlich, 1988). This phosphorus reflects a mobile sediment phosphorus pool, with increased phosphate desorption potential (Twinch & Breen, 1982)
- the combined phosphate immobilization ability of the microbial biomass and its associated organic matter acts as a phosphate buffer for surface sediments (Doremus & Clesceri, 1982)
- the temporary immobilization of phosphate due to rapid uptake of available phosphate by microorganisms followed by rapid excretion resulting in a mobile pool. Gross phosphorus transport in and out of bacterial cells far exceeds net transfers measured (Boström *et al.*, 1988a)
- the storage of phosphate in microorganism cells under aerobic conditions, for use and release when conditions shift to anaerobic (Boström *et al.*, 1988a)
- lastly, the authigenically formed phosphorus minerals in sediments have been examined as a possible phosphate buffering system for pore-water phosphorus, but they can hardly buffer changes which occur at a much faster rate than do these chemical precipitations (Enell & Löfgren, 1988).

The various mechanisms which keep phosphate available, despite its low solubility, indicate that phosphorus is a 'strategic' element for life in the sediment.

3.1.3.4. Phosphorus release from sediments

Phosphorus release and sediment characteristics

Transfer of phosphorus from sediments to the overlying water is prevented by the presence of the oxic/anoxic barrier at the interface.

Phosphorus dynamics, sorption and desorption, depend on sediment characteristics. With regard to desorption, three types of sediments exist: sediments which do not release phosphorus under any circumstances; sediments which release phosphorus at a given rate, not influenced by the addition of nitrate or acetate, the release rate increasing independently up to a certain level; sediments which release phosphorus when acetate is added. The addition of acetate stimulates bacterial activity, while the addition of nitrate stabilizes the redox potential and prevents release of phosphorus. The differences in behaviour are linked to the varying content in sediments of apatite-P, non-apatite-P and labile phosphorus (Boström & Pettersson, 1982).

Phosphorus sorbed to aluminium and iron is released when conditions of pH and redox potential of the sediments and phosphorus concentration in pore water change; phosphorus sorbed to calcium is released according to the solubility product constant of the calcium/phosphate compound formed; organic phosphorus releases phosphorus under the action of enzymes. Microbial activity releases acid fermenta-

Table 6. Principal buffering mechanisms of phosphorus in sediments

BUFFERING MECHANISMS OF SEDIMENTARY PHOSPHORUS	
PROCESS	TEMPORARILY IMMOBILIZED FORM
Adsorption/desorption on mineral particles	Labile phosphorus
Adsorption/desorption on organic particles	Biodetritus and humic compounds
Absorption/excretion (Phosphorus exchange)	Microorganisms
Absorption for stock constitution	Microorganisms

tion products which lower the pH of pore water. The consequent dissolution of carbonate minerals allows mobilization of calcium, magnesium and associated phosphorus. Increase of pH in pond water does not affect pore water pH, as the latter is buffered by the acidic products of bacterial metabolism (Marsden, 1989).

Phosphorus release and trophic level of a water body

Phosphorus immobilization in sediments is dependent on the saturation of fixation sites, while exchange of nutrients is dependent on the trophic level of the water body. Soil profiles in the pond mud give indications of the amount and type of accumulation occurring, and of the depth of oxygen penetration.

Eutrophication is the result of nutrient enrichment, especially phosphorus, which has accumulated on the pond bottom beyond the buffer capacity of the sediments. This excess loading entails a shift from aerobic to anaerobic metabolism in sediments, which further reduces the retention capacities of the latter.

Factors in sediments promoting desorption of phosphorus

Desorption is a readjustment of sorption equilibria and the factors which influence the sorption equilibria of phosphorus can be separated into those promoting removal of phosphorus from sorption sites and dissolution of phosphorus, and those promoting immobilization or sorption of phosphorus (Table 7). Amongst the factors promoting mobilization, different effects may be distinguished: those which modify the physical and chemical conditions required for sorption, and those which increase the concentration of soluble phosphorus, leading to saturation of sorption sites and thereby increase the concentration of soluble phosphorus.

Physical and chemical factors promoting sorption include temperature, pH, redox potential, dissolved oxygen concentration, and iron content of the sediments. They were discussed extensively above.

Pore water as a source of phosphorus

Interstitial phosphorus concentrations reflect the amounts of phosphorus sorbed onto sediments and the type of sorption. Loosely sorbed labile phosphorus and dissolved phosphorus are in a steady equilibrium. Soluble phosphorus is the phosphorus species most easily transferred out of sediments. The internal loading with regards to phosphorus refers to the amount of phosphorus immobilized by sediments, accrued from the pore water phosphorus. Saturation of sorption sites entails a saturation of pore water which brings about a concentration gradient appearing at the sediment-water interface and a directional diffusion flux from sediment to water.

Orthophosphate is the main form of phosphorus mobilized from sediments, but soluble forms of organic phosphorus, characterized by molecular weights of around 20,000 to 30,000, also ooze out of sediments enriched with organic matter (Enell & Löfgren, 1988). Physical mixing of sediment with water, as a result of turbulence acting on the surficial layers of sediments, promotes exchange of pore water with pond water and cause resuspension of sedimented particles (Gunatilaka, 1982). These aspects of phosphorus transport will now be discussed.

Aerobic and anaerobic transfers

Phosphorus release from sediment to water depends on prevailing conditions in sediments and in water: distinction is made essentially between anaerobic versus aerobic sediments and aerobic versus anaerobic water. Water is considered aerobic when dissolved oxygen concentration is higher than 2 mg l^{-1} (Marsden, 1989). Shifts between oxic and anoxic conditions may happen overnight when the supply of oxygen to the sediment is interrupted or following an algal crash. It results in an alteration of the patterns of phosphorus release and immobilization.

The sediment-water interface forms a physical and chemical barrier. Crossing this barrier involves a change of phase: the sediments are solids (or semi-solids), while water is liquid. Moreover, a chemical borderline is set by the redox-cline or oxycline, which is also often situated near the interface. The energy required to cross these borders comes from the establishment of concentration gradients, which promotes the diffusion of interstitial solutes which do not possess electrochemical charges. Molecular diffusion and gaseous ebullition, especially of methane,

Table 7. Factors affecting phosphorus release from sediments

FACTORS AFFECTING RELEASE:	promoting	impeding
Nature of the sediment	non-apatite minerals, (rich in aluminium and iron)	apatite minerals or calcareous rocks
Redox potential	negative redox values	positive redox values
pH	light acidic pH and slightly reducing conditions	high pH and presence of alkaline salts
Phosphorus loading or trophic level of the water body	saturation of the sediment with phosphorus	low total phosphorus concentration
Microbial activity	high microbial activity	low microbial activity
Factors affecting microbial activity	high temperature enhancing microbial activity	low temperature
Primary production, dissolved oxygen production	good oxygen supply to sediment ensuring continuous aerobic conditions in sediment	low oxygen supply to sediment and onset of anaerobic conditions in sediment
Supply of organic material to the sediment	high sedimentation rate of organic material of good nutritional value	low sedimentation rate

take place that way. Another way to cross the barrier is by the destruction of the protective oxidized surface. This microzone is destroyed by the onset of anaerobic conditions. Lastly, any turbulence, causing resuspension of sediments, insures transport of sediment material and pore water which mixes with pond water and permits release of any soluble phosphorus (Gunatilaka, 1982).

Importance of phosphorus release rates

Diffusion coefficients of sedimentary phosphorus across the interface of sediments are in the range of 1 10^{-6} cm^2 s^{-1} or 1 μmole PO_4^{-3} m^{-2} d^{-1} (Pomeroy *et al.*, 1965) or 0.6 10^{-6} to 2.2 10^{-6} cm^{-2} s^{-1} (Marsden, 1989). This transport is quite slow, but is considerably modified by ancillary processes, thereby being increased by several orders of magnitude. Release rates of phosphorus in aerobic sediments and in turbulent waters range from 1 to 2 mg m^{-2} d^{-1} to 40 mg m^{-2} d^{-1} and sometimes up to 550 mg m^{-2} d^{-1} (Marsden, 1989).

In laboratory conditions, recorded phosphorus effluxes from sediments, over defined time periods, were in the range of -7.7 to 1330 μmole PO_4^{-3} m^{-2} d^{-1}. They were not correlated to phosphorus concentrations, but to the NH_4 and CH_4 concentrations of pore water, which strongly suggests that phosphate flux is linked to mineralization processes, of which methane and ammonia are by-products (Sinke *et al.*, 1990).

Diffusion of phosphate is promoted by movement of groundwater; sediments release phosphate, when the movement of water is either upward or downward, even though in the latter case, the movement of water is in the opposite direction (Van Liere & Mur, 1982). Upward migration of phosphate might be hampered by adsorption and chemical precipitation.

Phosphorus is also transported by benthic organisms. Sorbed to biodetritus, it is temporarily immobilized; consumption of biodetritus by detritivorous organisms results in its exportation from the sediments. Part of the phosphorus ingested by these organisms is assimilated, the other part is excreted. This biodetritic phosphorus is therefore more mobile than when sorbed to colloids.

Gaseous ebullition

Methane, hydrogen sulphide, carbon dioxide, and molecular nitrogen are gases produced by microbial metabolism. When gas production and pressure conditions allow bubble formation, the gases rise and, in their upward movement, transport soil particles and dissolved substances. Spotte (1979) observed that about 90 % of the reactive phosphorus is removed from aquarium water in a time lapse of 24 hours by a normal aerator. Turbulence and disturbance induced by wind and benthic organisms cause the resuspension of sediment. Ions and solutes adsorbed to the disturbed material are desorbed and released to the water. On the other hand, soil particles with a high sorption capacity can, during their migration phase through water, sorb important amounts of dissolved phosphorus and increase phosphate removal from water (Boström *et al.*, 1988a).

Changes caused by resuspension of soil particles

Perturbations, such as turbulent mixing, bioturbation, and reworking of sediments caused by activities of benthic organisms, affect sediment characteristics and exchanges with water. With regard to animal activities, the depth to which sediments are affected varies. Turbulent mixing affects only the loose surficial strata. The water layer affected by sediment turbulences is sometimes called the lutocline, the interphase between the 'clean' water and the sediment-water mixture.

Effect of benthic organisms on sediment-water exchange

The activities of benthic organisms on and in the sediments increase the penetration of water into sediments and thereby the exchange between water and mud (phosphorus fluxes can be increased by a factor 5 to 10). The specific effects of organisms on sediments is termed bioturbation and involve transporting activities, including advective transport of both solutes and particulate matter (Andersson *et al.*, 1988).

Benthic activities include irrigation of the sediments by the burrow holes of benthic macroinvertebrates, and ventilation of the burrow holes by water currents caused by ciliary and muscular actions. These water movements are important in gas exchange, and in the removal of metabolites, egested wastes and detritus. Interstitial water is pumped in and out and mixes with the water in the burrow holes.

Benthic organisms (the composition of the benthic fauna is described in the paragraph on organic matter decomposition, p. 50) ingest and transport sediments (Rippey & Jewson, 1982). The transport of sediments from the surface to deeper strata removes organic rich (or easily degradable) sediment of high nutritional value from the surface, where decomposition proceeds in oxic conditions and is submitted to intense microbial activity, to deeper strata, where conditions are reducing and microbial activity diminished. The penetration of dissolved oxygen deeper into sediments (Rippey & Jewson, 1982) entails a higher sorption of phosphate onto the sediments; the release rate of phosphorus by sediments is decreased. The transport of sediment in the opposite (upward) direction has an opposite effect on phosphorus release, but the organic material carried from below to the surface is generally refractory (Kamp-Nielsen *et al.*, 1982). The effect of benthic organisms on the sediments varies according to their mode of life and their activities. Detritivorous organisms recycle phosphorus trapped in the biodetrital complex (Doremus & Clesceri, 1982). Their transport of particulate material is considerable.

In shallow ponds, the mixture of sediment and water by bioturbation and other disturbances allows contact between the sediments and the trophogenic layers of the water, which promotes exchange of nutrients, especially of phosphate, and permits a greater mobilization of sedimentary phosphorus (Boström *et al.*, 1988a) (see also p. 179).

The mass balance models of phosphorus in sediment and water generally do not take into account bioturbation and other sediment disturbances.

Removal of phosphorus from sediments by epipelic algae

Epipelic algae provide for their needs by removal of phosphorus from sediment. Superficial sediments in the euphotic zone are submitted to diel formation and destruction of an oxidized microzone. During photosynthetic activity, the surficial sediments are oxidized due to the release of oxygen, a by-product of photosynthesis. Phosphate is prevented from diffusion by the presence of this oxidized microzone. At night, this microzone is destroyed and phosphorus is taken up by the algae. On sediment samples from Lawrence Lake, an oligotrophic lake in Michigan, U.S.A., phosphorus efflux clearly showed these two distinct phases (Carlton & Wetzel, 1988). Advantage and disadvantage of periphyton and benthic algae over planktonic algae with regard to phosphorus uptake is discussed later (p. 106).

Transfer and phosphorus regeneration

Transfer of nutrients, especially phosphorus, from sediments to overlying water is often understood or depicted as a nutrient regeneration, benefitting the organisms of the water column. The importance of this regeneration in the supply of nutrients to the phytoplankton can be assessed by comparing the benthic fluxes to the uptake of nutrients in the water. This process contributes some nutrients, especially phosphorus, required for primary production. The regeneration of phosphorus is mainly effected by release of phosphorus sorbed to iron and by mineralization of organic matter. Interstitial phosphorus, assessed on the basis of C/N ratios, originates, to a major extent (50 %), from organic matter decomposition. The other part is supplied by the dissolution of iron(III)-phosphate compounds.

This regenerated phosphorus can diffuse, be sorbed, form complexes with organic or inorganic ions present, or precipitate as mineral. Significant correlations between methane, a final product of anaerobic organic decomposition, the dissolved reactive phosphorus, ammonia, calcium ions and iron existed in the eutrophic Loosdrecht lakes, Netherlands, and these concentrations increased with depth (Sinke *et al.*, 1990). This shows that available phosphate is supplied by decomposition.

Assessment of bioavailability of phosphorus

The amounts of total phosphorus found in sediments are often important. But this information is of little use to the fish farmer who wishes to know whether the stocks of phosphorus in the sediments can be mobilized and made available to primary production in his pond. As discussed earlier, phosphorus bound to different carriers can be determined analytically by sequential extraction, but, unfortunately, there is no concordance between the different fractions of phosphorus determined chemically and the bioavailable phosphorus, i.e. the amount available for primary production (Twinch & Breen, 1982; Klapwijk *et al.*, 1982; Logan, 1982). Bioavailability can be defined as the sum of phosphorus directly available, i.e. soluble reactive phosphorus and labile phosphorus, accrued from the phosphorus yielded by natural processes, such as desorption, dissolution, autolysis and hydrolysis. The specific rate of mobilization of each fraction of phosphorus depends on the total amount immediately available and on the processes by which phosphorus is released from its bonds. Distinction must be made between actual and potential or maximum availability.

Several methods have been used to assess available phosphorus. The first approach was to search for a concordance between chemical extractions and amounts of available phosphorus, but the correspondence was poor and unsatisfactory. Therefore, autotrophs (bacterioplankton and algae) were cultured on the sediments, directly or diluted with water, to determine how much, and from which fractions, the organisms were capable of extracting phosphorus. These bio-assays are time-consuming and tedious and are therefore not common. Also, they are biased by the fact that they do not exactly conform to reality, insofar that the algae commonly used for these assays are not the same as those in the pond; algal species have specific requirements for growth, and the response of a natural population might be drastically different from that of any given species. Another approach to bioavailability of phosphorus is through factors responsible for mobilization of phosphorus under natural conditions (Boström *et al.*, 1988b).

In the case of algal culture in the presence of sediments, the amount of phosphorus removed from the sediment is determined. The culture medium does not contain any phosphorus other than the sediment, and the algae have been phosphorus-starved before the beginning of the experiment. Algae in general use are *Scenedesmus* and *Selenastrum*. A quantitative budgeting of particulate and interstitial phosphorus indicates the amount of phosphorus removed by the algae. The phosphorus taken up by algae consists mainly of non-apatite inorganic phosphorus, i.e. phosphorus sorbed to aluminium and iron (Young & DePinto, 1982; van Eck, 1982) and clay (Klapwijk *et al.*, 1982). Little of the phosphorus linked to carbonates appears to be usable by the algae (Klapwijk *et al.*, 1982; van Eck, 1982). Phosphorus associated to apatite does not support any algal growth. Organic phosphorus is mostly biologically inert, though it may be hydrolysed by phosphatases. This is in contradiction with observations; in many situations, phosphates are supplied by organic matter decomposition. This apparent contradiction can be explained by the intermediary role of microorganisms in the transformation of organic matter and the regeneration of nutrients to the ultimate benefit of autotrophs (Boström *et al.*, 1988b).

Phosphorus compounds taken up by algae are orthophosphate and certain organic compounds, such as phosphate monoesters with an organic part of low molecular weight, and more complex molecules, such as complex organic phosphates. They cannot, however, make use of ATP or DNA. Portions of organic phosphorus of high or low molecular weight (HMWP and LMWP) are only used in an erratic way (Boström *et al.*, 1988b).

Effect of pond fertilization on pond bottoms

Fertilization of ponds will be discussed later, but the effects on the sediments can be outlined here. Phosphates in water will be sorbed by soil colloids and saturate sorption sites in the soil. Pond soils which are saturated with phosphorus are able to support higher phytoplankton populations for a long time. In Auburn, Alabama, U.S.A., Boyd & Musig (1981) observed that decrease in phosphorus concentrations in pond water during the first seven days after impoundment is negatively correlated to the initial phosphorus concentration of the soil. Large amounts of phosphorus immobilized in soil decrease the ability of sediment to remove inorganic phosphorus from the water and make each addition of phosphorus through fertilization more available for production.

Importance of phosphorus in sediments

1. Total phosphorus concentration in sediments is often high. This is the consequence of an internal loading process.

2. Sedimentary phosphorus is separated into interstitial or particulate. Interstitial phosphorus is, for the most part, the soluble reactive phosphorus (SRP), while particulate phosphorus is divided into different fractions, namely labile phosphorus, iron+aluminium associated phosphorus, calcium associated phosphorus and organic or biodetritic phosphorus. The relative proportion of each of these fractions can be determined by sequential extraction and chemical characterization.

3. Diagenetic transformation of phosphorus in the soil includes the mineralization of organic phosphorus, the sorption of phosphorus and eventually the authigenic mineral formation. Each of these transformations proceeds on its own time scale and specific kinetics; the former two are relatively rapid, while the latter occurs more slowly.

4. Most sediment phosphorus has an organic origin and the internal loading processes are of paramount importance for the saturation of sediments with phosphorus.

5. The most important sorption process in soil is the bonding of phosphate to ferric oxyhydroxides in aerobic sediments. The phosphate is released from its bond when the sediments become anaerobic and the iron is reduced to its ferrous form. All sorption processes result in a removal of soluble phosphorus, to a solid form, and are characterized by their reaction kinetics and

irreversibility or reversibility.

6. The sediment-water interface forms a barrier made of an oxidized microzone, preventing release of free phosphate from reduced sediments below.

7. The immobilization capacities of soil are linked to the sediment characteristics, to environmental conditions, and to the saturation of sediment with phosphorus. Sediments rich in iron have better 'labile' sorption abilities than sediments rich in calcium, which precipitate insoluble phosphate compounds. Environmental conditions which promote mineralization of organic phosphorus, without deterioration of the required living conditions (oxygen supply or availability of other relevant nutrients, for instance) for microorganisms, will lead partly to immobilization of phosphorus in biotic forms, which can be then exported out of the sediments by direct consumption by macro-detritivorous organisms. The phosphorus in excess of microbial needs will be immobilized by physical or chemical sorption. If environmental conditions deteriorate to anaeroby, the pace of organic decomposition will slow down and all accumulated phosphate in labile sorbed forms will be, little by little, released and diffuse to the overlying water. The sequence of diagenetic transformation is dependent on soluble phosphorus fluxes and kinetics of each sorption process, the reaction material and prevailing conditions.

8. The release rate of phosphorus by diffusion from sediments is generally low, but is increased by several orders of magnitude by ancillary phenomena, such as gaseous ebullition, and resuspension of sediments by wind induced turbulence and bioturbation. Bioturbation is the reworking of sediments due to activities of benthic organisms.

9. The portion of phosphorus in sediments available for primary production is called bioavailable phosphorus. This bioavailable fraction can be assessed with bio-assay methods and concords roughly with the portion of iron-aluminium associated phosphorus.

3.1.4. Sulphur

Importance and forms of sulphur present in sediments

Sulphur is an important element for living organisms, but is rarely in short supply. Sulphur is present in organic matter and released by decomposition. Sulphate is used as an alternative electron acceptor and plays a role in oxidative processes.

Organic matter decomposition generates sulphide compounds, which are transformed, for instance, into sulphate and hydrogen sulphide. In organic compounds, sulphur is bound either to carbon or to sulphate esters.

Sulphur is present in different oxidation stages, as thiol, sulphide or thioester (-2), disulphide (-1), sulphoxide (0), sulphone (+2), sulphonic acid (+4), and sulphate esters or sulphate (+6). All these compounds are intermediary products of decomposition. Not much is known about the different steps of the transformation processes of sulphur compounds in the sediments (Urban & Baker, 1989; Kelly & Rudd, 1989); sulphide ions eventually form iron sulphide and pyrite (Callender, 1982; Schuurkes & Kok, 1988).

Sulphate and organic matter decomposition

Microbial organic matter decomposition in sediments makes use of a sequence of available electron acceptors, as discussed earlier (*vide* p. 18). Sulphate is resorted to when sediments are fully anoxic, redox potential has reached a negative value of 60 to 100 mV (Boström *et al.*, 1988a) or -200 mV (Seitzinger, 1988), and other acceptors have been exhausted. Sulphate is partly supplied from the water column (Nembrini *et al.*, 1982). Sulphide ions, resulting from the reduction of sulphate, react with iron (II) to form iron sulphide (Nembrini *et al.*, 1982). Reaction is also possible with iron oxyhydroxide resulting in the same ferrous sulphide. Iron sulphide reacts further with elementary sulphur to form pyrite. Sulphate reduction is a proton-consuming reaction and therefore increases alkalinity in soils.

Sulphate oxidation and alkalinity increase

The loss of sulphate is accompanied by an equivalent loss of protons, which represents an alkalinity gain, but not a net production of new alkalinity. It is a transfer, from sediment to water, of acid neutralizing capacity. Ferrous ions in the sediments accumulate in the pore water of anoxic sediments and diffuse upward till they meet with oxygen, where they precipitate as ferric oxide. This oxyhydroxide gel is quite common and occurs in high concentrations at the sediment-water interface.

In sediment rich in carbonate, iron precipitates to form ferrous carbonate and consumes alkalinity. The iron oxyhydroxide can be reduced and constitutes a potential source for acid neutralization when coupled to a process of removal of soluble ferrous iron and irreversible fixation in a solid compound. This process is called the acid neutralizing capacity of iron. Iron oxyhydroxide is therefore assimilated into a solid-phase alkalinity (Anderson & Schiff, 1987).

Microorganisms responsible for decomposition

Bacteria responsible for sulphate reduction in free sulphide are not sensitive to pH changes and easily support a pH varying between 6 and 4 (Schuurkes & Kok, 1988). Mazumder & Dickman (1989) observe that purple and green sulphur bacteria require light and an absence of oxygen. They are found in the upper part of the hypolimnion or in the metalimnion.

3.1.5. Iron

As shown in the paragraphs on phosphorus and sulphur, iron holds a key position in the cycling of those nutrients. Iron is specially active when soluble and therefore only the portion of iron which is soluble and easily desorbed is to be considered. At high concentrations, iron (II) is toxic for most organisms. Iron activity is often discussed in the context of chelators.

Iron and manganese concentrations vary with sediment depth, as a result of cycling and transport. Their oxidation stage, as well as the position of the redox barrier, are important for their transport. This redox barrier is however not a sharp border, but more a discontinuous, sometimes diffuse line. When the soluble ferrous ion migrates and meets with oxic conditions, it is oxidized and precipitates. However, ferrous ions can also be included and occluded in bonds that resist oxidation, such as $FeOH^+$ or $FeCl^+$. The compounds most commonly found in hydromorphous soil are $Fe(OH)_3$, with a solubility product coefficient of 10^{-35}, $Fe(HCO_3)_2$ and $FeSO_4$. Iron carbonate and bicarbonate are used as energy sources by some microorganisms.

Iron (II) ion transport is achieved by various diffusion processes. The vertical concentration profile in the sediment gives an indication of the specific type of supply and transport. The two most common profiles are either concave or convex, and reflect different rates of transport in the sediments (Davison, 1982). Physical modifications at the interface affect transport and removal of products. Oxidation and transport kinetics of different ions vary; manganese ions are oxidized 50 times more slowly, but are more rapidly transported than iron (Davison, 1982).

The dissolution of iron in soil requires a drop in pH or in redox potential. Iron reduction is under microbial control.

Importance of sulphur and iron

1. Sulphur and iron are not often specifically considered because their cycles are not well understood. Sulphur forms many intermediary products. Soluble iron has great chelating capacities and forms part of many complexes. Soluble iron is toxic at high concentrations.

2. Sulphate and iron are both electron acceptors; iron oxyhydroxide is preferentially used to sulphate. Sulphate consumes protons when oxidized and, in combination with iron hydroxide, produces alkalinity in the soil. This is termed the acid neutralizing capacity of iron.

3. Sulphate and iron oxidation are microbial processes.

3.2. Organic matter and humus

3.2.1. Importance of organic matter

Origin of organic matter

Pond organisms which die in the course of the production period sediment to the bottom and become part of the organic loading of the sediments. Other contributions come from dead biomass of benthic organisms and unconsumed artificial feed. Depending on the location of the pond and the course of its feeder canal, some allochthonous organic matter can find its way into the pond and eventually to the sediments. This material is mostly leaves and plant debris (Cuffney *et al.*, 1990). The amount of sedimented organic matter is a function of primary production (Cappenberg *et al.*, 1982), and the sedimentation rate changes seasonally (Rippey & Jewson, 1982).

The superficial layers of the sediments are rich in organic matter, with a content decreasing with depth, according to the rate of mineralization (Rippey & Jewson, 1982). Organic matter is mainly made of organic carbon; the decomposition of organic matter is part of the carbon cycle.

The flocculent layer as the centre of organic matter decomposition

Organic matter settles on the pond bottom and forms a flocculent layer, a stratum of loose, uncompacted fluffy material, easily resuspended by water turbulence. This material is rapidly colonized by a microbial population (epipelic community) with a high respiration rate and a high heterotrophic, and sometimes even photosynthetic, activity, composed of bacteria, benthic algae, fungi and zoobenthos (Sweerts *et al.*, 1986). This flocculent layer is the centre of detritus decomposition in the aquatic environment (Doremus & Clesceri, 1982; Adams *et al.*, 1982; Schroeder, 1987).

3.2.2. Decomposition and nutrient regeneration

Decomposition

Decomposition is a generic term for a group of processes, which transform organic material into living matter and nutrients. It can also be viewed as a nutrient regeneration process, benefitting the water column. Processes differ according to the type of organic material, prevailing conditions and the type of microorganisms present. The aim of the microorganism is to benefit from the energy and nutrients contained in organic matter.

Dying and sinking phytoplankton cells settle to the pond bottom and their decomposition starts immediately after death, by autolysis during the sedimentation of the cell. Leafy and plant material needs shredding before it can be attacked by microorganisms; this is efficiently done by benthic macroinvertebrates (Cuffney *et al.*, 1990).

Various decomposition strategies can be chosen according to prevailing environmental conditions. They are either oxic or anoxic, oxidizing or reducing. The spectrum of organisms found in sediments is great and still, to a large extent, unexplored and unknown. The most active decomposition occurs at the sediment-water interface.

Organic matter is not a material of defined value; its nutritional value decreases with time, labile compounds being quickly hydrolysed, and with depth, as fresh sediments bury older ones. The ultimate forms are humic and fulvic compounds of no nutritional value.

Microorganism population growth is induced by the availability of organic material, and the size of the population is a function of the presence of material and suitable environmental conditions.

Decomposition is divided into three main steps, in which the end products of one process are inputs to the next process (Fig 15). This has been termed the trickle-down theory (Seitzinger, 1988). The first step is a depolymerization of organic polymers, e.g. polysaccharides, lipids, DNA, RNA, etc. into monomers, i.e. monosaccharides, amino acids, fatty acids, and nucleotides. The second step is a fermentation, releasing carbon dioxide and other fermentation gases, and yielding acetate, alcohols and methylamines. The last step is the terminal metabolism or the consumption of fermentation end products by terminal consumers, releasing methane,

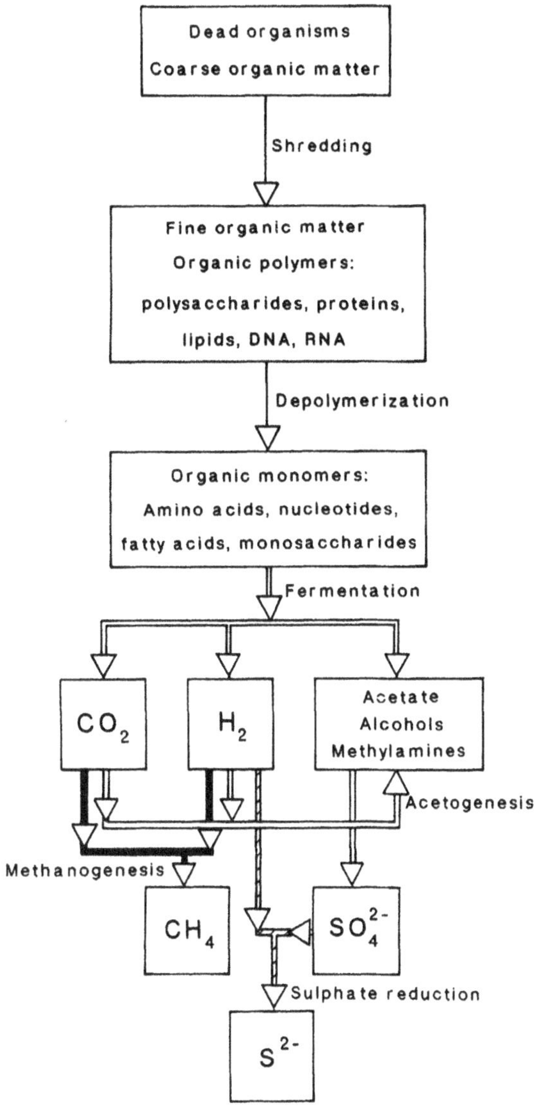

Fig. 15. Main steps of the decomposition process in sediment, including the shredding, depolymerization and fermentation of organic matter. Fermentation includes different alternative processes, depending on prevailing conditions and on the presence of electron acceptors (modified from Seitzinger, 1988).

carbon dioxide and reduced electron acceptors, by processes such as methanogenesis, acetogenesis, and sulphate reduction (Seitzinger, 1988). Each process requires specific conditions, with regard to, for example, pH, redox potential, or availability of some electron acceptor, and of course, availability of organic matter. The amount of energy released is a function of the type of reaction taking place. The aerobic condition of the sediment is largely determined by the input of organic matter. Planktonic production is the most important factor influencing sediment oxygen demand (Marsden, 1989). The amount of energy released by the transformation of organic matter has been established as follows (Seitzinger, 1988):

Electron acceptor	Energy released (in kJ mol^{-1} CH_2O)
O_2	475
NO_3^-	448
MnO_2	349
FeOOH	114
SO_4^{2-}	77
CO_2	58

Oxidation of organic carbon needs electron acceptors and this can be a limiting factor. Two groups of electron acceptors can be distinguished: those in soluble or diffused form, such as oxygen, sulphate and nitrate, and those in particulate form, such as iron and manganese oxides (Kelly *et al.*, 1988). In marine water, sulphate reduction is a more important process than in freshwater, where methanogenesis takes over, as salinity and sulphate concentration decrease. Energetic yield for the metabolization of organic matter through sulphate reduction is higher than by methanogenesis (Capone & Kiene, 1988).

Composition of the benthic microbial fauna

The benthic fauna is by and large made of detritivorous organisms, which live on the organic matter present. This organic matter is first ingested by the benthic macrofauna, which reduces the size of the organic material and makes it accessible to microbial attack. Along with the ingested material, the macrofauna consumes microorganisms which have colonized the material. The microorganisms living in the sediments are unfortunately not yet satisfactorily described and appear important in density and diversity. They have been classified in groups according to their basic metabolism. Often, they are difficult to separate from organic material; the term biodetritus encompasses the organic matter along with the microbial biomass living in it (Johnson *et al.*, 1989).

Bacterial populations of fish pond waters and sediments have not been inventorized and there is a dearth of information on them. A study on the bacteria present in the water and sediment of a carp pond in Japan identified 489 different strains, which were capable of decomposing substrates as different as casein, tributyrine, starch, cellulose and chitin. The first three substrata are also those most common in fish pond sediments and in unconsumed fish feed (Sugita *et al.*, 1987).

Aerobic decomposition and sediment oxygen demand

At the interface and in the surficial layer of sediments, conditions are aerobic: oxygen is provided by the overlying water. Decomposition processes are most active in the flocculent layer, where microbial respiration coincides with a demand for electron acceptors; the oxygen demand there is greatest, with the highest rate of oxygen depletion occurring at this interface (Adams *et al.*, 1982, on sediments of Lake Erie). Total oxygen requirements are termed the sediment oxygen demand (SOD), and indicate the importance of decomposition processes happening in the sediment (Jurkowska, 1987). Primary production is significantly related to annual sediment oxygen uptake. Nutrient-rich waters with high levels of primary production have anaerobic sediments; their rate of oxygen supply is not sufficient to satisfy the oxygen demand of mineralization (Rippey & Jewson, 1982; Marsden, 1989).

The oxygen exchange across the sediment-water interface depends on the renewal of the water in deeper portions of the pond. The transfer kinetics of oxygen have been studied by Hall *et al.* (1989). Oxygen supply to the sediments by diffusion processes has been estimated to reach a maximum of 4 g m^{-2} d^{-1} of oxygen. At Orissa, India, the oxygen demand of sediments in fresh water fish ponds was found to vary between 0.39 to 3.39 g m^{-2} d^{-1}, which represents some 5.1 to 29.5 % of primary production and 5 to 30 % of community respiration (Oláh *et al.*, 1987). This is in line with reported undisturbed sediment oxygen uptake rates in fish ponds in Israel peaking at 3 to 4 g m^{-2} d^{-1} (Schroeder, 1987). This represents about 2 g m^{-2} d^{-1} of organic carbon processed aerobically. Based on cellulose digestion in anaerobic conditions, the organic carbon processed would have reached 3 to 8 g m^{-2} d^{-1} (Schroeder, 1987), which is in the previously recorded range of oxygen consumption per molecule of methane transformed of 1:1.2-1.8 (Adams *et al.*, 1982). The relationship between sediment oxygen uptake rate and oxygen gradients in pore water argues against a molecular or eddy diffusion process operating alone (Rippey & Jewson, 1982). In sediments where the activity of benthos is high, vertical transport is best described by advection due to irrigation. Bioturbation has been found to increase the oxygen transfer rates by a factor 1.5 to 10, according to the nature of the sediments (Schroeder, 1987).

Sediment oxygen demand is divided into two components, biological and chemical. The biological SOD corresponds to heterotrophic respiration, i.e. the respiration of microorganisms and benthic macroorganisms. The chemical SOD pertains to the action of all reducing substances or electron-donors, such as iron (II), manganese (II) or sulphide, which rapidly oxidize in the presence of oxygen.

Sediment oxygen uptake is measured by oxygen profiles in the sediment; the oxygen concentration in sediments is considered to result from the difference between oxygen supply via diffusion across the interface, and oxygen consumption within the sediment. It is interesting to quantify the proportion of SOD due to the upward or benthic flux of dissolved reduced electron acceptors, such as Fe^{2+}, Mn^{2+}, NH_4^+ and CH_4, which are mobile and accumulate beneath the surface of sediments with low redox potential or with high decomposition activity. In the sediment of Lake Erie, the transport of these reduced constituents represented a sizeable oxygen demand, ranging from less than 1 % for ferrous iron and manganous manganese to as high as 26 % for ammonia and 30 % for methane (Adams *et al.*, 1982). Their transport depended on their concentration gradients. The reductant-oxygen ratio (in moles) for these substances is as follows (Adams *et al.*, 1982): for Fe^{2+}, 1:0.25; for Mn^{2+}, 1:0.5; for NH_4^+, 1:2 and for CH_4, 1:1.5.

Variation of dissolved oxygen concentration in the water at the interface causes changes in thickness of the aerobic layer, called the biofilm. The rate of decomposition is related to biofilm thickness. The diurnal exchange rate of the sediments versus the oxygen concentration and temperature in the water show a hysteresis effect due to the amplification of the discrepancy between exchanges under transient and under steady state conditions. In transient conditions, when oxygen supply increases, accumulated anaerobic end-products are oxidized, owing to the increase of the aerobic layer (Jeppesen, 1982). The diel oxygen fluctuations in the water due to phasic photosynthetic activity of autotrophs indicate that transient conditions are more realistic than steady state.

The importance of sediment oxygen demand increases with pond shallowness. Renewal of the water at the interface to increase benthic fluxes and to supply fresh dissolved oxygen may be of prime importance in fostering regeneration of nutrients in a pond. Practical means of implementing this water circulation will be discussed later (e.g. p. 200).

Anaerobic decomposition

Organic carbon mineralization in anaerobic conditions leads to methane as end product and, in aerobic conditions, to carbon dioxide (Cappenberg *et al.*, 1982). The various mineralization processes in water, and on and in the sediments, and the transfers across the interface constitute the carbon cycle. Carbon cycling in aquatic environments is the means by which the autotrophic populations are supplied with the carbon needed for production. Our knowledge of the various anaerobic breakdown pathways and turnover rates of carbon, and between its input and output in sediments is inadequate (Cappenberg *et al.*, 1982).

Nutrient enrichment experiments have shown that carbon can be limiting the growth of microorganisms. Growth of microorganisms attached to refractory organic matter, such as lignin, is not limited by nitrogen or phosphorus, as enrichment with these nutrients has been found not to enhance degradation. The limiting factor is the quality and the quantity of the organic carbon present. Carbon is limiting when the C:N ratio falls below 10:1 (Peters *et al.*, 1987). In sediments, the C:N ratio increases with depth (from 11:1 to 14:1) indicating that the proportion of organic carbon to organic nitrogen compounds increases too (Cappenberg *et al.*, 1982).

In waters with a high primary productivity, most organic matter comes from algal cell walls, which settle from the photogenic zone to the sediment (Cappenberg *et al.*, 1982; Rippey & Jewson, 1982). The decomposition is engineered by anaerobic bacteria and leads to the production of acetate, as an intermediary metabolite (Fig. 15). Acetate is further broken down either by anaerobic, sulphur-reducing, acetate-oxidizing bacteria in the upper layers of the sediment or by acetate-fermenting methanogenic bacteria in the lower layers. In the overlying water, acetate is broken down under aerobic conditions. The most active sedimentary transformation occurs in the lower sediments (Cappenberg *et al.*, 1982).

Acetate diffuses from the lower sediments towards the surface and high concentrations indicate an acetate point source localized at an intermediary depth. Acetate sinks exist both in the deeper layer (due to methanogenic activities) and at the sediment interface (due to aerobic, metabolic activities). Part of the acetate in sediment diffuses downwards, where it accumulates (Cappenberg *et al.*, 1982).

Mineralization rate is the combination of various simultaneous processes and their kinetics: production of acetate, methanogenic transformation, oxidative acetate transformation, and downward migration. To this should be added all other transformations discussed earlier, involving transformation of nitrogen and phosphorus. All these processes are drastically modified by bioturbation and physical and chemical disturbances of the sediments, such as gaseous ebullition and wind-induced turbulence.

Acetate is not the only intermediary product in organic carbon decomposition: lactate and formate are produced as well. Lactate, which is produced on a smaller scale, is a more voluminous molecule and therefore diffuses more slowly. Formate is unstable and transforms easily into hydrogen and carbon dioxide, which serve as precursors for the reactions of the hydrogen/carbon dioxide fermenting methane-producing bacteria in the upper sediments. These last intermediates, which are less oxidized than acetate, play a less important role (Cappenberg *et al.*, 1982).

The breakdown of complex organic matter into low-molecular weight intermediates such as acetate is a rate-limiting step. The ultimate transformation of acetate to carbon dioxide or methane has a rapid turnover rate. In organic matter decomposition, the major part of the input is returned to the overlying water by gaseous diffusion and complementary processes of gas transport (Cappenberg *et al.*, 1982).

The anaerobic decomposition processes lead to an accumulation of reduced electron acceptors, which accumulate in deeper horizons and migrate upwards to be oxidized. These aspects have been discussed earlier (p. 9).

Decomposition rate and nutritional value of organic matter

Sedimented organic matter is often of poor nutritional quality and detritus decomposers may be nutrient-limited. Nutrient requirements of surface sediment decomposers can be satisfied in two ways: diffusion from lower sediments or removal from the overlying water column. In the sediment, pore water is the intermediary compartment, concentrating and conveying soluble nutrients (Doremus & Clesceri, 1982).

Decomposition first involves an enzymatic attack, which detaches the most labile radicals from the organic compounds. The first to be removed are the nitrogen radicals, by ammonification. These transformations are reflected in the evolution of the C:N ratio, which increases with depth.

Mineralization is confined to the surficial layers of sediments, where acid-soluble and hydrolyzable compounds are metabolized, with removal of carbon and nitrogen atoms. The C:N ratio reaches its maximum value in the lower portion of this layer. The bacteria active here are anaerobic, belonging to the group of sulphate-reducers in marine and brackish waters, or to the acetate-fermenters or acetate-oxidizers in freshwater. They are almost always present, though their activities can be latent, depending on the volume of substrate present.

In deeper strata of the sediments of a tropical lagoon (Biétri bay, in the urban zone of Abidjan, Côte d'Ivoire), decomposition was slowed down by geochemical complexation processes and by immobilization of organic matter. Organic matter is sequestered in clay-humate complexes within fine mineralogical particles. Humic acid undergoes polymerization, in which compounds of molecular weight below 100,000 make up 70 % of the total. In anoxic sediment, the mineralization processes by heterotrophic organisms diminish in favour of humification processes and immobilization of organic detritus (Guiral, 1984). This humic material is composed of organic complexes, largely uncharacterized chemically, in a reduced and dispersed state, so that they are not easily separated from the mineral and the other undecomposed organic material present. Humic matter is a gel-like product, black or dark-brown in colour, synthesized by microorganisms. Being a colloid, this matter is able to sorb solutes.

In eutrophic waters (Ebrié Lagoon, Côte d'Ivoire), the resuspension of sediment in well oxygenated water reactivates heterotrophic bacteria, which drastically increases the oxygen consumption in the overlying water. This may have serious consequences for life in a pond (Carmouze & Caumette, 1985).

Effects of decomposition processes on sediments

The decomposition of organic matter in sediments modifies the conditions of the latter, inducing a decrease in pH and an increase in the concentration of exchangeable aluminium, which affects phosphate sorption and benthic fluxes (Twinch & Breen, 1982). The rate of decomposition of organic matter is seemingly reduced by acid conditions, although studies in acidified Swedish lakes do not support the hypothesis of decreased microbial activity in acidified freshwater sediments (Hasselrot & Hultberg, 1984).

The surficial flocculent layer prompts changes in the redox potential profile beneath it, which are essentially a shift from aerobic to anaerobic conditions.

Effects of benthic organisms on decomposition and benthic fluxes

Based on a survey of Polish lakes, Kajak (1988) stated that the abundance of the benthic fauna reflects the trophic state of ponds. The presence and diversity of the fauna is linked to the importance of aerobic decomposition and to pH. The number of taxa of benthic macroinvertebrates present in sediments is related to sediment pH; below 5.5, molluscs, crustaceans and larvae of mayflies are absent and the number of taxa of caddis-fly larvae (Trichoptera) and chironomids is low. This may be linked to the non-availability of calcium ions and high concentrations of free aluminium ions, which are toxic for these organisms (Friday, 1987).

The macrofauna belongs to few main groups only: tubificids (Oligochaeta Tubificidae, e.g. *Pomatothrix*), chironomids (Diptera Nematocera, e.g. larvae of *Chironomus* and *Chaoborus*), amphipods, and burrowing bivalves. Other organisms found include copepods, ostracods, nematodes, rotifers and protozoans (Andersson *et al.*, 1988). Species assemblages of benthic fauna are incompletely known, due to a lack of identification keys.

Oligochaetes of the tubificid family feed at shallow depths (3 to 6 cm below the surface) and deposit their faeces on or near the surface of the sediment (Andersson *et al.*, 1988). In lake Esrom, Denmark, their daily ingestion rate was estimated at four times their biomass (Kamp-Nielsen *et al.*, 1982). Their transport of particulate material is therefore considerable. The renewal of water inside their burrow holes, required for gaseous exchange, removal of metabolites and egested wastes (ventilatory movements), is provided by undulatory muscular contractions of the posterior end. Oligochaeta do not construct permanent galleries.

Tubificids promote exchanges at the water-sediment interface, in particular the diffusion of ferrous iron, which is transported to the surface, where it is subjected to oxidation and forms clouds of colloidal ferric oxide at the interface.

Ammonium concentration in the upper layers of a sediment can reach saturation levels. This is the result of intense ammonification by microorganisms, as a consequence of benthic macrofaunal activities, such as ingestion, egestion and mixing of the sediment with excreta.

Nitrate supply to sediment causes an increase in the disappearance of carbon, which indicates an increase in respiration or heterotrophic activity, and an intensification of microbial denitrification. Tubificids are responsible for anaerobic decomposition of organic matter, as denitrifying microorganisms are present in the intestine of these worms as well as in sediments.

Tubificids affect phosphorus cycling by increasing the redox potential and causing stronger binding of phosphate, depending on sediment type (Andersson *et al.*, 1988). They increase the mineralization rate of organic matter, promote exchange by bioturbation, especially the benthic flux of nitrogen compounds (Fukuhara & Sakamoto, 1988, in Lake Suwa, a shallow eutrophic lake in Japan), stimulate the growth of microbial biomass and of their predators, and add to the organic load of the sediment by their excreta (Kurihara, 1989, in ricefields in Japan).

Chironomids are surface filter or surface deposit feeders, which transport sediment from the interface to subsurface layers. They live in straight or U-shaped tubes. Water is pumped inside these tubes by body movements (Andersson *et al.*, 1988).

Chironomids are important with respect to biomass, production and recycling of organic matter. They influence water quality, as they drastically modify oxygen circulation above sediments, as well as oxygen penetration to sediments. By their burrowing activities, they greatly enhance the transport of dissolved substances from sediment to water. In Lake Balaton, Hungary, the presence of chironomids accelerates organic matter transformation, and organic material is also made more accessible to bacterial attack (Dévai, 1990).

Amphipods feed at the sediment-water interface, while bivalves burrow inside sediments (Andersson *et al.*, 1988).

The various effects of bioturbation can be summarized as follows: bioturbation transports sediment downward, i.e. buries organic rich sediment and prevents rapid degradation, and carries sediment upward, i.e. reexposes it to aerobic decomposition. At the same time, burrow holes facilitate penetration of aerobic water, electron acceptors and bacterial decomposers, which leads to a more efficient sediment mineralization. Other changes include redox potential increase and modifications in phosphate sorption. By ingesting material, deposit feeders increase the rate of particulate matter mineralization, though most detrital material ingested is not assimilated because of its low nutritional value (Andersson *et al.*, 1988). The shredding of fibrous organic material by macroinvertebrates makes it available to further bacterial attack (Cuffney *et al.*, 1990). However, for macrobenthic fauna, aerobic conditions are required. Under prolonged anoxic conditions, most organisms cannot survive. Under low dissolved oxygen conditions, the presence of chironomids and tubificids increases the dissolved oxygen flux.

Exchanges with the water column

The transformation of organic material is essentially a process involving the recovery of nutrients and energy for decomposer metabolism. Hence, it transforms dead organic material into living organic material, along with release of some chemical by-products. Regeneration of organic matter benefits the biotic communities of the water column primarily via consumption of benthic organisms by pelagic organisms. Besides this, and under specific conditions, excess soluble nutrients are released by the sediments (benthic fluxes). As discussed above, the most active stratum involved in decomposition is the flocculent layer, which is an entangled mixture of microorganisms and organic matter. In lake Valencia, Venezuela, this flocculent layer is intensively browsed upon by tilapias (*Oreochromis mossambica*) (Bowen, 1981). In Awassa Lake, Ethiopia, 10 cm sized tilapias (*O. niloticus*) were found to feed actively on benthic epipelic diatoms, belonging to the genera *Synedra*, *Cymbella*, *Neidium*, and *Gomphonema*. The food of tilapia of 25 to 30 cm size (standard length) has been found to consist of chironomid larvae and adult insects. On increasing in age and size, these fish change to a more herbivorous diet, though their regime may still be diversified. They consume phytoplankton, loose benthic detritus rich in bacteria, epiphytic diatoms, and periphyton (Tudorancea *et al.*, 1988).

The chemical by-products are essentially reduced electron acceptors, which migrate downwards and/or upwards and constitute benthic fluxes. They confirm the importance of the mineralization processes.

Importance of organic matter and its decomposition in the sediment

1. Organic matter settles from the water column to the sediment; this sedimentation rate is related to the primary productivity of the pond. The organic matter forms the flocculent layer on the sediments, which is a biodetrital layer, very active in decomposition and made of an entanglement of bacteria, fungi, algae and detritus.

2. Decomposition processes are varied, and depend on environmental conditions. The main two groups of decomposition processes are aerobic and anaerobic.

3. The sediment oxygen demand is an indicator of the oxygen consumption of the sediment, i.e. community respiration, especially by decomposers (microorganisms and others), and chemical oxidation by the reductants. The SOD is a measure of the importance of aerobic decomposition occurring in the sediment.

4. The rate of organic matter decomposition depends on the nutritional value of the organic matter, i.e. the nutrient content, expressed by the C:N:P ratio, and the energetic value.

5. The presence of benthic macroorganisms completely modifies the decomposition processes, in particular the sedimentation stratigraphy, the penetration of oxygen inside the sediments, and the benthic fluxes.

6. Regeneration to the water column of nutrients resulting from decomposition occurs via consumption of benthos by pelagic organisms and benthic fluxes.

Chapter 2.

Water and productivity

1. Water characteristics

Water is the culture environment for fish or other aquatic organisms. It is the physical support, in which they carry out their life functions, such as feeding and excretion. Water properties can be divided into two main categories: physical characteristics, like density and temperature, and chemical characteristics, the concentration of solutes and nutrients.

The water column is bordered below and at the sides by the pond sediment, and above by the atmosphere. Exchanges between water and sediment have been discussed in the previous chapter; exchanges with the air will now be discussed.

One of the main problems facing aquatic organisms is the low concentration of oxygen in water as compared to the aerial environment. The chemical composition of water is also much more variable than that of air and values of acidity or alkalinity can vary drastically, deeply modifying the living conditions of the organisms. The difference in density of water as compared to that of air has also influenced the shape and locomotion of aquatic organisms.

Overall water characteristics are summarized in the concept water quality, but, unfortunately, there is no universally accepted definition of this concept (see Couillard & Lefevre, 1985, for 20 different indices used). Because the specific requirements of cultivated species vary over such wide ranges, a general definition of suitable ranges of different water characteristics, or water quality criteria applicable to fish farming, is not possible.

2. Physical characteristics of water

2.1. Temperature, density and viscosity

The sun is the primary energy source on earth. The sunlight brings solar energy to ponds, which is absorbed, warming up the water mass. The low thermal conductivity of water restricts the heating to the upper water layers. Evaporation of water and diffusion of gases in solution in the water cause heat losses.

Air and water temperature depend on latitude, altitude and mean daily number of sunshine hours. Water temperature is also affected by specific characteristics, such as turbidity, wind force, plant cover, and microclimatic conditions of the immediate vicinity of the pond. Variations in water temperature at the surface closely follow changes in air temperature.

Temperature variations in shallow ponds are wider and affect the water mass to a greater extent than in deep ponds. In water bodies less than 1 m in depth, water temperature can reach 40° C or more under direct sunlight. In deeper ponds, heating is restricted to the upper layers and water stratifies thermally. Temperatures of 30 to 35° C are tolerated by fish, but above these temperatures, aquatic life is threatened. During periods of high temperatures, fish seek refuge in deep, cool water.

Increase in water temperature results in evaporation, which may entail considerable losses of water from a pond, depending on degree of humidity of air and wind velocity. Edwards *et al.* (1981) recorded evaporation rates of 25 l m^{-2} d^{-1}, in a humid tropical climate (Bangkok, Thailand).

Turbid water increases more rapidly in temperature than non-turbid water, regardless of whether turbidity is mineral or organic. Under the same sunshine conditions, the increase in temperature is more rapid in water rich in plankton than in water poor in plankton.

Temperature is a factor of great importance for aquatic ecosystems, as it affects the organisms as well as the chemical and physical characteristics of water. For example, the metabolism of cold-blooded organisms or poikilotherms is dependent on temperature, as are the solubility of dissolved oxygen, and the density and viscosity of water. There is, therefore, a need to keep pond water within some range, avoiding temperatures which are either too high or too low, and in agreement with the temperature requirements of cultured organisms. Excessively high water temperatures can be controlled by water renewal; water temperature in feeder canals is generally lower because it is not standing and poor in plankton.

Water density and viscosity decrease with temperature, while the former increases with salinity. These water properties are not commonly taken into consideration or measured, but do affect stratification (see p. 130) and motility of the aquatic organisms and seston.

2.2. Water movements and stratification

Water movements are either horizontal or vertical convective movements. Water movements connected with transport are not discussed here. They are treated in details by Coche & Van der Wal (1981).

Most ponds are impounded before stocking and no major water renewal occurs, except for occasional additions to restore a correct water level.

Effects of wind and waves

Waves are produced by wind. Their height is proportional to wind speed and fetch. Waves move a thin, sometimes even thick, layer of surface water, and homogenize water temperature. As referred to earlier (see p. 41), wind-induced turbulences even affect the flocculent layer on the pond bottom. In Lake George, Uganda, wave induced water movements cause resuspension of the sediment layer at the water-sediment interface to a depth of 5 to 14 cm (Viner & Smith, 1973). Wave action also inflicts long-term damage on dykes, situated perpendicular to the predominant wind direction.

Wind blowing across a water surface induces, besides waves, circulatory currents known as Langmuir cells. These are horizontal tubes of rotating water, their axes aligned approximately parallel to wind direction. Adjacent tubes rotate in opposite directions. Tubes of varying diameter can be present at the same time. Both wind and waves must be present to generate these roll vortices, but even a light wind over small-amplitude waves can set them up. Cells have diameters ranging from a few centimeters to several meters: at a wind speed of 5 m s^{-1}, a typical cell might have a diameter of 10 m and a surface speed of 1.5 cm s^{-1} (Kirk, 1986).

Wind action in tropical countries is responsible for a daily destruction of thermal stratification of ponds and shallow water bodies. This mixing of water affects all production parameters of a pond, and specifically photosynthetic activities (Kirk, 1986; Barbosa & Tundisi, 1989, in a shallow tropical Brazilian lake).

Convection currents

Vertical convection currents occur in all deep water bodies and appear at the change of season, when daily temperature regime changes. Water masses at higher temperature, and therefore lighter, move upward and displace other water masses, causing a seasonal overturn of the entire water mass. The superficial water temperature cools down as a consequence of a drop in air temperature, leading to the creation of warmer water layers at greater depth. This overturn mixes water thoroughly, carrying upwards nutrients released from the sediment by benthic flux and supplying deeper layers with surface water rich in dissolved oxygen. Upward flux of nutrients during overturn may be partly prevented by sorption, complexation or trapping reactions occurring during transport, as observed in Biétri Bay, in the urban zone of Abidjan, Côte d'Ivoire. As a matter of fact, organic colloids and humic matter readily sorb nutrients and solutes which come in their vicinity (Guiral *et al.*, 1989). Positive and negative responses to mixing exist for diatoms and blue-green algae respectively, though a simple relationship between mixing and gross estimates of the heterogenous phytoplankton populations is unlikely. Mixing may result in the dispersion of diatom populations to diatom-bare water (Talling, 1969).

Different types of stratification of ponds and other water bodies

Thermal stratification has been abundantly described in limnology manuals. The upper zone, in which wind homogenizes the water, is called the epilimnion. This stratum rests on a water mass in which water temperature decreases rapidly with depth. This is called the thermocline or metalimnion. The deepest water mass, termed hypolimnion, is defined by a stable temperature. Thermal stratification of water bodies has been studied extensively in temperate climates.

In tropical waters, thermal stratification exists as well, but to a minor extent. There is no seasonal overturn, as temperatures change little over the months. In shallow lakes, such as Lake Chad and Lake George, with mean depths 4 m and 4.5 m, respectively, thermal stratification is destroyed by wind action. These lakes are polymictic. In some cases, especially deeper lakes, overturn is only partial and does not affect the whole water mass; these lakes are called meromictic, such as Lake Malawi (Talling, 1969). In other lakes, such as Lake Upemba (1500 ha and 2 m mean depth), on the Lualaba river, Zaire, overturn occurs daily. These lakes are polymictic and holomictic.

A fish pond resembles a shallow lake, and daily thermal stratification occurs (Boyd, 1979; Chang & Ouyang, 1988). During daytime, sunlight heats up the water and heat is accumulated in the upper 1 m layer. Beneath this superficial layer, water temperature of the intermediary strata changes drastically over a narrow depth, located between 1 m and 1.75 m. Below this, water temperature remains stable (Fig. 16).

This thermal stratification is labile and may change at any moment, subject to water movements or eddies, caused by wind or other disturbances, such as inflow of water from a feeder canal, or heavy rain (Boyd, 1979). Chang & Ouyang (1988) observed a steady stratification during daytime and distinguished in the tropical fish pond (in the Pearl River Delta, Guangdong Province, China) the three main thermal zones defined above. In the late evening or at night (between 8 to 10 pm), a convective

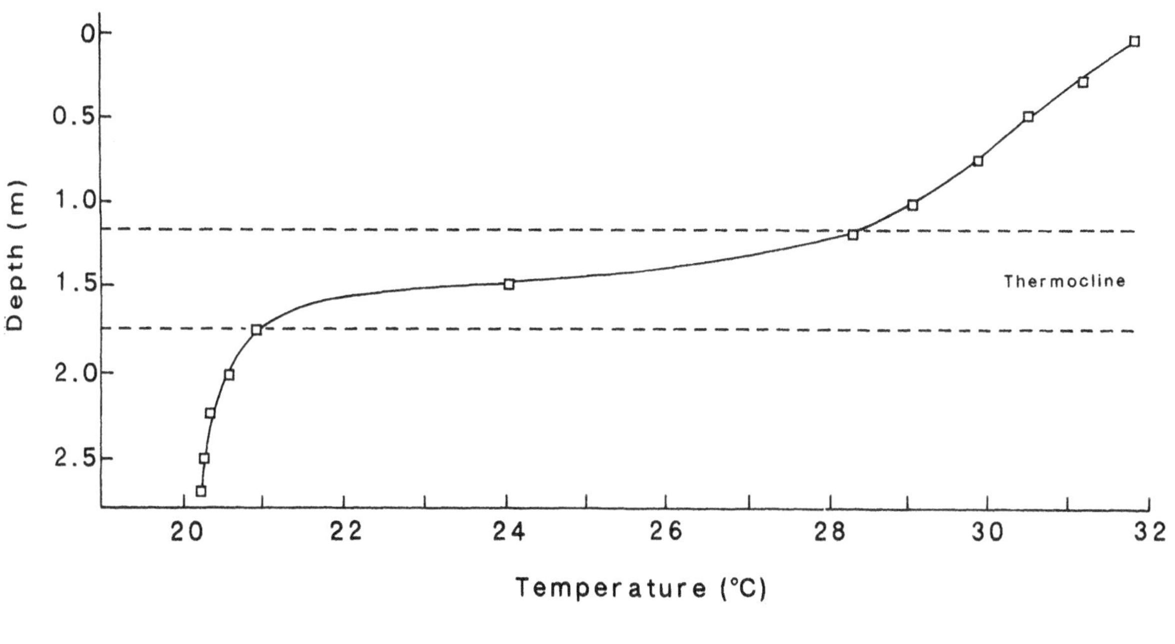

Fig. 16. Thermal stratification in a fish pond. Changes of water temperature with depth, showing the presence of a thermocline (after Boyd, 1979).

water movement caused destruction of the daytime stratification, mixing the upper 0.7 to 1.0 m water layer. The cooling off of the superficial layers was the result of a breeze (less than 100 cm s^{-1}), which is common in the late evening in tropical countries.

In shallow ponds in Auburn, Alabama, U.S.A., overturn homogenized the whole water mass, while deeper ponds remained stratified over the warmer months of the production season (Boyd, 1979). The accumulation of heat is greater in more turbid ponds, with stratification being more marked. Disappearance of plankton in ponds may cause disappearance of thermal stratification.

2.3. Colour, transparency and turbidity

Water colour

The colour of an object results from absorption of wavelengths in visible light. Absorption is the sum of partial absorptions, by water itself, by particles in suspension (seston), by dissolved matter in the water, and by phytoplankton.

A distinction between soluble and particulate matter is drawn arbitrarily by filtration through a membrane pore filter of 0.45 μm pore size. Dissolved matter ranges over a wide variety of organic and inorganic material, such as pigments released by plankton, humic matter, mineral salts, and nutrients. Large quantities of colloidal iron in suspension provide water with a brown opalescent colour. Particles in suspension, termed seston, are either non-living or living. The living ones are, in fact, the plankton. Sometimes, particles in suspension are brought by water fed to the pond, especially at flood times. An example is fine clay, which gives a milky appearance to the water and takes long to settle. In the Niger, water loaded with montmorillonite-kaolinite clay is called white flood, and these particles appear to be coated with bacteria and algae (John, 1986).

The abundance of phytoplankton and zooplankton in suspension is mainly responsible for the colour of pond water. Green, blue-greenish or brown-greenish water indicates a good plankton population. Turbidity caused by clay is brown, beige, or milky in appearance. Turbidity modifies the penetration of light and directly affects photosynthesis and primary productivity. Seston is also responsible for light scattering in water. Such scattering is stronger in the range of short waves than in that of long waves.

Light propagation in water and attenuation of light by plankton

The solar rays reaching the water surface are partly reflected, and partly refracted, according to their angle of incidence. The closer to the perpendicular the rays fall, the deeper they penetrate; if rays fall outside a certain angle, they are totally reflected. When the water surface is calm and smooth, the light penetration is better than when it is agitated.

The light penetrating the water covers a spectrum of wavelengths, including the visible spectrum. Each wave is absorbed and scattered according to its wavelength and specific absorption coefficient (Kirk, 1986). About 53 % of the incident light is absorbed within the first meter of water (Yentsch, 1980; Kirk, 1986: 105, Fig. 6.1). Fong & Ju (1987),in shallow water ponds located in Tainan county, Taiwan, established a relation between incident light intensity and depth:

$$\text{Intensity (in log of lux)} = 5.517 - 0.052 \times \text{depth (in cm)},$$

and between the extinction coefficient and turbidity:

$$K = 0.0084 + 0.0020\ T,$$

where turbidity is expressed in mg.l^{-1}.

The zone receiving more than 1 % of the incident light is called the euphotic zone. There phytoplankton can develop. Beneath this zone, respiration exceeds photosynthesis.

Phytoplankton algae possess photosynthetic pigments adapted to the light available in water (Yentsch, 1980). The various pigments cover almost the full width of the spectrum (Fig. 17), so as to make best use of the light available at any depth. In the water, the vertical zonation of algae is determined by the specific pigments they possess.

Water colour is rarely measured. Transparency or turbidity is more commonly recorded. Transparent water indicates an absence of plankton, while turbid water signifies presence of plankton and/or suspended particles.

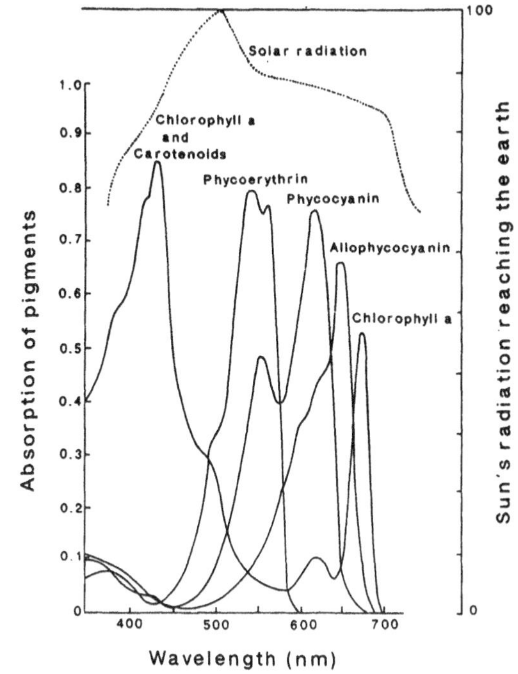

Fig. 17. Absorption spectra of different photosynthetic pigments and their relation to solar radiation reaching the earth (from Stockner & Antia, 1986) (with kind permission of Prof. Stockner on behalf of authors & Can. J. Fish. Aq. Sci.).

Water turbidity

Turbidity results from the presence of seston and from the resuspension of sediment (see p. 180). Most abiotic matter is colloid or non-colloidal material, which eventually settles downwards, as soon as water movements permit. The abiotic particles in suspension can be separated into mineral and organic material. Mineral particles consist of silt and clay, brought in by feeder canals, while organic material originates from the pond or from feed or manures added.

Movement and maintainability of suspended particles

The movements of particles in suspension are dependent on water movements, caused by waves, roll vortices, convective currents, upwellings, and currents generated by living organisms, such as filtrating zooplankton. Such particles are further affected by bacterial mineralization, which changes their shape and density, and by the ionic composition of the water, which causes flocculation or dispersion of colloids (see p. 13) (Weilenmann *et al.*, 1989). They are nevertheless subject to gravity.

Sedimentation of particles in suspension depends upon their size and electrostatic charge. Small sized particles take more time to settle. Most particles carry surface charges, usually negative, which originate from carboxylic or phenolic radicals, in the case of organic material, or from hydroxyl radicals, in the case of mineral particles. The ability to remain in suspension depends on surface chemistry, mutual electrostatic repulsion, or formation of aggregates.

The surface charge of a particle attracts a cloud of ions of opposite charge, and determines the dimension of the adsorbed cloud and the possibility of particles to come close to each other. Water conductivity, i.e. the total amount of electrolytes in solution, reflects the size of the ionic cloud. Electrophoretic mobility measures the size of the ionic cloud, or the stability of the particles, owing to electrostatic repulsion. The mobility of particles is altered by changes in conductivity, dissolved organic matter, and pH (Gerritsen & Bradley, 1987). Increase of cation concentration, such as calcium, in the water entails coagulation of the aggregates, and causes sedimentation. Increase of dissolved organic matter enhances dispersion and prevents sedimentation, and can be considered as a eutrophication factor (Weilenmann *et al.*, 1989). Particles coated with a film of humic matter are more stable than those without. The presence of blue-green and green algae, and of bacteria, confers a greater homeostasis to pH gradients and prevents flocculation (Gerritsen & Bradley, 1987).

The mobility of particles in water is particularly important for filtering organisms, as it affects the suspension of phytoplankton, bacterial adherence to substrates, and algal flocculation (Gerritsen & Bradley, 1987).

The role of coagulation in particles sedimentation

After collision, particles can remain together and form aggregates. The stability of such aggregates is a function of the concentration equilibria of ions in solution, and of surface properties of the particles.

The aggregation process takes place in two distinct steps: transport of the particles, and their attachment. Transport of particles is subject to movement of fluids (brownian diffusion, fluid shear) and to differential sedimentation, a consequence of differences in size and densities. The tendency of particles to form aggregates is essentially a chemical process (see p. 112). Aggregation increases the size of particles and modifies sedimentation rate. Even zooplankton grazing can be considered a coagulation process. However, it reduces sedimentation rate, rather than accelerating it (see p. 164).

Effect of turbidity on production

Organic particles in suspension are important for life in a pond, especially for decomposers and filtering organisms living in water, with a heavy organic load, such as that of ponds receiving organic manures.

Mineral particles reduce light penetration and, thus, photosynthetic rate. Mineral particles also interfere directly with life. In experiments in limnocorrals in a small lake in North Carolina, Cuker (1987) observed that clay in suspension reduced net community productivity, chlorophyll a concentration and algal cell numbers. The composition of the community changed to the benefit of flagellates and at the expense of blue-green algae.

Coarse clay particles (less than 2 μm), at high but not at low concentrations (50 to 100 mg l^{-1} and 10 mg l^{-1}, respectively), and fine particles (less than 1 μm), at both high and low concentrations, have been found to differently affect various species of cladocerans, by reducing their growth, especially under borderline conditions of food availability. Rotifers in the same experimental conditions, were not affected (Kirk & Gilbert, 1990). As we shall explain later (p. 162), low turbidity reduces the degree of vertical migration of cladocerans and copepods, become they become less vulnerable to predators in conditions of reduced visibility (Cuker, 1987).

Among effects on the benthos, one should note that large amounts of sedimenting material tend to bury life at the sediment-water interface.

Suspended material also affects fish. It deposits on the gill filaments and in the opercular cavity, forming a film of fine particles, which impede respiration and cause death by suffocation. Eggs, larvae and fry are much affected. High turbidity, such as 200,000 mg l^{-1}, causes death of fish. In natural conditions, turbidity seldom exceeds 200,000 mg l^{-1}, with values around 2,000 mg l^{-1} common in muddy waters (Boyd, 1979). Tilapia do not fare well in turbidities above 1300 mg l^{-1}. Many species living in bogs and shallow waters became adapted to turbid conditions by reduction of eye size and by mucus secretion. Mucus helps to precipitate particles in suspension, surrounding the fish with an envelope of clarified water. Some aquatic plants also clarify water.

Turbidity measurement

Methods for estimating turbidity are: measurement of the amount of settleable matter, or of total solid matter, Secchi disk measurement, and nephelometric measurement. The first two are analytical and quantitative, while the other two are qualitative.

Procedures for measuring amounts of settleable matter and total solid matter are detailed in Boyd (1979).

The Secchi disk, first used by Angelo Secchi in 1866 during a scientific expedition, is painted black and white on alternate quadrants, weighted and fixed to a graduated rope. The disk is immersed in water and the depth at which it disappears from sight is noted as P_1. The disk is lowered further into water and then slowly hauled up again; the depth at which the disk reappears is noted as P_2. The Secchi disk measure is the mean of the two measures:

$$P = \frac{1}{2} (P_1 + P_2)$$

The disappearance depth is a visual measure of water clarity (Preisendorfer, 1986). The Secchi disk measurement should be made between 10 am and 2 pm on clear days and with a non-overcast sky. The observer should stand perpendicular to the disk and not wear sunglasses. The sun should preferably be at his back (Boyd, 1979).

The depth of the euphotic zone corresponds roughly to 2.5 times Secchi disk visibility (Fig. 18). Minimum visibility requires about 15 % of total daylight and the vertical extinction coefficient of incident light is about 1.7/P to 1.9/P.

Almazan & Boyd (1978) established a correlation between Secchi disk visibility and phytoplankton concentration, and between the former and organic particulate matter concentration in ponds (Boyd, 1979: 278, Fig. 13.8 & 13.9). Measurements made by different observers do not vary in a significant

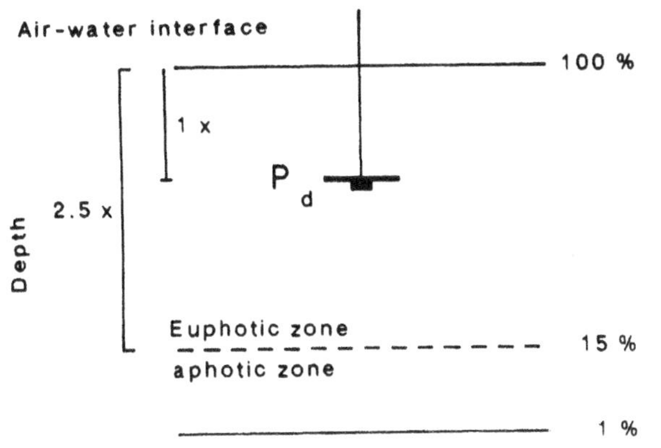

Fig. 18. Absorbance of light in water with depth and Secchi disk visibility according to light absorbance.

way, which indicates that the measures are objective. A detailed study of the conditions for use of the Secchi disk was made by Preisendorfer (1986).

The use of the Secchi disk to assess the abundance of plankton in ponds is a current practice. An acceptable plankton productivity in a pond corresponds to a Secchi disk visibility of 0.4 to 0.8 m (Boyd, 1979), and coincides with an organic particulate matter density of 7.5 to 14 mg l^{-1}.

The nephelometer is an optical instrument used to measure turbidity in comparison to standard solutions. Absorption in the water, of which the turbidity is to be measured, is compared to absorption of a solution of known turbidity. The measure is expressed in nephelometric turbidity units, NTU. This instrument is an attachment to the spectrophotometer, but can be used for field work. Nephelometric measurements give a good indication of absorption of light in water. Sigler *et al.* (1984) established an equation for transforming nephelometric units into amounts of suspended material, expressed in mg l^{-1}:

$$NTU = 10 + 0.178\ (\text{mg l}^{-1}) \text{ with } r^2 = 0.764.$$

Turbidity control

Methods to control turbidity include the use of lime (see p. 79 & p. 112) and alum, $Al_2(SO_4)_3.14H_20$, at the rate of 20 mg l^{-1}. Alum has an acidifying effect on water, 1 mg l^{-1} of which destroys 0.5 mg l^{-1} of alkalinity and produces 0.44 mg l^{-1} of carbon dioxide. Alum is not toxic for fish at these concentrations. The acidifying effect of alum is neutralized by the use of lime, 0.37 mg l^{-1} of which, in the form of $Ca(OH)_2$, neutralizes 1 mg l^{-1} of alum (Boyd, 1979).

Importance of physical characteristics of water

1. Optimum water temperature is essential for fish farming, as temperature should remain in a range suitable to the requirements of cultivated organisms (temperature preferendum); water temperature affects a series of other variables of which dissolved oxygen concentration is certainly the most important, but which includes also, for instance, the activities of decomposer microorganisms.

2. Water temperature is governed by thermal exchange with air and by direct heating by insolation. The temperature exchange within the water mass is low and thermal stratification in fish ponds does occur. However, thermal stratification is destroyed daily.

3. Control should be exerted on water temperature, when the latter exceeds values which may directly affect fish and cause their death. An influx of cooler water helps to reduce high temperature.

4. Turbidity of water is caused by mineral and/or organic particles in suspension. Mineral turbidity is composed of clay and silt particles, and is detrimental to living organisms, while organic turbidity is made up of plankton organisms, though organic colloids can also be involved, especially in ponds supplied with liquid manure. Colour of the water characterizes the type of seston load of the water.

5. Water turbidity impedes light transmission in water and reduces photosynthetic rate.

6. Control of abiotic turbidity is done by precipitating particles, using flocculating agents, such as lime.

3. Chemical variables of water

Factors determining water chemistry

The chemical composition of water depends on the ions and the gases in solution. Three main sources provide water with these substances: exchange with sediments, exchange with overlying air, and cycling of elements within the water column. To this should be added the supply of feed and fertilizers to ponds during production time. *Sensu stricto*, only ions participate in the chemical composition of the water, as the properties of gases in water are strictly part of its physical properties. However, gases and ions are often interrelated, like for example, the transformation of gaseous carbon dioxide into carbonate and bicarbonate ions.

The amounts of gases dissolved in water, such as oxygen, carbon dioxide, and molecular nitrogen, depend on atmospheric pressure and on specific solubility of each gas. Solubility increases with a decrease in temperature. Compared to the composition of atmospheric air, the gas composition of water is fundamentally different: oxygen concentrations are much lower, carbon dioxide concentrations are much higher, and ammonia, hydrogen sulphide and methane, almost non-existent in the atmosphere, are present in varying amounts, according to specific anaerobic decomposition processes going on. The concentrations of these gases depend on production and exchange processes prevailing in the water, on agitation and movement of the water, and on temperature. Moreover, their concentrations vary with depth. Water is usually aerobic in the upper layers, but may be anaerobic at greater depths.

Water is a solvent. Owing to its polar nature, the water molecule attracts ions, surrounding itself with an envelope of molecules. The ions are said to be hydrated. Ionization of salts in water gives a certain mobility to the ions, which can be set in motion under the influence of electrochemical forces. The electrolytic mobility of an ion depends on its charge, its diameter and on electrochemical potential. The concentration of ions in solution depends on the solubility coefficient and the reaction equilibria with the solid compound, generating these ions.

3.1. Dissolved oxygen

Importance of dissolved oxygen

Dissolved oxygen is one of the key factors of life. Oxygen is much more important to aquatic than to terrestrial life, as oxygen has a low solubility in water and is often a limiting factor for life in water. Some species are able to breathe atmospheric oxygen, by surfacing and gulping air, and do not rely entirely on oxygen dissolved in water. However, oxygen is also needed for all oxidation, nitrification and decomposition processes. Oxygen concentration in water is controlled by four factors: photosynthesis, respiration, exchanges at the air-water interface, and supply of water to the pond (Krom *et al.*, 1989b; Erez *et al.*, 1990).

Oxygen transfer at the air-water interface and within water

The quantity of oxygen dissolving in water depends on atmospheric pressure and on the partial pressure of oxygen in contact with the water surface. Between the two media, an equilibrium is established, responsible for exchange of oxygen between air and water, according to the degree of saturation of the water. At the peak of production in the early afternoon, oxygen saturation can reach 250 %. In these cases, the exchange is towards release of oxygen into the air. Oxygen is then not the only gas diffusing out of the water; other gaseous products, such as nitrogen are also carried along. If the water is below saturation levels, oxygen dissolves in water.

Transfer at the interface is proportional to the ratio of saturation level to actual concentration, to the surface area of contact, to the volume of water, to the period of contact, and to the transfer coefficient. The transfer coefficient depends on the degree of agitation and turbulence and ranges between 0.01 to 0.03 g O_2 m^{-2} h^{-1} in calm water, to about 0.75 g $m^{-2}.h^{-1}$ in rough water (Boyd, 1979). These numbers have been determined under laboratory conditions and give an indication of the range of values in controlled conditions, but are an underestimation of the real importance of the phenomenon in nature.

The diffusion of oxygen to deeper strata within the water column decreases with the logarithm of depth. It, therefore, rapidly becomes an insignificant process as depth increases. In Lake George, the surface water was strongly supersaturated (216 %), but an oxygen depletion (49 % of saturation) was observed at a depth of only 1.8 m (Talling & Talling, 1965). A change in atmospheric pressure modifies the saturation level, i.e. the maximum concentrations of oxygen soluble in water. These changes are, however, modest, being around 7 % for a variation of the atmospheric pressure equivalent to 50 mm of mercury at 25 to 30° C. Atmospheric pressure decreases with altitude.

Solubility of oxygen in water

The maximum amounts of oxygen soluble in water (the saturation level) depend on temperature and/or salinity. With increasing temperature or salinity, the solubility of oxygen decreases (Table 8).

Oxygen sources and consumption in water

Only a small part of the oxygen dissolved in water originates from air, the major part coming from photosynthetic activities. In Israel, Milstein *et al.* (1989) have shown that water quality of reservoirs used for fish farming is primarily the result of algal activities, then decomposition processes, and, lastly, wind action.

Oxygen concentrations in water between the surface and the euphotic limit are mainly the result of photosynthetic activities, and cannot be explained solely by diffusion.

Heterotrophic respiration is one of the principal sources of oxygen consumption in a pond. Heterotrophs include all animals contained in the pond, from aerobic microorganisms and zooplankton to fish. At night, the action of the autotrophs has to be added. Decomposition of organic matter also consumes much oxygen, though this is actually a part of the heterotrophic respiration (being consumption of oxygen by microorganisms).

Dynamics of change in dissolved oxygen in ponds

The dissolved oxygen dynamics in a pond are still largely unknown, particularly the vertical dynamics, i.e. interactions between the supersaturation at the surface and oxygen depletion in deeper water.

The photosynthetic rate varies in the course of day, and with it, the rate of oxygen release to the water. In the epilimnion, algae produce oxygen, while in the hypolimnion, they consume oxygen. In the metalimnion, consumption and production are balanced. The total production of O_2 per unit surface area can be considered fairly stable. Surface production is compensated by a reduction of production with depth, or a reduction of compensation depth, i.e. the depth at which production and consumption of O_2 are balanced (Chang & Ouyang, 1988).

In the epilimnion of ponds in Guangdong Province, China, the ranges of diurnal variations of dissolved oxygen were important and reached up to 10 mg l^{-1}. The 20 to 30 uppermost centimeters were frequently supersaturated with oxygen, which consequently involved a loss, by diffusion, of oxygen from water to the atmosphere. In hypereutrophic ponds, the amounts of nutrients available daily are sizeable and favour the growth of blue-green and green algae. Blue-green algae possess positive buoyancy properties; they rise to the surface, to form thick algal mats. This upward migration is promoted by the supersaturation of gases in the water and gaseous ebullition. Along with them, they pull to the surface plankton, microorganisms, and detritus. At the surface, oxygen produced by photosynthesis does not dissolve in water, but the major part diffuses into the air. Their presence at the surface not only reduces the impact of the wind on the water surface, but also prevents exchanges with the water beneath; in particular, the diffusion of gases is inhibited.

In the hypolimnion, dissolved oxygen concentrations usually do not exceed 2 mg l^{-1} and diurnal changes are not greater than 1 mg l^{-1}. The lowest oxygen concentrations are recorded at mid-day, when solar radiation is most intense and thermal

Table 8. Dissolved oxygen concentration in relation to temperature and salinity (after APHA, 1976)

Temperature (° C)	Dissolved Oxygen (in mg l^{-1})			
	Chloride concentration in water (g l^{-1})			
	0	5	10	15
15	10.2	9.7	9.1	8.6
16	10.0	9.5	9.0	8.5
17	9.7	9.3	8.8	8.3
18	9.5	9.1	8.6	8.2
19	9.4	8.9	8.5	8.0
20	9.2	8.7	8.3	7.9
21	9.0	8.6	8.1	7.7
22	8.8	8.4	8.0	7.6
23	8.7	8.3	7.9	7.4
24	8.5	8.1	7.7	7.3
25	8.4	8.0	7.6	7.2
26	8.2	7.8	7.4	7.0
27	8.1	7.7	7.3	6.9
29	7.9	7.5	7.1	6.8
30	7.6	7.3	6.9	6.5
31	7.5			
32	7.4			
33	7.3			
34	7.2			
35	7.1			

stratification is at a maximum. During daytime, the oxygen supplies are reduced, but consumption never stops. At night, vertical circulation of the water reinjects new quantities of dissolved oxygen into these layers and homogenizes the dissolved oxygen concentrations of the epilimnion and hypolimnion. The oxygen consumption rates are dependent on the amount of oxidable organic matter present. A relationship between oxygen consumption and dissolved oxygen has been established (Chang & Ouyang, 1988):

$$\text{dissolved oxygen consumption (in mg } l^{-1} h^{-1}) = 0.149 \times [OD]^{0.668}$$

Oxygen consumption rate is dependent on temperature and pH, being maximum at a pH of around 8. In the case of a regularly resuspended sediment, consumption has been found to double (Chang & Ouyang, 1988).

Supersaturation of the upper water layers should help to improve the conditions of the hypolimnion, if circulation of the pond water occurs during the hottest hours of day (Chang & Ouyang, 1988).

Factors controlling dissolved oxygen

The photosynthetic activity of algae changes in the course of the day. Factors modifying photosynthetic activity also affect dissolved oxygen concentration. They are, among others, temperature, light, cloud cover (Fig. 19), turbidity, nutrient concentration, and abundance of autotrophs. Photosynthetic rate decreases with depth, though the vertical distribution of phytoplankton is more complex than a simple decrease of abundance with depth (see p. 146).

In the presence of a phytoplankton bloom, daily oxygen fluctuations are high: the phytoplankton releases much oxygen during the photosynthetically active period, but, at night, consumes more oxygen than was released during daytime. In ponds with an abundant benthic flora, the pattern of dissolved oxygen stratification is reversed. The oxygen produced by benthic algae is released at the sediment level, and diffuses upwards towards the surface.

The dissolved oxygen regime in ponds is heavily dependent on the total organic matter (dissolved, particulate, on the pond bottom) in the process of being decomposed. Decomposition, aerobic or anaerobic, is a result of the availability of dissolved oxygen and organic load. Almost 90 % of the recycling of organic matter in deep lakes, such as Lake Titicaca, occurs in the euphotic zone, supplying the major part of the inorganic carbon for photosynthesis. The euphotic zone is characterized

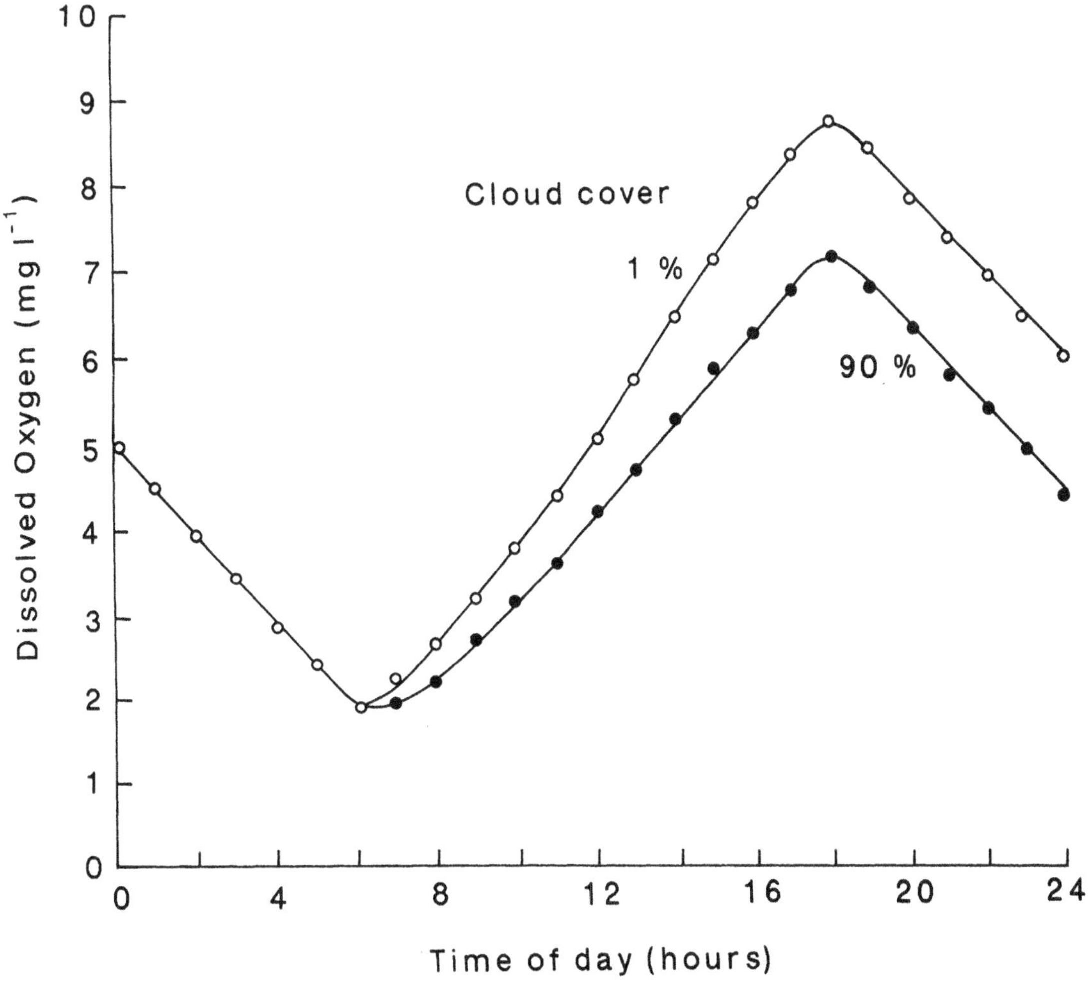

Fig. 19. Diel changes of dissolved oxygen according to a cloud cover of 1 and 90 % (from Piedrahita *et al.*, 1984) (with kind permission of Prof. Piedrahita & J. World Maric. Soc.).

by intense heterotrophic activity. The metabolism of the water body is tightly correlated to oxygen supply (Carmouze *et al.*, 1984).

In catfish production ponds in the United States, Cole & Boyd (1986) found a negative correlation between the fish feeding rate (amount of feed supplied to the pond per hectare per day) and dissolved oxygen concentration at dawn. The most important factors which influence aerobic bacterial activity are temperature, pH of the substrate, and the nature of the organic material. One criterion used to assess the nutritional value of organic matter is the C:N ratio. The closer this ratio in organic matter is to the ratio of the bacteria themselves, the more nutritive is the substrate, the faster organic matter is decomposed and the higher the rate of oxygen consumption (see p. 198 for bacterial oxygen consumption in ponds). Fibrous or structural material, such as lignin and chitin, takes a long time to decompose (Boyd, 1979).

The dissolved oxygen concentration (DO) varies in the course of the day, being at minimum in the morning, slightly before dawn, and at maximum in the early afternoon. The minimum DO concentration in the morning is the result of the night respiration of all heterotrophs and autotrophs combined. By early afternoon, photosynthesis has attained its optimum rate, the release of oxygen is highest and the water reaches its highest DO level. Oxygen decrease during the night can render the water anoxic, causing problems for the fish. Supersaturation of the water by oxygen may also be a problem, as it can provoke gaseous embolies to fry and fingerlings. The influence of low DO concentrations has a greater impact when other water quality factors are unfavourable, and the fish is already stressed. There is thus a need to reduce daily oxygen fluctuations, i.e. the difference between the daily maximum and minimum, and to contain the nightly drop in DO.

Boyd and associates (Romaire *et al.*, 1978; Romaire & Boyd, 1979; Boyd *et al.*, 1978) tried to predict nocturnal decreases in DO. Of the night consumption of oxygen, 75 and 85 % was found to be accountable to the plankton, in non-fertilized and fertilized ponds respectively, with the rest attributed to fish and benthic respiration (Romaire *et al.*, 1978). Oxygen production is directly and significantly correlated to the intensity of solar radiation, concentration of chlorophyll a, and to percentage O_2 saturation at dawn. Similarly, a model has been developed, combining all factors, allowing prediction of the DO concentration at dawn, using chemical oxygen demand, COD, and temperature, these two factors being correlated. COD is estimated from measurements of turbidity or Secchi disk visibility (Boyd *et al.*, 1978). The most important factors affecting dissolved oxygen concentrations at dawn are biotic or algal turbidity, planktonic respiration, inorganic nutrient depletion, and algal biomass (Smith & Piedrahita, 1988).

Measurement of dissolved oxygen

The measurement of dissolved oxygen is usually done by a modified Winkler method or by oxygen electrodes. Dissolved oxygen is expressed as mg l^{-1}, ml l^{-1}, or ppm. The factor to convert mg l^{-1} into ml l^{-1} is 1.43 mg ml^{-1} at normal pressure and a temperature of 0° C.

Dissolved oxygen concentration is sometimes expressed as percentage saturation, i.e. the ratio between the amount of oxygen per litre and the maximum amount soluble at the prevailing temperature.

Example: the saturation value at 35° C is 6.98 mg l^{-1},
the recorded oxygen concentration at 35° C reads 3.5 mg l^{-1}
the percentage saturation is:

$$\frac{3.5 \times 100}{6.98} = 50.1 \%$$

Methods for controlling dissolved oxygen

Interventions of the fish farmer to prevent oxygen depletion are of two types: ensure aeration of the pond, or control phytoplankton production. Solutions proposed to control phytoplankton production include the use of various chemicals, such as simazine, copper sulphate or potassium permanganate. Planktivorous fish (biological control) can also be used. Among the species used are the silver carp (*Hypophthalmichthys molitrix*), the bighead carp (*Aristichthys nobilis*), and different species of tilapia (*Oreochromis aureus*, *O. niloticus*, and their hybrids). Plankton consumption by fish is discussed on p. 184.

Artificial aeration of ponds is discussed on p. 199.

Sudden death of a great amount of plankton, such as the dying off of a plankton bloom, results in a dramatic decrease in dissolved oxygen in the water. It is therefore of great importance to keep a vigilant eye on phytoplankton density and, if necessary, to reduce it (preventing mass destruction of phytoplankton). A discussion on factors promoting the appearance of phytoplankton blooms is given later. Filamentous algae at excessive densities can be removed with the help of a seine (Heinen *et al.*, 1988). Boyd *et al.* (1975) observed the dying off of blue-green algae of the genus *Anabaena*, induced by a change in light intensity; this resulted in mats of algae floating at the surface, constituting a light screen. Any such observation should lead to the removal of this floating material. Another method is to replace part of the pond water, thereby diluting the plankton (see p. 142).

Simazine is a herbicide toxic to phytoplankton, but apparently not toxic to fish. A treatment of 0.5 mg l^{-1} or 0.25 mg l^{-1}, in repeated applications, kills off blue-green algae. Simazine is commercialized under the trade name Aquazine. Tucker & Boyd (1978) used a treatment comprising three applications: initially 0.8 mg l^{-1}, followed by two doses, of 0.25 mg l^{-1}, after 3 and 1 months. This gave a good control of phytoplankton, but was accompanied by a decrease in dissolved oxygen, unfavourable for production. Copper sulphate ($CuSO_4$) is much more toxic for fish and should be used at a dose of 2 mg l^{-1} or 0.84 kg ha^{-1}. However, at this concentration, Tucker & Boyd (1978) did not observe an effect on *Microcystis* blooms.

Potassium permanganate is a strong oxidizing agent and is used in fish ponds to reduce biological oxygen demand, by oxidizing surplus organic matter. Tucker & Boyd (1977) studied the effects of applications of 2, 4, and 8 mg l^{-1} in ponds and observed that the product decreased biological oxygen demand, reduced phytoplankton density and photosynthetic rate, but did not affect dissolved oxygen concentration. *In vitro*, permanganate is bactericidal, but this property has not been confirmed in pond use. To increase its bactericidal effect, it has been proposed to treat ponds with a dose of 4 mg l^{-1}, in addition to the amount determined by the so-called '15-minute $KMnO_4$ demand'. The quantity of permanganate reduced after 15 minutes by organic matter present in the fish pond is determined and, to this amount, 4 mg l^{-1} is added.

Tucker (1984) discussed the oxidation processes which go on during the 15 minute period. He concluded that it is essentially the soluble organic matter which is oxidized, and this represents only a variable part of total organic matter. He recommended the extension of this period to 180 minutes, to achieve a more thorough destruction of excess organic matter. Tucker (1989) proposed multipling the 15-minute demand by 2.5. The concentrations of potassium permanganate used in these ranges are not harmful to American catfish.

The use of these products should be restricted and a better approach would be to investigate ways of combatting the causes of algal proliferation, such as by decreasing nutrient supply, in the form of fertilizers or feed, or by controlling algal biomass mechanically. Toxic products may indeed eventually reappear in fish (Boyd, 1979; Smith, 1988).

Importance of dissolved oxygen in ponds

1. Dissolved oxygen is essential for pond life, being required by all organisms for heterotrophic respiration. All aerobic chemical oxidation processes require oxygen as well. Oxygen find its way into a pond either by diffusion from the air or by photosynthesis. It is this latter process which provides the bulk of oxygen to the water mass.

2. The dynamics of photosynthesis, involving changes in the course of the day and with depth, are the major determinant of the spatial and temporal partitions in oxygen concentration within the pond, imposing concentration gradients and a vertical stratification. This stratification is destroyed by homogenization processes, such as periodic overturns of the water, and wind and wave action.

3. Control of dissolved oxygen dynamics in a pond, entails, firstly, management of the phytoplankton mass and the photosynthetic rate, and, secondly, making proper use of the over-turn process, with the aim of facilitating the transport of oxygen to deeper levels, where dissolved oxygen can easily become depleted.

3.2. Conductivity and salinity

Definition

The conductivity of water is its capacity to transport an electrical current. The transport is ensured by the ions present. The measure gives an indication of the total amount of ionizable salts in solution and is expressed in $\mu S\ cm^{-1}$. Earlier, one spoke of resistivity, which is the reverse of conductivity.

Distilled water has a conductivity of 0.5 to 2 $\mu S\ cm^{-1}$, which increases to 4 $\mu S\ cm^{-1}$, when stored. During conservation, carbon dioxide and ammonia from the atmosphere dissolve in water. The conductivity of fresh water varies between 50 to 1500 $\mu S\ cm^{-1}$ (Boyd, 1979), but some polluted waters reach 10,000 $\mu S\ cm^{-1}$. Seawater has a conductivity around 35,000 $\mu S\ cm^{-1}$. The degree of mineralization or total load of salts of a water is in direct relation with its conductivity.

Factors responsible for mineral load of water

The total amount of mineral salt in water is the result of several processes (Golterman & Kouwe, 1980):

- the solubilization of salt, in particular, calcium bicarbonate, originating from soil and rocks, with which the water comes in contact during percolation and run-off
- dilution by rainfall
- precipitation due to interaction with other salts or to evaporation of water.

The last two processes lead to a dominance of sodium and chloride ions (Golterman & Kouwe, 1980). African waters are largely dominated by rock erosion, concentration by evaporation, and precipitation of calcium carbonate. Calcium carbonate begins precipitating when the saturation concentration of calcite is reached, which corresponds to an alkalinity of approximately 2.5 $meq\ l^{-1}$ of solid matter in solution. When dissolved matter reaches beyond 1 $mg\ l^{-1}$, most calcium ions are removed from solution. The evaporation-crystallization process leads to a substantial loss of carbon at each step of concentration (Kilham, 1990).

Dominant ions

To characterize a water, its ionic composition needs to be analysed. The amounts of nutrients present give an indication of the potential production of the water. The chemical environment is important, though it does not give sufficient information to assess the production capacities of an environment. It is not the total amount of nutrients present in any form, but their availabilities, seasonal fluctuations and their minimum concentrations that matter. Also, nutrients appearing in limiting concentrations have a direct impact on life in a pond.

Temperate freshwaters are characterized by an abundance of calcium, magnesium and sodium, and bicarbonate, sulphate and chloride. The ionic composition of African lake waters is distinctive, being characterized by high contents of sodium bicarbonate and carbonate, and a wide range of salinities (Talling & Talling, 1965; Welcomme, 1972). In some cases, magnesium is present in large amounts (Golterman & Kouwe, 1980). Among the minor constituents, a high proportion of silica and low concentration of nitrate are features probably widespread in tropical lakes (Talling & Talling, 1965).

Types of water based on conductivity

According to conductivity, three main groups are distinguished: waters with conductivities below 600 $\mu S\ cm^{-1}$; between 600 and 6000 $\mu S\ cm^{-1}$; and beyond 6000 $\mu S\ cm^{-1}$ (Talling & Talling, 1965).

Waters with low conductivities are found in marshland areas. They are usually of dark colour and have high concentrations of organic matter, in particular humic matter. Waters with intermediate conductivities contain mainly sodium, chloride and bicarbonate ions. Waters with high conductivities result from high evaporation rates and salt concentrations (Talling & Talling, 1965).

With increasing conductivity, the influence of major ions changes: at low conductivities, increase of conductivity is accompanied by an increase in calcium ion concentration; in waters of intermediary conductivities, bicarbonate and carbonate ions play a more important role; at high conductivities, calcium is replaced by sodium (Talling & Talling, 1965). The predominance, in African lakes, of carbonate and bicarbonate accounts for the correlation between conductivity and alkalinity (Talling & Talling, 1965).

Interactions between ions occur, such as between calcium, carbonate and phosphate ions. Calcium carbonate is dissolved by carbon dioxide. If carbon dioxide is removed from solution, calcium precipitates. In some cases, supersaturation of calcium and carbonate beyond the solubility constant of calcite ($K_{sp} = 1.10^{-8}$) occurs (Golterman & Kouwe, 1980). Such supersaturation of carbonates may be related with organic load, organic compounds sorbing calcium ions, and preventing precipitation.

Conductivity is used to assess the expected yield of fisheries reservoirs, with the use of the morphoedaphic index (MEI), establishing a relationship between conductivity (in μS cm^{-1}) and mean depth (in meters), and between MEI and yield (kg ha^{-1}) for African water bodies (Welcomme, 1983):

$$MEI = \frac{\text{conductivity}}{\text{mean depth}} \quad \text{and} \quad Y = 14.3136\ MEI^{0.4681}$$

Modes of representation of ionic composition

The chemical composition of ions in solution can be represented by the Kufferath ionic diagram. To draw such a diagram, four concurrent axes, equidistant by the same angle, are drawn. Each axis is graduated in percentages, representing calcium, sodium, potassium, magnesium, bicarbonate, carbonate, chloride and sulphate. The concentration of each ion is calculated in milliequivalents and expressed as a percentage of the sum of the ions of each sign. Each percentage calculated is placed on scale on one half-axis; the cations are arranged on one side, opposite to the anions, and the different points joined. Sometimes, only three axes are drawn, sodium and potassium, and carbonate and bicarbonate, being grouped together (Fig. 20).

Salinity and chlorinity: definition

Salinity is the sum of all solid substances in solution in 1 kg of water, when all carbonate ions have been converted to oxide ions, all bromide and iodide ions replaced by chloride ions, and all organic matter oxidized (Spotte, 1979). Salinity is expressed in g kg^{-1}. Sea water normally has a salinity of 35 g kg^{-1}. Salinity is difficult to measure by direct chemical methods and is often measured refractometrically (salinometer). It is also possible to measure chlorinity (S) by titration, and establish salinity with the help of the relation:

$$S\ (g\ kg^{-1}) = 1.80655 \times Cl\ (g\ kg^{-1}).$$

Chlorinity measures the concentration of halide ions, expressed in g kg^{-1}. Chlorosity is the same measure, expressed in g l^{-1}.

Water with a salinity below 0.5 g kg^{-1} is considered fresh; salinities between 0.5 and 30 g kg^{-1} indicate brackish water; when salinity is beyond 30 g kg^{-1}, the water is either seawater, or hypersaline.

3.3. Alkaline cations

(1) Calcium and magnesium

Calcium and magnesium are the most abundant ions in temperate freshwater. However, in African waters, sodium and magnesium are often dominant. According to Weninger (1985), in humid tropical areas, waters are commonly dominated by calcium and bicarbonate ions and are alkaline. In areas where soil erosion is important, owing to heavy seasonal rains, dominant ions are magnesium and bicarbonate. In waters rich in humics, calcium is retained by a strong ion exchange capacity system. In tropical rainforests, ammonium and iron are present in high concentrations.

The ratio of monovalent to divalent ions (M/D), which is equivalent to the ratio of sodium and potassium to calcium and magnesium, has, for a long time, been considered useful for predicting the phytoplankton assemblage in a given water body

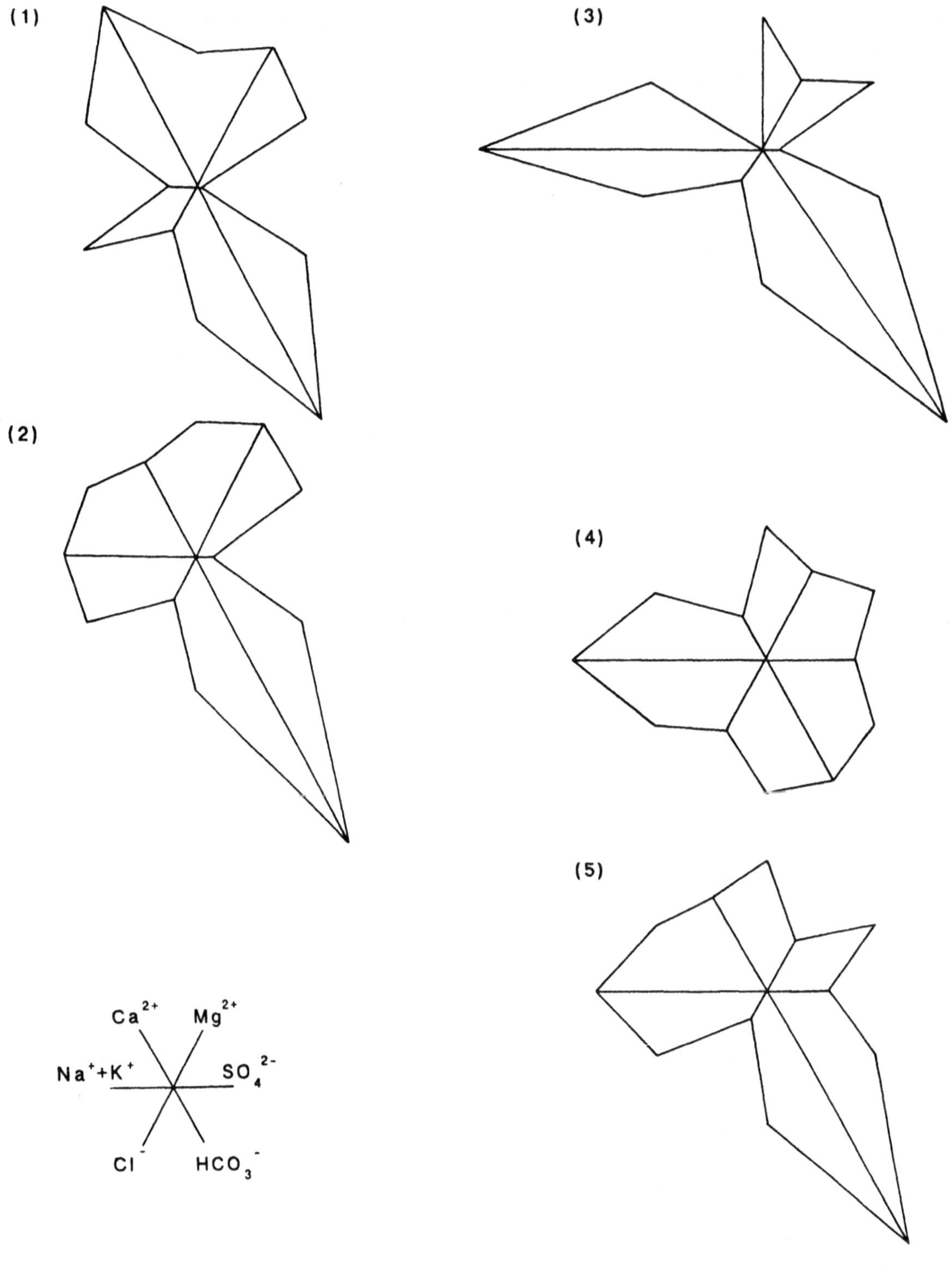

Fig. 20. Ionic diagrams for African waters belonging to the four main types of water, drawn using data of Weninger (1985): (1) Calcium-bicarbonate water, from Madagascar, groundwater (85, Table 2); (2) Magnesium-bicarbonate waters, from Reunion, Cascades (86, Table 3); (3) Sodium-rich waters, from Madagascar, spring pool (88, Table 4,); (4) and (5) Water enriched with humic acids, ammonia and iron, from Kinshasa, lake and S. E. Nigeria, streams (89, Table 5)

(Pearsall, 1924). In a study of 99 Scottish freshwater lochs, Shoesmith & Brook (1983) showed that a M/D ratio below 1 was favorable for diatoms and blue-green algae, and a M/D ratio above 1 suitable for desmids.

This ratio alone cannot explain the dynamics of phytoplankton; ions, such as nitrate, phosphate and silicate, must be considered individually. The importance of the M/D ratio is difficult to disentangle from the usual correlation between increasing M/D ratio values and decreasing amounts of phosphate, nitrate and ammonium salts (Lund, 1965). Correlations between a phytoplankton ratio, using the proportion of the number of species of blue-green algae, Chlorococcales, centric diatoms and euglenids to desmid species numbers, and the concentrations of calcium and magnesium have been established, illustrating the role of these ions in the biology of these species. This ratio also gives information on the trophic state of the water body. Values below 1 indicate oligotrophy, values from 0.0 to 0.3, dystrophy, and values above 1, eutrophic conditions (Shoesmith & Brook, 1983). However, there seems to be little evidence for limitation of production by a direct lack of calcium. Calcium usually occurs in excess of requirements, even in soft waters (Lund, 1965). The importance of calcium is largely in relation to pH and the carbonate-bicarbonate buffer system, and to microbial processes.

The ratio attempts to distinguish between calciphobe and calciphile algae, but there seem to be relatively few planktonic forms, whose distribution is restricted by calcium. In some examples commonly quoted, such as desmids, other factors may be equally important, such as pH or alkalinity (Talling & Talling, 1965). In lakes with different calcium contents, phytoplankton assemblages can be grossly the same. Only a few algae seem to have strict requirements (Lund, 1965).

Calcium and magnesium concentrations are interdependent with those of carbonate, through the carbonate-bicarbonate buffer system and chemical reactions. Biological processes, such as photosynthesis, respiration, and decomposition, which modify the concentrations of carbon dioxide, act on the availability of these cations. Moreover, these cations are involved in many physiological reactions, alone or with phosphates, such as the formation of bones and shells. Calcium is quite abundant in macrophytes and algae, being some 2.5 and 0.5 mg g^{-1} in dry weight, respectively.

(2) Sodium and potassium

Sodium is generally predominant in waters of high salinity, and its concentration increases fairly regularly with conductivity and total ionic concentration (Talling & Talling, 1965).

Potassium is rarely considered important for the ecology of algae (Lund, 1965). This ion is apparently always present in smaller amounts than sodium, the difference tending to increase with salinity.

Importance of conductivity and alkaline ions

1. Conductivity is a measure of the amount of electrolytes dissolved in water, or the total ionic charge of the water.

2. Dominant ions in African lake waters are the cations sodium and magnesium, and the anions carbonate and bicarbonate. In comparison to temperate waters, the predominance of calcium is decreased; this may involve several other profound changes in water quality.

3.4. Alkalinity, pH, hardness and carbon dioxide

3.4.1. pH

Definition of pH and pH regulation

pH is a measure of concentration of hydrogen ions in water and, as such, determines its acid, alkaline or neutral character. Any water is a mixture of acids and bases and is the result of acid-base titration processes. The compounds in solution come from geochemical and biological sources. The most important acid in freshwater is carbonic acid, originating from carbon dioxide dissolving from the air into the water, and from carbonaceous rocks. Respiration and photosynthesis releases and takes up, respectively, carbon dioxide, thus modifying the pH of the water. Some microbiological processes, such as denitrification and sulphate reduction, generate acidity. Finally, the pH of water is strongly influenced by soil pH.

Equilibria exist between the concentrations of carbon dioxide, bicarbonate and carbonate. The reactions between the different ion species are rapid and act as a pH buffer system. There are, however, other buffer systems, though not of the same importance.

pH is an intensity factor, while the acid neutralizing capacity (ANC) and the base neutralizing capacity (BNC) are capacity factors. The buffer system is an implicit function of both intensity and capacity (Stumm & Morgan, 1981).

Optimum pH values

pH is a general measure, its value being determined by a number of processes occurring in ponds. Little is known about the actual way in which pH affects organisms. pH values below 4 and above 11 are lethal for most fish. Generally, pH values ranging between 6.5 and 8 are considered as optimum.

Productivity studies in Canadian acid lakes have shown that the relationship between primary productivity and concentrations of total phosphorus are not modified by pH; biological production stays high if the nutrient supplies remain steady (Schindler *et al.*, 1985: in a Canadian Shield lake after a multiyear experimental acidification; Kerekes *et al.*, 1990: in twenty lakes across Nova Scotia, Canada). Productivity limitation due to acidification can result from inhibition by elevated trace metal concentrations, particularly inorganic monomeric aluminium, and/or increased phosphorus limitation as a consequence of adsorption and/or co-precipitation of phosphate with aluminium (Mulholland *et al.*, 1986).

3.4.2. Alkalinity and hardness

Definition of alkalinity

Alkalinity, or total alkalinity, is defined as the net negative charge of all ions which react with the hydrogen ion. This definition corresponds with that of the acid neutralizing capacity (ANC), which measures the net deficit in protons, by titration of all bases with a strong acid to a preselected reference point (Stumm & Morgan, 1981). It is the sum of all anions, but mainly carbonate, bicarbonate, and hydroxyl ions. The latter are less important:

$$Alk = [HCO_3^-] + [CO_3^{2-}] + [OH^-] + [A^-] - [H^+],$$

where A stands for the proteolytic anions.

Alkalinity is expressed in meq l^{-1} or mg l^{-1} equivalent $CaCO_3$. The factor of conversion is as follows: 1 meq l^{-1} of alkalinity is equal to 50 mg l^{-1} equivalent $CaCO_3$. The concentration of anions determines the buffer capacity of water.

The base neutralizing capacity (BNC) is the equivalent sum of all acids that can be titrated with a strong base to an equivalent point. Conceptually, it is a measure of the concentration of the species containing protons in excess minus the concentration of the species containing protons in deficiency of the proton reference level. It is a measure of the net excess of protons over the reference level of protons. The reference level is defined by the composition of a pure solution, where the proton concentration balances the anion concentration (Stumm & Morgan, 1981).

Processes modifying alkalinity

Equilibria between the different carbonate ions, and

their impacts on alkalinity, have been summarized by Spotte (1979) as follows (Fig. 21):

- when carbon dioxide is added or removed from water, alkalinity does not change, as carbon dioxide does not bear any electrical charge
- when bicarbonate ions are added or removed, alkalinity increases or decreases according to 1 meq l^{-1} per meq l^{-1} of bicarbonate
- addition or subtraction of carbonate entails a change of 2 meq l^{-1} of alkalinity per meq l^{-1} of CO_3^{2-}
- addition of a strong base or acid causes a modification in alkalinity according to the amount of meq l^{-1} added, without, however, changing the total concentration in carbon dioxide.

In his reasoning, Spotte does not take into account equilibria between the amounts of carbon dioxide and carbonic acid in solution. The relations between these equilibria and alkalinity are discussed in Stumm & Morgan (1981).

Various biological and chemical processes directly affect alkalinity. Photosynthesis, denitrification, sulphate reduction, and dissolution of lime or other calcareous rocks increase or produce alkalinity (consume protons), while respiration, nitrification, and sulphide oxidation decrease or consume alkalinity (release protons) (Stumm & Morgan, 1981; Cook *et al.*, 1986). The relationship between alkalinity and pH is illustrated in Fig. 22.

Alkalinity can be exchanged between water and sediments. The loss of alkalinity equivalents in the

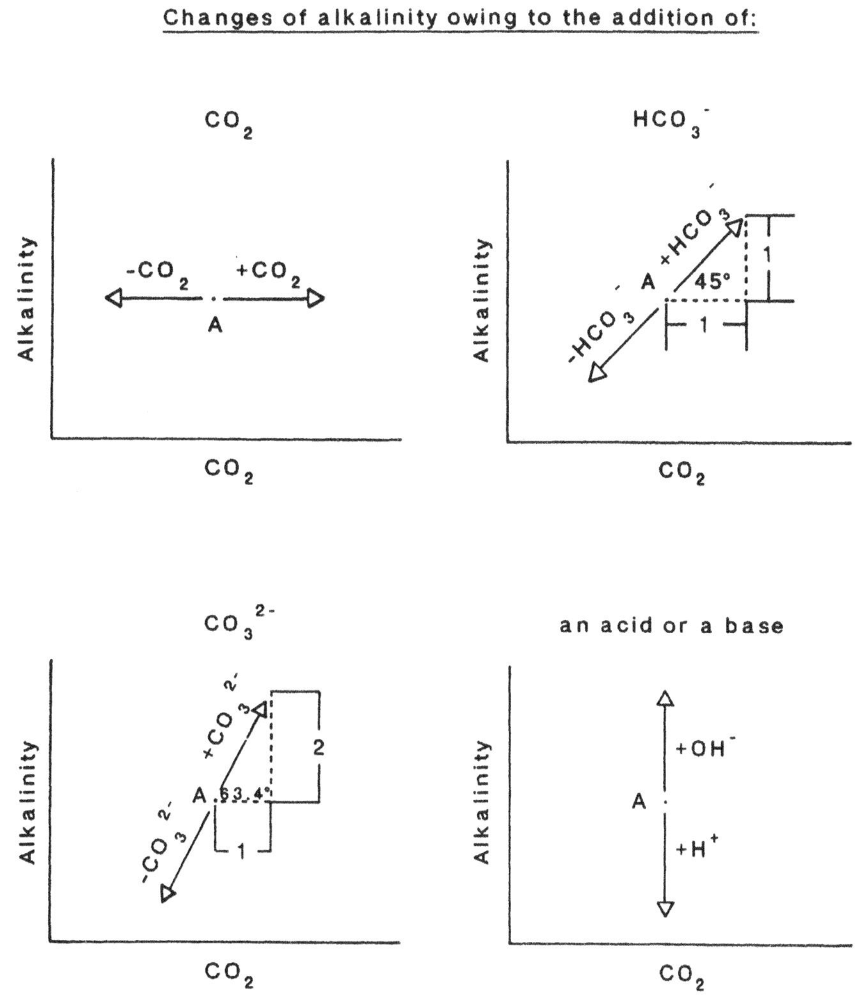

Fig. 21. Impact of different alkaline ions on alkalinity (from Spotte, 1979) (with kind permission of Dr. Spotte & Wiley).

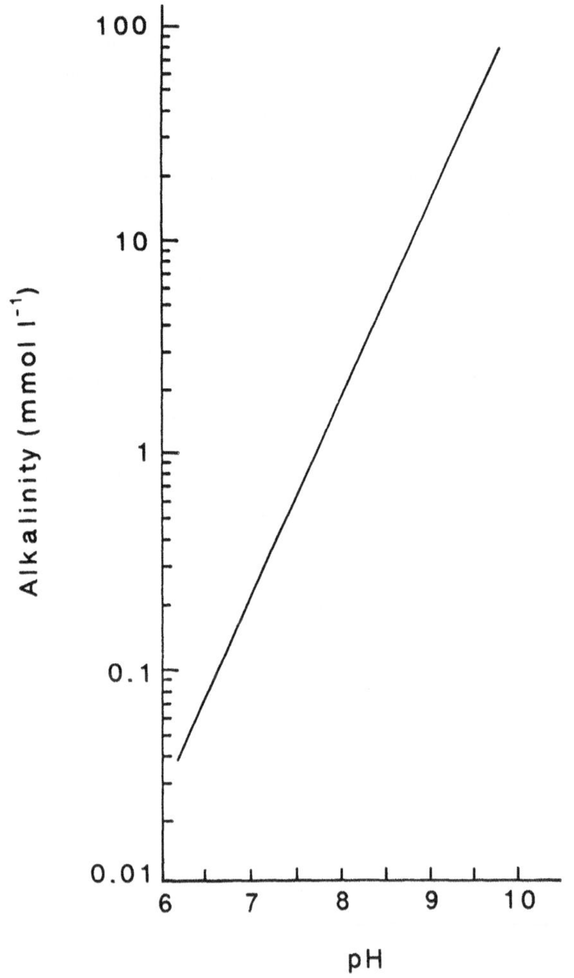

Fig. 22. Relation between alkalinity and pH (after Golterman & Kouwe, 1980, with kind permission of Dr. Golterman on behalf of the authors).

water column can be compensated by supply from the sediments, and vice-versa (see p. 45).

Total alkalinity of natural waters varies between 5 to several hundred mg l^{-1}. Water is considered to be hard when total alkalinity is above 40 mg l^{-1} of $CaCO_3$, while waters with values below this are said to be soft (Boyd, 1979). Hard water is usually more productive than soft water.

Other definitions of alkalinity have been used in the past: carbonate alkalinity, phenolphthalein alkalinity, and carbon dioxide alkalinity.

Definition of hardness

Hardness is a measure of the amount of calcium and magnesium ions associated with carbonate. Total hardness is expressed in mg l^{-1} equivalent $CaCO_3$ and measures the total concentration of calcium and magnesium. As for alkalinity, other definitions have been used in the past, such as temporary hardness and permanent hardness.

3.4.3. Carbon dioxide

Importance of carbon dioxide

Carbon dioxide takes part in the carbon cycling of the ecosystem, forming an input to organic carbon production and other processes related with primary productivity. Moreover, it appears as an output or byproduct of respiration and organic matter decomposition. Dissolution of carbonaceous salts in water is also a source of carbon dioxide (Fig. 23). Decomposition of organic matter in water will be discussed later (p. 115, and also p. 100).

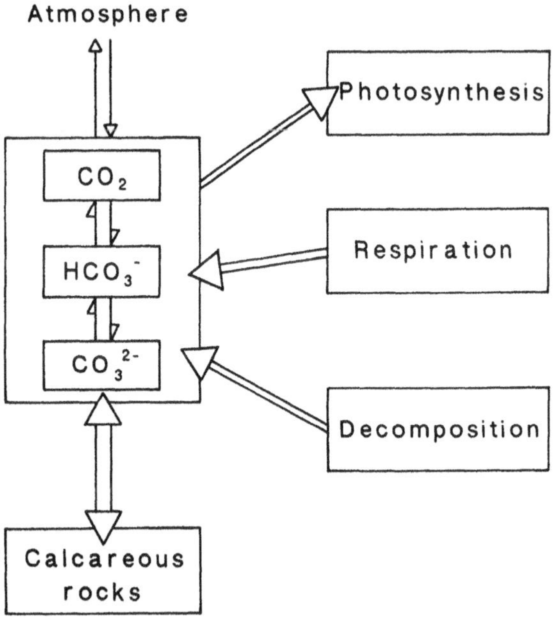

Fig. 23. Relationship between the different carbonate ions within the aquatic environment *sensu lato*.

Carbon dioxide forms ion species which play an important role in controlling pH and restricting its fluctuation range in water. Carbon dioxide dissociates into bicarbonate and carbonate ions; their respective concentrations are maintained in steady equilibrium, so that a change imposed on the concentration of one species is spread and 'absorbed' by the different ions, owing to equilibrium interactions, which buffer changes in concentration.

Carbon dioxide penetrates into water across the soil-water and water-air interfaces. The exchange with the sediments involves carbon dioxide of geochemical (limestone from the soil or added by liming) and biological origin (microbial and benthic activities). The different carbon species in equilibria in water are as follows:

$$CO_{2\,(g)} \rightleftharpoons CO_{2\,(aq)} + H_2O \rightleftharpoons H_2CO_3 \rightleftharpoons H^+ + HCO_3^- \rightleftharpoons 2H^+ + CO_3^{2-}$$

Carbon dioxide is well soluble in water. Its solubility is modified by temperature and salinity; a rise in either decreases solubility (Table 9). The higher the pH, the more the equilibrium shifts to the right.

Most carbon dioxide is continuously recycled in the water column: removal of carbon dioxide during daytime by autotrophs, and release of carbon dioxide by all the organisms at night. Community respiration depends on the biotic load of a pond. If total biomass is high, the decomposer population is significant, with microbial respiration representing a great part of the community respiration. The carbon dioxide which dissolves in water makes a small contribution, (some 0.2 g m^{-2} d^{-1}, less than 5 %) to the dioxide demand of primary productivity in productive ponds (Schroeder, 1987). The carbon dioxide content of the air is about 0.03 %.

Carbon dioxide affects fish respiration. High concentrations of carbon dioxide interfere with the capacity of hemoglobin to fix and transport oxygen and even cause release of oxygen already fixed. This phenomenon is called the Bohr effect, and it slows down the metabolism of fish. Fish tolerate, without too many problems, concentrations of carbon dioxide up to 60 mg l^{-1}, provided that dissolved oxygen concentrations are high. It is recommended that the concentration of carbon dioxide remains below 70 mg l^{-1}; however, in hard waters, concentrations as low as 10 to 15 mg l^{-1} cause nephrocalcinosis.

Table 9. Solubility of carbon dioxide in relation to temperature and salinity (according to Spotte, 1979).

Temperature	Solubility of CO_2 (10^{-4} mol l^{-1} atm^{-1})		
(° C)	Concentrations in chlorides ($mg.l^{-1}$)		
	0	15	20
16	440	385	367
18	413	364	347
20	389	344	328
22	367	325	310
24	347	308	294
36	328	291	278

Carbon dioxide and buffer capacity

The buffer capacity of water is its ability to maintain pH within a narrow range of values, resisting changes in pH upon addition or production of acids or bases. Water is characterized by an acid neutralizing capacity and a base neutralizing capacity

(Stumm & Morgan, 1981). Among the ions in freshwater, carbonate ions play a key role in the buffer capacity of water.

Carbon dioxide is the most important input to photosynthesis. The consumption of carbon dioxide (primary production) in a productive pond in Israel is around 10 g m^{-2} d^{-1} of carbon, of which only about 5 % is supplied from the air (Schroeder, 1987). This means that 95 % of the carbon dioxide originates from autochthonous production, i.e. respiration (about 71 %) and decay processes. Respiration is a continuous process, while photosynthesis is restricted to the daytime. Therefore, the removal of carbon dioxide occurs roughly in half the time needed to supply it. The stocks of carbon dioxide are not limited to the amounts needed for photosynthesis: they are always much larger and, in the above cited example of Schroeder, amounted to three days' gross primary productivity demand. The unused carbon dioxide must be prevented from acidifying the water to such an extent as to cause damage to living organisms. The changes in concentration of such large amounts of carbon dioxide associated with primary productivity and respiration can induce strong fluctuations of pH if the buffer system controlling such variations is weak. This is the case in waters with low or high alkalinities. In waters of high alkalinity and high pH, carbonate is the dominant ion and little carbon dioxide is made available by the buffer system, thus reducing photosynthesis (Boyd, 1972). pH variations are subject to photosynthetic activities during the day and to respiration at night; pH increases during the day and decreases at night to reach a minimum at dawn.

The ionization constants, K_1 and K_2, of carbonic acid H_2CO_3 are, respectively, $10^{-6.35}$ and $10^{-10.33}$. With these values, the concentrations of carbonate ions can be calculated (Fig. 24). Carbonate is only present in measurable amounts when pH is above 8.34, when all carbon dioxide is ionized. The two ionization constants decrease with temperature and salinity.

Carbon dioxide in water reacts with carbonate salts to form bicarbonate:

$$CaCO_{3\,(solid)} + CO_2 + H_2O \rightleftharpoons Ca^{2+} + 2HCO_3^-$$

Changes in the carbon dioxide concentration induce changes in bicarbonate ion concentration. If carbon dioxide is removed from solution, the bicarbonate concentration decreases; this happens when the concentration of carbonate and carbonic acid increases, which implies an increase in pH, as carbonate ions are more alkaline than bicarbonate. In the case of carbon dioxide addition, the reactions are reversed and pH decreases. The combination of the two ionization constants enables the calculation of pH:

$$[H^+] = \sqrt{K_1 \cdot K_2 \frac{[CO_2\,tot]}{[CO_3^{2}-]}}$$

pH is directly proportional to the concentration quotient of carbon dioxide to carbonate. It is, therefore, the bicarbonate ions which maintain the balance in the concentrations of carbon dioxide and carbonate and buffer changes in pH.

Carbon dioxide and photosynthesis

Carbon dioxide and light are required for photosynthesis. Their availability may limit the photosynthetic rate. The photosynthetic rate varies at any given light intensity with carbon dioxide concentration, approximately in accordance with the Michaelis-Menten equation:

$$P = \frac{P_m\,[CO_2]}{K_m(CO_2) + [CO_2]}$$

where P is the photosynthetic rate,
P_m is the maximum photosynthetic rate,
K_m is the substrate concentration which gives half the maximum rate.

Observed curves of photosynthetic rate vs. CO_2 concentration are approximately fitting this relation; *in situ* dioxide concentrations may indicate to which extent dioxide availability limits photosynthesis in natural waters. The K_m values for the first intracellular enzymatic reaction of CO_2 (carbon dioxide reduction and combination with ribulose 1,5-diphosphate, in the presence of ribulose carboxylase (Lehninger, 1965)) is about 20 μM in the isolated state. This indicates the maximum rate at which CO_2 can be processed within the cell. Values of K_m determined for freshwater phytoplankton range between 20 to 170 μM (determined at pH 5.5 and 560 μE PAR m^{-2} s^{-1} are for *Pediastrum boryanum*,

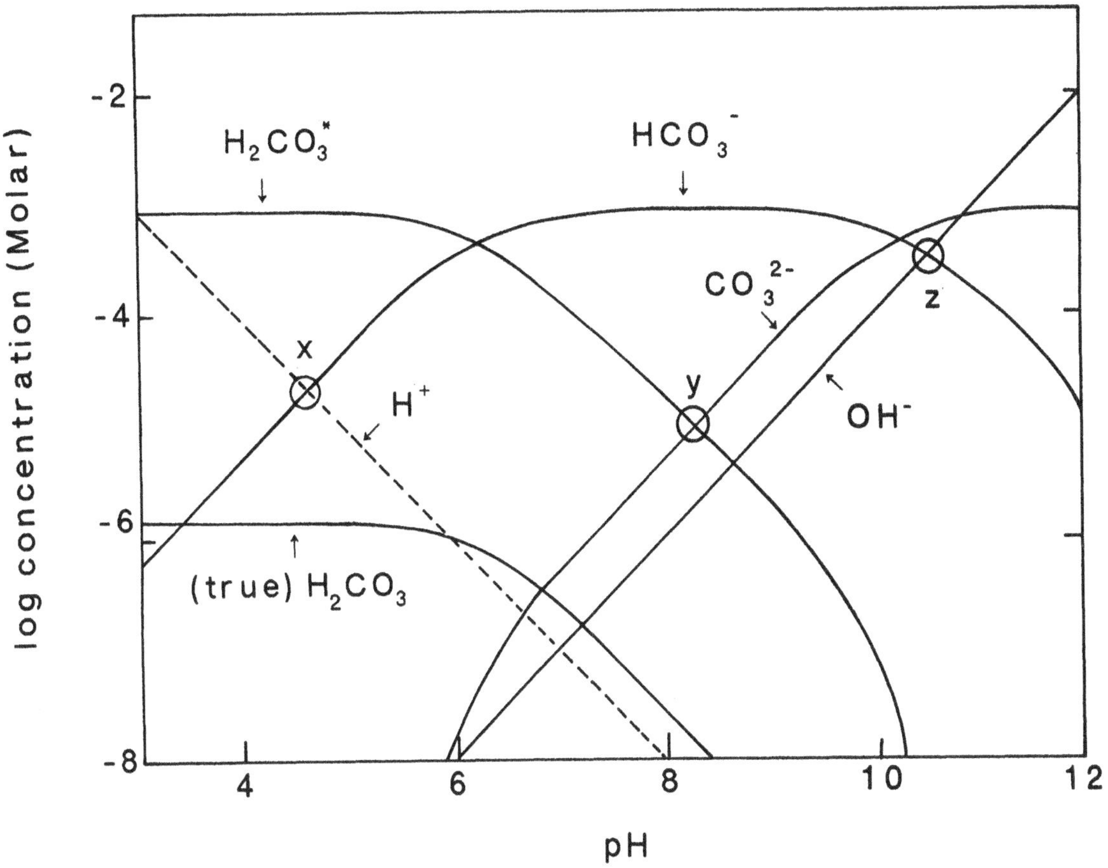

Fig. 24. Distribution of solute species in aqueous fresh water carbonate systems at 25 °C. Logarithmic equilibrium diagram for fresh water. $[H_2CO_3^*] = [H_2CO_3] + [CO_{2\,aq}]$, but here equivalent to $CO_{2\,aq}$, as $CO_{2\,aq}$ is much more abundant than H_2CO_3. H_2CO_3 is a much stronger acid than $CO_{2\,aq}$. The equivalence points marked x, y, z, correspond to pure solutions (molar concentrations) of $H_2CO_3^*$, $NaHCO_3$ and Na_2CO_3 respectively (from Stumm & Morgan (1981), with kind permission of Prof. Stumm & Wiley).

(Chlorococcales), 40 μM; for *Cosmarium botrytis*, a desmid, 170μM; for *Anabaena cylindrica*, a blue-green alga, 60μM. A full noon summer sunlight is about 2000 μE PAR m^{-2} s^{-1}) (Kirk, 1986).

Fresh water in equilibrium with the atmosphere at 15° C contains dissolved dioxide at about 14 μM. It could therefore be expected that aquatic plant photosynthesis in natural waters is undersaturated with respect to dioxide concentration, and should respond to an increase in dioxide concentration with an increase in photosynthetic rate. This does not only apply at saturating light intensity, but even at low light intensities. Most inland water contain more inorganic carbon in the form of bicarbonate ion, HCO_3^-, than dioxide, and increases with pH (Kirk, 1986).

It would clearly be an avantage for aquatic plants to use bicarbonate as a carbon source for photosynthesis. In fact, many do, but not all. (The interconversion reaction is catalysed by a chloroplast enzyme, the carbonic anhydrase; OH^- accumulates and is excreted to counterbalance the uptake of bicarbonate). Most higher plants can utilize bicarbonate, but amongst freshwater phytoplankton this ability varies widely. A possible reason why bicarbonate is less effectively used is that energy must be used for its active transport into the plant, whereas dioxide diffuses in freely. Carbon is not obtained from bicarbonate, except at high pH, above 9. The ability of blue-green algae such as *Anabaena*, *Microcystis*, *Spirulina* to continue photosynthesizing in water of high pH, and effectively zero free carbon dioxide could be one of the main factors contributing to their frequent domination of eutrophic lakes in late summer, and their dominance of highly alkaline waters such as the African soda lakes (Kirk, 1986).

The 14 μM concentration applies only to waters in equilibrium with the atmosphere. But any productive water is likely not to be in equilibrium so far as carbon dioxide is concerned. Carbon dioxide concentration varies with time and depth in accordance with processes of consumption and production.

Photosynthesis decreases the concentration of free dioxide, by increasing pH. Sinking rate of diatoms increases when carbon dioxide becomes limiting (Kirk, 1986).

Influence of alkaline ions on buffer capacity

In waters with a low proportion of calcium, sodium and potassium ions, i.e. soft waters, production of carbon dioxide leads to the formation of sodium and potassium carbonate, which are much more soluble than calcium carbonate, and the increase of bicarbonate ions entails a decrease in pH. In the presence of calcium ions, the precipitation of $CaCO_3$ buffers the pH. Waters with low calcium and high sodium concentrations are much more frequent in Africa (see p. 66).

The importance of bicarbonate ion concentration in the buffer system is expressed by the Henderson-Hasselbach equation:

$$pH = p_{K1} + \log \frac{[HCO_3^-]}{[CO_{2\,tot}]}$$

p_{K1} is the negative logarithm of the value of the ionization constant, $-\log 10^{-6.35} = 6.35$.

The effects of the ionization of carbon dioxide to form carbonate species are modified by other ions, such as sodium, magnesium, and sulphate (Table 10 & Fig. 25). Cations are able to bond with carbonate and sulphate ions, but the bicarbonate ions are always the most important.

Table 10. Reactions which take part in the bicarbonate-carbonate buffer system (according to Spotte, 1979).

Type of reaction								p_K
Ca^{2+}	+	H_2CO_3	$\rightleftharpoons$	$CaHCO_3^+$	+	H^+	(1)	5.450
Mg^{2+}	+	H_2CO_3	$\rightleftharpoons$	$MgHCO_3^+$	+	H^+	(2)	5.502
H_2CO_3			$\rightleftharpoons$	HCO_3^-	+	H^+	(3)	6.124
Na^+	+	HCO_3^+	$\rightleftharpoons$	$NaHCO_3$	+	H^+	(4)	6.745
Mg^{2+}	+	HCO_3^+	$\rightleftharpoons$	$MgCO_3$	+	H^+	(5)	7.665
Ca^{2+}	+	HCO_3^+	$\rightleftharpoons$	$CaCO_3$	+	H^+	(6)	7.913
Na^+	+	HCO_3^+	$\rightleftharpoons$	HCO_3^-	+	H^+	(7)	9.186
HCO_3^+			$\rightleftharpoons$	CO_3^{2-}	+	H^+	(8)	9.776
H_2O			$\rightleftharpoons$	OH^-	+	H^+	(9)	13.791

In the case of a dying algal bloom, the bicarbonate-carbonate buffer system reacts in a surprising way: the production of ammonium and carbon dioxide, originating from decomposition of the algae, in water with a weak buffer system, leads to a drastic drop in pH. In alkaline waters, with a strong buffer system, a large fraction of the ammonium ions forms gaseous ammonia, which is toxic and can cause fish kills (Krom *et al.*, 1989b) (see also p. 148).

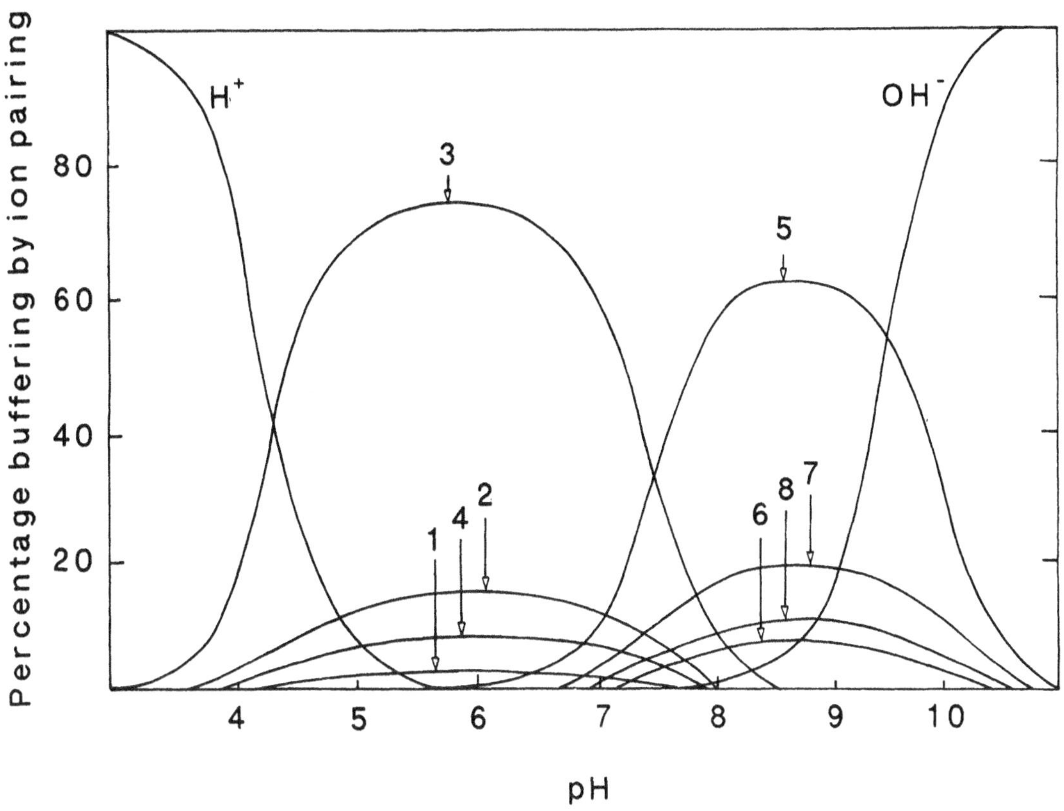

Fig. 25. Contribution of the different ions in solution in water to the buffering capacity of water, expressed as a percentage of buffering by ion pairing, according to the pH (from Spotte, 1979) (with kind permission of Dr. Spotte & Wiley).

Importance of alkalinity, hardness, pH and carbon dioxide

1. pH is a measure of the acidity, alkalinity or neutrality of water. Alkalinity measures the concentration of all species capable of reacting with hydrogen ions, especially carbonate and bicarbonate ions. Hardness is a measure of the amount of calcium and magnesium ions in water.

2. Carbon dioxide is essential for life in a pond as it provides the inorganic carbon needed for photosynthesis. Carbon dioxide dissociates in water to form carbonate and bicarbonate, and participates in the carbonate-bicarbonate buffer system, which regulates pH.

3. Most of the carbon in the water is continuously recycled but some exchanges with air and with sediments exist as well.

3.5. Liming

3.5.1. pH control of water by liming

pH variations in water result from the interaction of various abiotic and biotic processes, as discussed above. Biological processes are responsible for the most rapid modifications, as well as for cyclic variations. In ponds with high plankton biomass and productivity, pH rises slowly from dawn to early in the afternoon. These are the two most important values for a fish farmer to record, as they represent the daily minimum and maximum. If the range between these two values is wide, buffer capacity is weak. If the values are close to each other, buffer capacity is strong. Most fish do not tolerate large pH changes. pH values between 6.5 to 8 are considered optimum.

In areas most commonly available for fish farming, such as marshlands and swampy areas, waters are generally acidic. Natural waters with alkaline pH are less frequently found.

The most common aquacultural practice to control pH is liming, which consists of spreading lime on the pond soil or onto the water.

3.5.2. Effects of liming

Liming is a management practice, aimed at improving the abiotic and biotic conditions of the environment. The effects of lime on soil and water are as follows (Huet & Timmermans, 1970):

a) increase of sediment pH

Lime neutralizes the acidity in soil and water, by adding alkalinity. Proton and hydroxyl ion concentrations decrease and increase respectively, as $CaCO_3$ reacts with protons to form bicarbonate and calcium ions. Laboratory studies of sediment taken at 10-11 m depth in lake Hovvatn, in southern Norway showed effects of lime up to a depth of 6 cm (Smayda, 1990). Thus, modification of soil pH affects water pH. The reaction between lime and acid water is first order and is controlled by the concentration of protons on the surface of the lime particles, and not by mass transport of protons to these particles (Compton *et al.*, 1989).

The calcium ions of lime react with and neutralize the free acid ions of actual acidity, and acid ions sorbed (potential acidity). The decrease in acidity also reduces the solubility of toxic iron (II), aluminium and manganese (II) ions, which precipitate.

The neutralizing action of lime is more important in soil than in water. Lime, even when used in fine particles, eventually settles down to the pond sediment. Here, the concentration of particles is denser, carbon dioxide is more abundant, the actual and potential acidity is higher, and reaction kinetics are more rapid.

b) decrease in redox potential of sediments

The redox potential and pH control the transfer of ions across the water-sediment interface. In a fish pond, sediment is usually oxidized on the surface and reduced underneath. Liming significantly decreases the redox potential, E_7, of the sediment to a depth of 5 to 6 cm, with recorded changes in the order of 100 to - 66 mV. Oxidized conditions persisted at the surface in a Norwegian lake, but, below this, conditions were highly reductant (Smayda, 1990). This pronounced reduction is perhaps due to a significant mineralization within the sediment, which makes use of all electron acceptors available to oxidize organic matter.

c) increase of the base unsaturation of soil, hardness and alkalinity of water, and buffer capacities

Lime reacts with the free carbon dioxide present in water, thus competing with phytoplankton. As a result, the photosynthetic rate decreases (Fig. 26). This transformed carbon dioxide is, however, not lost from the environment; it increases the alkalinity and the buffer capacity of the water.

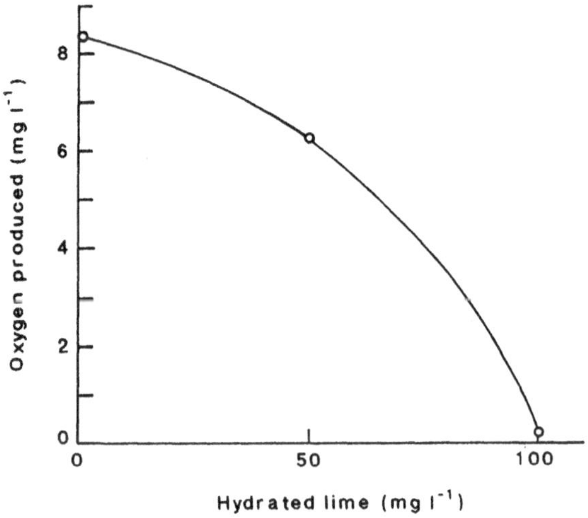

Fig. 26. Effect of different doses of hydrated lime on oxygen production (by-product of photosynthesis) in fish ponds (average value of three ponds), owing to the fixation of carbon dioxide by hydrated lime.

The calcium and magnesium ions from lime are adsorbed onto colloids, thereby increasing the concentration of alkaline ions and the base reserve and base saturation of the sediment. This addition of calcium and magnesium increases the water hardness.

d) acceleration of the decomposition rate of organic matter

Decomposition rates increase with liming. Turnover times of free glucose in surface sediments become shorter, indicating great, persisting increases in microbial activity after neutralization. Easily degradable organic compounds, which accumulate in sediments under acidic conditions, are rapidly decomposed by invertebrates and microorganisms (Hasselrot & Hultberg, 1984).

Lime increases the decomposition rate of organic matter (Fig. 27), especially cellulosic material in sediment, and is responsible for the crumb structure of dried pond mud. This decomposition reduces the risk of propagation of bacterial and parasitic diseases, and decreases the sediment oxygen demand. Acid products resulting from fermenting processes of decomposition are neutralized by lime, thereby dissolving carbonate salts and allowing the mobilization of calcium, magnesium and associated phosphates (Marsden, 1989). This nutrient mobilization stimulates bacterial activities.

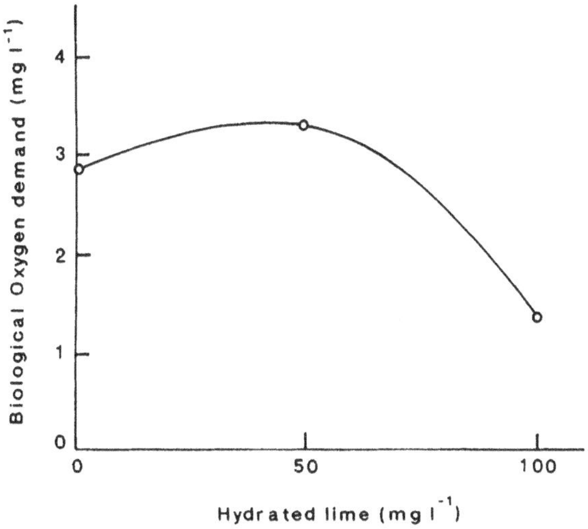

Fig. 27. Effect of different doses of hydrated lime on the biological oxygen demand in fish ponds (average value of four ponds).

Decomposition of organic matter in water is also stimulated by liming, probably as a result of flocculation of dissolved organic matter by calcium ions. Further, dissolved organic matter is adsorbed onto calcium carbonate and precipitates (Saunders, 1980). In three small lakes in Michigan, the precipitation of dissolved organic matter after liming considerably increased light penetration, resulting in the stimulation of green photosynthetic sulphur bacteria in deep, anoxic waters; bacterial blooms resulted. The stimulation of organic matter decomposition had the additional effect of reducing hypolimnetic oxygen levels, with the consequence of hydrogen sulphide entering the photic zone (Leavitt *et al.*, 1989). Though the presence of sulphur bacteria is unlikely in fish ponds, better penetration of light in ponds where, for example, liquid manure is used, can be achieved with lime.

e) precipitation of suspended organic matter

In waters with high concentrations of humic matter, lime reacts with it to form stable aluminium-humate complexes at the surface of the limestone, thus inactivating the lime (Hasselrot & Hultberg, 1984). But, in low concentrations, humic matter, which gives a brownish colour to water and affects light penetration and productivity, precipitates. Precipitation of calcite is prevented by the presence of organic matter. Polyphenols, derivatives of phosphonic acid, polyphosphate, fulvic acids, magnesium and inorganic phosphate, at high concentrations, also hinder the precipitation of calcite, while, at low concentrations, they participate in adsorption and coprecipitation (Stabel, 1986).

Calcite coprecipitates with some algae, by surface nucleation occurring between solutes and cell walls, or by physiological reactions. A possible explanation of this phenomenon is as follows: the photosynthetic assimilation of bicarbonate and release of hydroxyl ions increase the pH of the microenvironment of the cell, so that supersaturation concentrations of calcite are reached, which lead to crystal formation on the cell wall and to precipitation. Algae can also act as surface catalysts for precipitation (Stabel, 1986).

f) acceleration of nitrification processes

Nitrifying bacteria react strongly to hydrogen ion concentration. Although a wide range of pH optima have been reported, there is almost general agreement that, as pH moves into the acid range, nitrification rates decline (Kaiser & Wheaton, 1983). In the light of present information, pH should be above 6.5. Thus, the alkalinizing effect of liming stimulates nitrification.

g) detoxifying and sterilizing effect

Liming promotes aerobic decomposition, limits fermentation and, thus, prevents accumulation of intermediary toxic products. Quick lime and calcium cyanamide kill parasites and pathogens. Spread on soil, they also eradicate aquatic macrophytes with superficial roots.

h) increase of the availability of phosphate and molybdate ions

In water or soil of acid pH, the acid ions aluminium and iron, react with phosphate to form precipitates; in alkaline conditions, phosphate combines with calcium to form apatite. The highest availabilities of phosphate, as well as molybdate, are at intermediate pH. Lime in excess has a deleterious effect, not only on the availability of phosphate, but also on that of iron, manganese, copper, zinc and boron.

i) increase in planktonic and benthic species

In acid waters, most studies revealed a reduced species richness, biomass, and productivity (Almer *et al.*, 1974; Hendrey & Wright, 1976; Kwiatkowski & Roff, 1976; Yan, 1979; Bleiwas *et al.*, 1984), though others observed a shift in phytoplankton groups composing the community, without a decline in biomass or productivity (Schindler, 1985). Liming induces a shift in community structure, with the appearance of species characterized by faster growth and greater nutrient requirements than most acid tolerant species, suggesting a subsequent increase in primary productivity. With regard to the zooplankton communities of acid waters, cladocerans are lacking, while rotifers and copepods are reduced in numbers. Liming of Swedish lakes resulted in increased rotifer density and species number, due likely to increased availability of food for filtering species (Hasselrot & Hultberg, 1984). Changes in the benthic fauna also occurred.

3.5.3. Types of lime used and their equivalences

Different types of lime are on the market and it is useful to know the differences between the products and to establish equivalences (Brady, 1984):

- agricultural lime, crushed limestone, marl, and oyster shells are all calcium carbonate compounds, mixed with varying amounts of magnesium carbonate. They consist, on average, of 75 to 95 % $CaCO_3$, and are almost insoluble in water, dissolving slowly under the influence of carbon dioxide. Dissolution depends on the size of the particles; fine particles of less than 1 mm in size are most efficient and dissolution extends over 1 to 2 months. Amongst the above-cited products, agricultural lime is the most easy to use.
- Quick-lime is actually calcium oxide, CaO, obtained by the burning of lime. It also contains varying amount of magnesium oxide. The action of quick lime is fast, caustic and toxic and, when the base saturation is low, the induced pH changes are great. It is used to kill off parasites and disinfect ponds, tanks, etc. It contains about 85 to 95 % CaO.
- Pulverized, hydrated or slaked lime, $Ca(OH)_2$, is obtained by mixing quick lime with water. The product is caustic and toxic for fish. Slaked lime hardens in air, due to the presence of other oxides as impurities. It contains about 95 to 96 % $Ca(OH)_2$.
- Calcium cyanamide, $Ca(CN)_2$, is toxic, with a residual effect lasting 2 to 3 weeks and is in powdered or pelletized form. The cyanide radical contains nitrogen, which is released after chemical transformation and acts as a nitrogen fertilizer.

Lime is assessed on its chemical composition and the equivalences of the different products are expressed according to three different, but similar, criteria (Brady, 1984):

- the calcium oxide equivalence or the conventional amount of oxide. The different concentrations of calcium and magnesium compounds are converted to their calcium oxide equivalents. Specific conversion factors are used for each compound and the results of the conversions are added to give the total equivalent amount in calcium oxide. The conversion factors are established using the molecular weights of the calcium or magnesium compounds. For example, the conversion factors of calcium and magnesium carbonate are 0.65 and 0.56, respectively
- the neutralization power or calcium carbonate equivalence. The principle is the same as for the calcium oxide equivalence, the calcium and magnesium compounds present (the chemical composition is usually listed by the manufacturer on the bag) being converted into calcium carbonate equivalents. Factors of conversion are established by using the molecular weight of the compounds and are, for example, 1.786 and 1.186, respectively, for calcium oxide and magnesium carbonate
- the percentage of calcium and magnesium, expressed as a percentage in weight of each of these elements.

The calculation of chemical equivalences should not mask the fact that differences between different lime products exist. Each product has its specific reaction and none are totally equivalent.

3.5.4. Liming methods

If the pond has been drained, lime is cast on the bottom and dykes, while, after impoundment, it is sprinkled on the water surface or fed into the pond by the feeder canal. Each of these practices has specific aspects. Liming of soil before impoundment accelerates decomposition processes in the sediment, kills off parasites and eradicates weeds. To obtain this effect, it is better to use quick lime or calcium cyanamide. Huet & Timmermans (1970) recommended a dose of 1000 to 1500 kg ha^{-1} for quick lime or 1000 kg ha^{-1} for calcium cyanamide. The total dose is divided into several applications separated by 8 to 15 days, preferably during periods without rain. The impoundment and stocking of the ponds should be given a supplementary 10 to 15 days lead time. In liming pond water, care should be taken to prevent the pH from becoming too alkaline, especially from reaching values beyond 9.5. Liming of the water fed into the pond is most efficiently carried out by the rotary drum system (Zurbuch, 1984).

When handling lime, protective clothing, especially gloves and goggles, should be worn, and it is recommended to rub some protective ointment on exposed skin.

3.5.5. Identification of the need for lime

The procedure for establishing the need for lime in a fish pond has been extensively described by Boyd (1979). In summary, the hardness of water, which gives a reflection of soil pH, and the base unsaturation of the soil, are measured. Base unsaturation is estimated by adding a soil sample to a buffer solution at pH 8, and measuring the resulting decrease in pH. If the hardness of the water is above 20 mg l^{-1}, no liming is necessary. Also, if hardness is low, but alkalinity high, liming would not be useful. Lime requirements are read from a chart (Table 11), based on initial sediment pH and the decrease in pH of a buffer solution, caused by the soil sample (Boyd, 1976c).

Table 11. Lime requirements, expressed in kg ha^{-1} of calcium carbonate (neutralization power = 100), to increase the total hardness or the alkalinity of the pond water to 20 mg l^{-1} or more (Boyd, 1976c)

Initial pH of the soil	Amount of calcium carbonate (in kg ha^{-1}) needed in connection with the pH decrease of the buffer solution									
	7.9	7.8	7.7	7.6	7.5	7.4	7.3	7.2	7.1	7.0
5.7	91	182	272	363	454	544	635	726	817	908
5.6	126	252	378	504	630	756	882	1008	1134	1260
5.5	202	404	604	806	1008	1210	1411	1612	1814	2016
5.4	290	580	869	1160	1449	1738	2029	2318	2608	2898
5.3	340	680	1021	1360	1701	2042	2381	2722	3062	3402
5.2	391	782	1172	1562	1953	2344	2734	3124	3515	3906
5.1	441	882	1323	1765	2205	2646	3087	3528	3969	4410
5.0	504	1008	1512	2016	2520	3024	3528	4032	4536	5040
4.9	656	1310	1966	2620	3276	3932	4586	5280	5980	6552
4.8	672	1344	2016	2688	3360	4032	4704	5390	6048	6720
4.7	706	1412	2116	2822	3528	4234	4904	5644	6350	7056

The number of samples to be taken to determine lime requirements depends on pond size. Soil properties vary according to its depth. The sediments in the deeper portion of the pond, having higher CEC, are usually more anaerobic, have a more acid pH, and, therefore, require more lime. According to Boyd (1976b), the number of samples should be 12 in ponds of less than 1 hectare, and 25 in ponds between 1 to 4 hectares. The effect of liming lasts for a period of about 10 impoundments and drainages (Boyd & Cuenco, 1980). After each draining of the pond, a dose one fourth to one fifth of the initial dose should be supplemented.

3.5.6. Other methods of neutralization

Neutralization of sediments with sodium bicarbonate

Sodium bicarbonate (soda) has been injected into sediments enriched with organic matter, to saturate them with sodium. The impact of these sodium ions on the neutralization of sediments was compared to that of calcium, used in liming. Sodium ions were exchanged with hydrogen ions, involving consumption of alkalinity, as the protons reacted with bicarbonate to form carbonic acid. Phosphorus was released from the sediments, stimulating algal growth. This gave rise to the idea of using soda, by adding solutions of alkaline sodium, to control adsorption and exchange processes in acid sediments and create a long-lasting neutralization capacity. Sediments act as reversible cation exchangers. Lime is inactivated by humic matter and calcium-humate complexes have a restricted reversibility. This signifies that some of the calcium involved in these complexes is lost to further exchanges; this is not the case with sodium (Lindmark, 1982). In some other experiments, in two small lakes in eastern Norway, soda was deposited onto lake sediments, but here results were unsatisfactory. Because of low levels of water turbulence and mixing, the soda and neutralized water were trapped in a stable layer, formed by differences in density (Rosseland & Skogheim, 1984). This subject is clearly in need of further studies.

Use of nitrogen fertilizers

In waters with high pH and low hardness, one solution to reduce the pH is the use of ammonium fertilizers. Ammonium fertilizers set protons free during the process of nitrification:

$$NH_4^+ + \frac{1}{2} O_2 \longrightarrow NO_2^- + 2H^+ + H_2O \quad (1)$$

$$NO_2^- + \frac{1}{2} O_2 \longrightarrow NO_3^- \quad (2)$$

The drop in pH has, however, no extended effect. The presence of nitrate induces plankton blooms, which bring about wide pH fluctuations. Moreover, all ammonium ions are not nitrified and those which subside are transformed into ammonia. This ammonia, at high pH values, forms toxic gaseous ammonia.

Other products

Alum, an aluminium sulphate, has the immediate effect of diminishing pH (1 mg of alum decreases the alkalinity by 0.5 mg l^{-1} and produces 0.44 mg l^{-1} of carbon dioxide), but without alleviating the causes for the high pH. It is therefore only a temporary measure against critically high pH.

Gypsum, $CaSO_4.2H_2O$, is a substance which contributes calcium ions to water and increases hardness in a proportion of 2.15 mg l^{-1} per mg l^{-1} of total hardness. It is, however, not comparable to a liming product.

Importance of liming

1. Liming is a common practice in aquaculture, aimed at keeping soil pH in an optimum range. The beneficial effects of liming include a long-lasting increase of water and sediment pH, an increase of organic matter decomposition rates in the sediments, precipitation of dissolved organic matter in the water column, and a sterilizing effect on the sediment.

2. A few products are available for liming, but agricultural lime is the most suitable for routine use.

3. Other non-calcareous products have also been tested, some hopes arising from the use of soda, sodium bicarbonate, but a proper methodology has yet to be outlined.

4. Nutrients in water

Relation between nutrients, water fertility and fertilizers

Nutrients represent the natural fertility of the water, on which primary productivity and, ultimately, fish production depend. This natural production is achieved with minimum operational costs, as no feed or fertilizer are added to the pond. In many cases, it is financially rewarding to take maximum advantage of natural productivity, especially plankton production. The increase of production beyond this base level requires knowledge of nutrient requirements and their cycles. Unfortunately, there is as yet no standard procedure for the use of fertilizers, and recommendations found in the aquacultural literature are empirical.

Nutrients are supplied to ponds in the form of inorganic or organic fertilizers. Inorganic fertilizers are characterized by their composition of primary nutrients, i.e. nitrogen, phosphorus and potassium (N:P:K), in mineral form and in finite amounts. Organic fertilizers are either animal excrements or products of organic decomposition, such as compost. Compared to inorganic fertilizers, the composition of organic fertilizers is never well defined, the respective amounts of nitrogen and phosphorus varying according to the type of material used to feed the animal, in the case of excrement, or used to make the compost, and to the time and conditions of stocking. Organic fertilizers contain, in addition, decaying organic material. This material is essentially an energy reservoir and a 'mine' of organic carbon. Potassium is seldom considered, as it is usually accepted that the needs for this element are satisfied by the amounts naturally contained in water. Therefore, potassium fertilizer is usually not added; it is the N:P proportion that matters.

The amount of autotrophic producers is a function of the concentrations of the various nutrients. This relationship has often been considered as the sole determinant of the dynamics of aquatic systems and, on that basis, mathematical models have been constructed. However, persisting presence of algae in great numbers, despite conditions of low phosphorus and nitrogen concentrations, were observed in White Lough, a hard water lake, Northern Irland, and implied that biological factors alter, in a significant way, the relationship between algae and phosphorus (Foy & Fitzsimons, 1987). Actual concentrations of nutrients have sometimes no direct relationship with the physiological state of the autotrophs, while past concentrations do. An improved knowledge of the ecology of algal species and their interactions is also needed to better understand the factors involved in production.

Recycling and regeneration of nutrients

In a steady dynamic system, inputs or supplies of nutrients have to be balanced by outputs. If the supplies of nutrients from outside are less than the rate of utilization, nutrients have to be regenerated or mobilized from the pool of non-available nutrients or storage compartment, by recycling. The importance of the recycling process depends not only on the deficit of nutrients or the proportion of available to non-available forms, but also on production and utilization rates of regenerated nutrients, described in terms of instantaneous changes in available forms (Howard-Williams, 1985).

The regeneration rate is the rate at which a nutrient re-enters the system and is re-utilized; this rate decreases as the proportional amount of the available form increases and vice-versa. Each nutrient has a maximum loading capacity for its available forms and if the nutrient concentration is close to the maximum load, recycling and regeneration rates are low. In running waters, the continuous unidirectional water flux carries away nutrients and their utilization is only possible during a short time. This gave rise to the concept of nutrient spiralling, defined as the distance a nutrient will travel before being taken up; this will depend on the upstream supply rate and the utilization or exchange rate (Howard-Williams, 1985). Similarly, in a production system, there is exportation of nutrients in the form of fish at harvest; the supply flux of nutrients must be regulated with regards to utilization rates.

4.1. Rules for the use of fertilizers

Some general rules for the assessment of fertilizer needs should be followed based on their modes of action.

The limiting factor(s) of natural productivity should first be determined. If they are physical factors, such as light or temperature, which are difficult to modify drastically, there is not much one can do. However, other physical and chemical factors can, by proper management, be adjusted to suit production needs.

Empirical determination of fertilizer needs, by comparative trials, requires a careful choice of ponds. Firstly, the experimental ponds have to be of intermediate size, not less than one fifth of the area of the target ponds. They should have been previously used for fish farming for at least three years, have the same fertilization history, and their natural productivity should be known. The stocking rate of fish should be the same as in the target fish ponds. The feeding or non-feeding of the fish should be considered. If the actual contribution of fertilization to production is to be assessed, no feeding should be carried out. But if feeding and fertilization are supplementary, a comprehensive sequence of the conditions should be tested: no fertilizer and no feed; fertilizers only; fertilizers and feed. Concerning general pond conditions, the pond should be weeded, and pH of the water should be neutral or within the optimum pH range and the hardness and alkalinity within the normal range (above 20 mg l^{-1}). If liming is required, fertilization should not take place prior to a period of 8 to 15 days after liming. Lime precipitates with phosphate and therefore liming and fertilization should be applied separately.

Before discussing the use of inorganic fertilizers in fish farming in Africa, it is important to mention that intensive fish farming never depends exclusively on inorganic fertilizers; their use is supplementary to artificial feeding. Inorganic fertilizers, because of their fixed chemical composition, are a useful experimental tool to determine the nitrogen or phosphorus needs in specific ponds, and serve especially in the first steps of the empirical determination of fertilizer needs.

Fertilizers stimulate primary productivity and therefore some questions need to be asked beforehand, such as how much plankton is required, what role this plankton should play, and what proportion of natural feed to artificial feed is desirable. The production of natural feed is also justified by the fact that most artificial feeds in common use (in Africa) are not nutritionally balanced, so that natural feed can provide missing nutrients, such as essential amino acids, vitamins and minerals.

The cost of inorganic fertilizers argues strongly against their exclusive use, without any other feed or manure, in intensive culture. Any use of inorganic fertilizer should therefore take production costs into account. This criterium alone will indicate whether use of inorganic fertilizer makes sense.

Organic fertilizers should be used in small amounts, and preferably at short intervals. The pond has no way to store, in the long term, the nutrients (at least nitrogen) supplied; a consequence of this is a rapid decline of plankton biomasses, as soon as fertilization stops (Shearer *et al.*, 1987). In ponds with sediments of low exchange capacities, it is especially important to apply fertilizers frequently, but in small amounts.

Importance of nutrients for the production of natural feed

1. Nutrients are the primary inputs for autotrophic production, along with light. Knowledge of the nutrient needs is essential for achieving high primary production and, eventually, high fish production.

2. No standard procedure for the determination of fertilizer needs has yet been established, partly because nutrient cycling in ponds is not well understood. Mathematical modelling would require quantitative knowledge of the various processes involved.

3. Inorganic and organic fertilizers are available for use in ponds. Nutrients are present in defined amounts in inorganic fertilizers. The latter are rather costly. Organic fertilizers are cheap, and provide, besides nitrogen and phosphorus, organic carbon, which is an energy source. However, the supply of nitrogen and phosphorus, from organic fertilisers, is usually not properly controlled as the concentrations of these nutrients are variable. Potassium is usually not required in ponds.

4. The determination of fertilizer needs requires identification of the factors which limit production.

4.2. Nitrogen and nitrogen compounds

Nitrogen, along with carbon, oxygen and hydrogen, is one of the four major elements, of which organic molecules are composed, and is therefore quantitatively important in living matter. Nitrogen needs of animals are met principally by proteins, while plants use mineral nitrogen compounds. Nitrogen assimilation occurs simultaneously with that of other nutrients, in accordance with a given proportion between nutrients. The carbon-nitrogen-phosphorus ratio (C:N:P) most commonly used with reference to algal needs is the Redfield ratio of 106:16:1, though other ratios are mentioned, such as a C:N:P of 100:12:1 (Blažka *et al.*, 1980). This proportion is not fixed, as the levels of nitrogen and phosphorus in algal cells change constantly, through assimilation and other metabolic activities.

4.2.1. Nitrogen cycling in water

Nitrogen compounds in water

Nitrogen is found in water as gaseous molecular nitrogen, (N_2), NO_3^-, NO_2^-, NH_4^+, gaseous ammonia, NH_3; and organic compounds: urea, amino acids, and polypeptides. The nitrogen compounds found in natural waters are of various origins:

- nitrate ions come from run-off water, especially from fertilized agricultural soils, and from nitrogen fixation in soils. They are not firmly sorbed onto soil particles and are easily washed away with heavy rains. They are present in sediments and polluted water as well
- nitrite ions, NO_2^-, are produced by bacterial oxidation of ammonium or by reduction of nitrate. They are only present in water with low dissolved oxygen
- ammonium concentrations are usually low, but may fluctuate widely in time and space. Ammonium is produced by decomposition of organic matter or animal excrements, by deamination (McCarthy, 1980). High concentrations of ammonium ions inhibit phytoplankton development (Pańczakowa & Szyszka, 1986)
- organic nitrogen substances include urea, released by animal excretion, free amino acids and peptides, which are of importance for microorganisms, and a number of unidentified organic compounds.

The transformation cycle of the different compounds is shown in Fig. 28.

Nitrogen fixation in water

Nitrogen enters the cycle as atmospheric molecular nitrogen, which dissolves in water and is fixed by nitrogen-fixing bacteria and blue-green algae. In freshwater, blue-green algae are responsible for most of the planktonic nitrogen fixation. In mesotrophic and eutrophic waters, benthic nitrogen fixation is done mainly by heterotrophic bacteria, while, in oligotrophic waters, blue-green algae are important. In oligotrophic and mesotrophic waters, nitrogen fixation does not represent an important nitrogen supply, whereas in eutrophic waters, where phosphorus is not limiting, nitrogen fixation contributes significantly to primary production (Howarth *et al.*, 1988a).

Nitrogen fixation is significant when the N:P supply ratio is below the Redfield ratio of 16:1 (Horwarth *et al.*, 1988b). Nitrogen fixation is carried out in the algal heterocysts and is proportional to their density (McFarland & Toetz, 1988). In some African lakes, like Lake George and L. Tanganika, nitrogen fixation accounts for about half of total nitrogen input (Hecky *et al.*, 1991). The fixation rate of atmospheric nitrogen in Lake George, Uganda, (mean depth of 2.5 m) was 0.44 to 1.69 mg m^{-2} h^{-1} (Horne & Viner, 1971) and 11 to

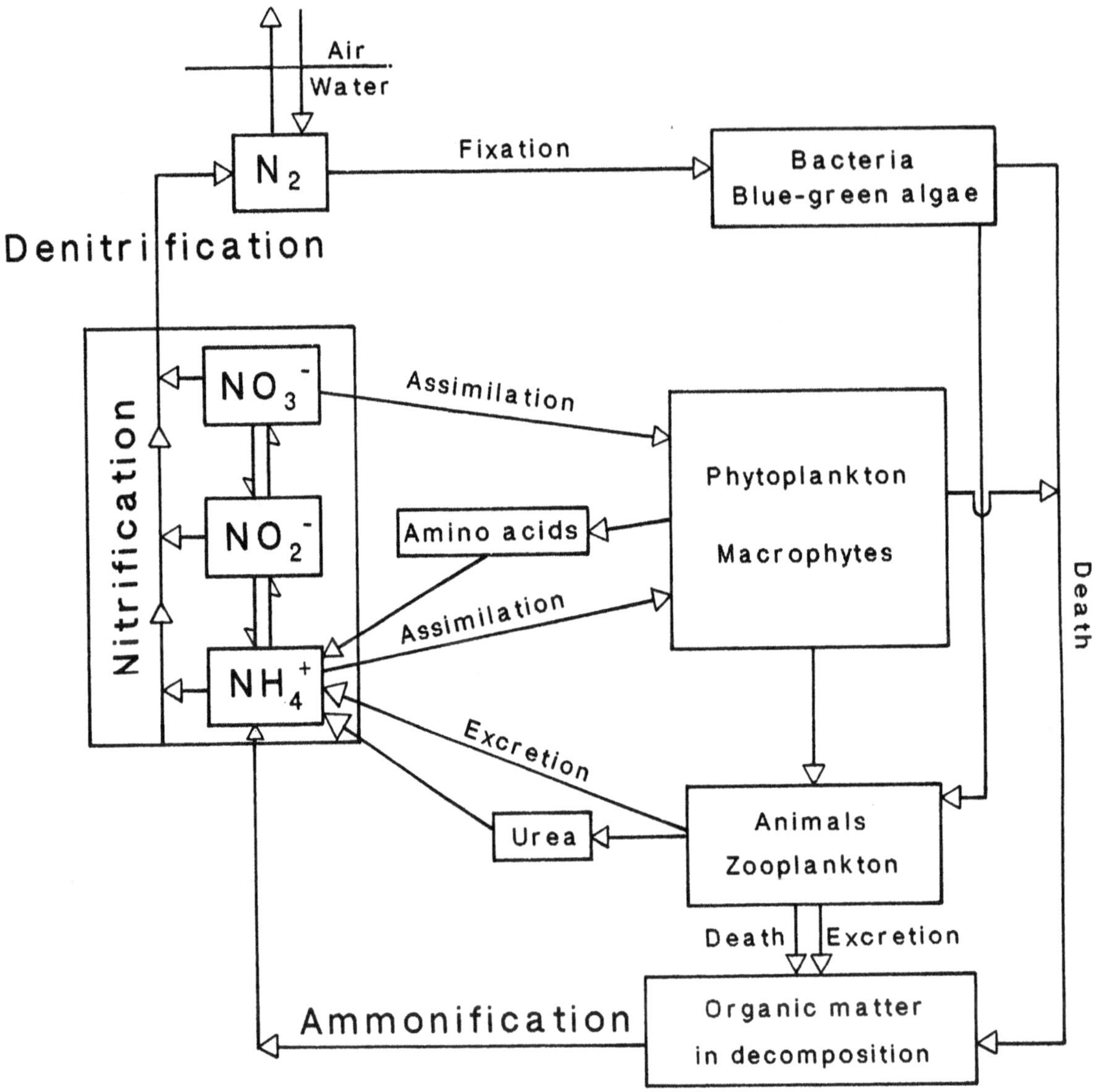

Fig. 28. Nitrogen cycling within ponds.

58 mg m^{-2} d^{-1} of N (Ganf & Horne, 1975). Nitrogen fixation was maximum in Lake George at the surface and at a depth of about 40 cm and minimum at depths of 15 cm and 150 cm (Horne & Viner, 1971). Nitrogen fixation in Hungarian fish ponds amounted to 4.7 mg m^{-2} d^{-1} or 0.33 mg m^{-2} h^{-1} of N, and was temporarily inhibited by the use, in the pond, of nitrogen fertilizers (El Samra & Oláh, 1979).

Factors inhibiting nitrogen fixation

Considerable N-fixation occurs in the dark. Oxygen inhibits the nitrogenase, as photorespiration and nitrogenation compete for reductants (Ganf & Horne, 1975). The activity of nitrogenase depends more on the previous history of the cell than on prevailing nutrient concentrations (Wynne & Berman, 1990).

As nitrogen fixation is inhibited by oxygen, the

occurrence of the process in water has to be explained. According to Paerl (1985), calm conditions in the water allow the creation of anoxic microzones around cells, conducive to fixation. These microzones are also conducive to the extracellular reduction of microelements, such as molybdenum and iron.

Ammonium and ammonia (NH_{4+} and NH_3) are strong inhibitors of nitrogen fixation (McFarland & Toetz, 1988), their effect being less in water than in sediment (Howarth *et al.*, 1988b).

Favourable conditions for nitrogen fixation include calm waters of low mixing depths and long stratification periods, light intensity, and, probably, high temperatures (Horwarth *et al.*, 1988b).

Ammonification

Ammonification is a transformation of organic matter (Fig. 29), yielding ammonia. It transforms into the ammonium ion in the presence of water or is released from water in gaseous form, when the solubility of ammonia is exceeded.

Ammonification is performed by heterotrophic bacteria, whose density is strongly dependent on organic substrate available (Barat & Jana, 1987). Their activity is more important in the presence of light and when dissolved oxygen is abundant (Jana *et al.*, 1983). High oxygen concentrations are related to active metabolism of phytoplankton, which find, because of the activities of ammonifying bacteria, high concentrations of available nutrients; this relationship benefits both the bacteria and phytoplankton. Harrison (1978) established a correlation between ammonification and ammonium and carbon dioxide assimilation, suggesting a coupling between nitrogen recycling and photosynthesis. High concentrations of calcium, magnesium and phosphate ions also have a positive effect on ammonification.

Drying up of the pond bottom between production periods reduces, by ammonification, the amounts of organic matter in the sediment and, as a consequence, the amount of ammonium in solution after the dying off of planktonic blooms, in subsequent productions. Thus, Krom *et al.* (1989a) observed that ammonium concentrations in water were higher

Denitrification: $NO_3^- \rightarrow NO_2^- \rightarrow NH_3$

$NO_2^- \downarrow$ $N_2O \rightarrow N_2$; $NO_2^- \searrow N_2$

Ammonification: Org N $\rightarrow$ HN_4^+

Nitrification: $NH_4^+ \rightarrow NO_3^-$

Fig. 29. Chemical transformations involved in the main processes of the nitrogen cycle.

after the dying off of algal blooms in ponds which had not been dried during the production interphase as compared to those which had. This would mean that the sediments of undrained ponds were almost or fully saturated with respect to ammonium, and could not adsorb ammonium from the water.

Nitrification

Ammonium ions are absorbed directly by plants or phytoplankton, or first transformed, by nitrification, into nitrite and, later, nitrate. The oxidation of ammonium is done by chemoautotrophic bacteria, to which this reaction provides energy, as shown in equations (1) and (2) (p. 82).

The complete oxidation of 1 mg of $N\text{-}NH_4^+$ requires 4.6 mg of oxygen. The bacteria responsible for reaction (1) belong to the genus *Nitrosomonas*. This reaction produces 13.9 to 20.1 kJ per mole of ammonium oxidized. The bacteria of the genus *Nitrobacter* perform reaction (2), in which 3.68 to

5.0 kJ per mole of nitrite are liberated. Optimum pH are 7.8 for reaction (1) and 6.6 to 6.7 for reaction (2), at temperatures of 25 to 35° C (Kaiser & Wheaton, 1983).

This process is largely dependent on the availability of a substrate, to which bacteria can attach. The process is inhibited by high light intensity (Lipschultz *et al.*, 1985), but presence of oxygen is a prerequisite for this photoinhibition, which is maximum at 420 nm, i.e. blue light. The substrate or analogous compounds (ammonium, methylamine or methanol, and nitrite or citrate) partially protect the enzymes. The temporary accumulation of nitrite in the water column is sometimes due to excessive light (Diab & Shilo, 1988) or to a paucity of phytoplankton (Krom *et al.*, 1989b).

Gaseous ammonia and nitrous acid also inhibit nitrification. At high concentration, ammonia inhibits the oxidation of ammonium and nitrite. Nitrification releases protons and is therefore a source of acidification of the water, consuming alkalinity. This can be controlled by use of lime (Hunt & Boyd, 1981).

Denitrification or dissimilation

In the absence of oxygen, some bacteria use nitrate ions instead of oxygen as electron acceptors for respiration, releasing molecular gaseous nitrogen as a byproduct (Jørgensen & Sørensen, 1985). This process is called dissimilation or nitrate respiration. The products released are nitrite, ammonium, nitrogen oxide or molecular nitrogen. The gaseous products tend to escape from the water and are lost to the environment. When the end-product is nitrogen oxide or molecular nitrogen, the process is called denitrification. The number of bacteria capable of denitrifying are numerous and not all are obligate anaerobes. They can also make use of alternative energy metabolisms (Kaspar, 1985).

Dissimilation is inhibited in the presence of oxygen, which affects the synthesis of nitrate and nitrite reductases, the enzymes which perform this transformation.

Assimilation of nitrogen by plants and phytoplankton

Nitrogen ions are physiologically required by autotrophs, especially for amino acid synthesis. Among the different nitrogen species, ammonium is preferred to other forms, i.e. nitrate, nitrite and urea, because the uptake of these other forms requires a supplementary energy expenditure to transform them into ammonium, NH_4^+, for example, 18.5 kJ mole^{-1} at 25° C for nitrate (Thompson *et al.*, 1989). The uptake of ammonium increases with light intensity and is maximum at a light intensity which inhibits nitrification (Lipschultz *et al.*, 1985). Nevertheless, ammonium uptake also takes place in the dark, and can reach 70 % of the diurnal maximum.

Nitrate or nitrite absorption requires the presence of nitrate or nitrite reductase, yielding ammonium as an end product. The enzymes depend on the intracellular substrate pools which reach concentrations much in excess of those in the medium. Intracellular nitrite never appears in detectable amounts, which suggests the existence of small pools, compared to those of intracellular nitrate and ammonium (McCarthy, 1980). Nitrite reductase activity is inhibited in the presence of nitrate. Enzyme activities vary in the course of the day and absorption is a process which happens within minutes (Krom *et al.*, 1989a). Nitrate reductase is more active during daytime, with a maximum activity at mid-day, while nitrite reductase activity increases during nighttime, to reach a maximum which peaks at around five times the diurnal minimum. Glutamate dehydrogenase is also more active at night; this enzyme regulates the incorporation of the amine radical (amination) in the α-ketoglutarate. Assimilation is maximum during the first half of the day, and minimum during the first half of the night (McCarthy, 1980).

When the concentrations of NH_4^+ in the water are above 5 μg-at l^{-1} of nitrogen, phytoplankton absorbs ammonium in preference to nitrate, even when nitrate concentrations are high (McCarthy, 1980: 213, Fig. 5.4). Ammonium accounts for 92 % of the nitrogen uptake of the cells, with nitrate contributing not more than 7 %, when ammonium ($N\text{-}NH_4^+$) ion concentrations are above 1 μg-at l^{-1} (McCarthy, 1980). The importance of ammonium uptake is therefore 4 to 30 times greater than that of nitrate; ammonium recycling is effected in 0.3 to 1.23 days.

Nitrogen uptake affects water quality. The uptake of mineral nutrients, such as nitrogen and phosphorus ions, which are metabolites, cleanses the water. At the same time, plants and phytoplankton release amino acids and vitamins into the water, improving its quality.

Relative importance of the various processes

Budgeting the amount of nitrogen recycled in a pond is difficult: measurements of the amounts involved in the various processes, such as nitrogen fixation, denitrification and ammonification, are cumbersome and time-consuming and need to be considered simultaneously to give a real picture of the nitrogen fluxes.

A considerable amount of nitrogen is supplied by the organic loading to the pond. This material is continuously recycled; for some compounds, this occurs rapidly as they leach from the excreta and are almost immediately taken up again, while others go through the decomposition process, being gradually released. Some of the material released, such as the faecal pellets of the zooplankton, for example, are small quanta suspended in the water column, and thus not easily discriminated.

Bacteria and zooplankton make use of large amounts of amino acids for their energetic needs, and release much ammonia into water. This ammonia supply is difficult to assess (Golterman & Kouwe, 1980). The turnover time of ammonia is short, varying between some tens to some fractions of hours, and is important in sustaining production.

Nutrient enrichment experiments on a natural seawater plankton assemblage gave following results. In conditions of low concentrations, most ammonia uptake was bacterial. Increase in dissolved nitrogen concentration benefitted phytoplankton. Its increased uptake resulted in increased primary production. This trend was similar as regards phosphate, where the uptake could also be partitioned between bacteria and phytoplankton, according to availability. Availability of a nutrient determines the outcome of exploitative competition between bacteria and phytoplankton, and between species of phytoplankton. As more of a limiting resource is directed into larger cells under a patchy nutrient regime, these cells should do better when nutrient supplies are in pulses, than when present at a low, homogeneous concentration (Suttle *et al.*, 1990). Availability does not imply persisting concentrations, but includes short-lived high concentrations or pulses of nutrient supply. Pulsed supplies permit a higher diversification of the assemblage, where species with different requirements can coexist.

An example of a nitrogen budget in a fish pond in Israel is given by Schroeder (1987). The different inputs of nitrogen into the water, and outputs and losses to the sediment and to fish were quantified. The principal sources of nitrogen in the pond were phytoplankton respiration (670 mg m^{-2} d^{-1} of N) (51.6 %), followed by manure and fertilizers added to the pond (29.7 %), and fish metabolism (15.6 %). Oxic microbial processes at the sediment surface and zooplankton metabolism were insignificant, bioturbation processes could not be quantitatively assessed, and nitrogen fixation was not mentioned. The total quantity of nitrogen supplied to the water column was estimated to be 1.3 g m^{-2} d^{-1} of N. The gross primary productivity removed nitrogen from the water at a rate of 1.31 g m^{-2} d^{-1} of N, representing 89.7 to 96.3 % of the daily withdrawals. Of this algal production, 38.9 % sedimented to the bottom, out of which 70 % accumulated in the sediments and 20 % was returned to the water by oxic decomposition and benthic fluxes. Of the nitrogen added by manuring and fertilization (386 mg m^{-2} d^{-1}, i.e. 30 kg N ha^{-1} wk^{-1}), 18 % was incorporated into fish, and 52 % was lost through seepage, diffusive flux to the atmosphere, and denitrification.

Ammonia toxicity for aquatic organisms

Ammonia concentrations together with pH need to be monitored, especially when ammonia-containing fertilizers are added to the water. With increasing pH, the ammonium ion-ammonia equilibrium tends to shift towards ammonia:

$$NH_4^+ + OH^- \rightleftharpoons NH_3 + H_2O$$

The dissociation constant of the reaction is $10^{-4.74}$ at 25° C. The proportion of gaseous ammonia at different pH and temperature is shown in Table 12 and Fig. 30.

Gaseous ammonia is toxic to fish; the fish farmer should make sure that concentrations do not reach lethal levels. There is, however, disagreement between authors on the lethal dose.

The maximum acceptable ammonia concentration is 0.1 mg l^{-1} of N-NH_3; the conversion factor for N-NH_3 to NH_3 is 0.8235 (Meade, 1985). Ammonia toxicity causes osmoregulatory imbalance, kidney

Table 12. Percentage of ammonia in solution at temperatures between 16 and 30° C and pH values between 6 and 10 (according to Emerson *et al.*, 1975)

Temper-ature (° C)	pH								
	6.0	6.5	7.0	7.5	8.0	8.5	9.0	9.5	10.0
16	0.0295	0.093	0.294	0.925	2.87	8.54	22.8	48.3	74.7
17	0.0318	0.101	0.317	0.996	3.08	9.14	24.1	50.0	76.1
18	0.0343	0.108	0.342	1.07	3.31	9.78	25.5	52.0	77.4
19	0.0369	0.117	0.368	1.15	3.56	10.5	27.0	53.9	78.7
20	0.0397	0.125	0.396	1.24	3.82	11.2	28.4	55.7	79.9
21	0.0427	0.135	0.425	1.33	4.10	11.9	29.9	57.5	81.0
22	0.0459	0.145	0.457	1.43	4.39	12.7	31.5	59.2	82.1
23	0.0493	0.156	0.491	1.54	4.70	13.5	33.0	60.9	83.2
24	0.0530	0.167	0.527	1.65	5.03	14.4	34.6	62.6	84.1
25	0.0569	0.180	0.566	1.77	5.38	15.3	36.3	64.3	85.1
26	0.0610	0.193	0.607	1.89	5.75	16.2	37.9	65.9	85.9
27	0.0654	0.207	0.651	2.03	6.15	17.2	39.6	67.4	86.8
28	0.0701	0.221	0.697	2.17	6.56	18.2	41.2	68.9	87.5
29	0.0752	0.237	0.747	2.32	7.00	19.2	42.9	70.4	88.3
30	0.0805	0.254	0.799	2.48	7.46	20.3	44.6	71.8	89.0

failure, suppressed excretion of endogenous ammonia, resulting in neurological and cytological failure, and damage to the gill epithelium, which leads to suffocation. Establishment of safe concentrations of ammonia is based on gill histology, ammonia poisoning giving rise to hyperplasia of the gill epithelium, separation of the epithelium from the underlying membranes, and hypertrophy and fusion of gill lamellae, leading to suffocation (Meade, 1985).

As ammonia molecules are small and do not carry electrical charges, they are transported by passive diffusion and penetrate cells by crossing the cell membranes. They have to be eliminated by active excretion. A correlation between ammonia concentration in water and blood has been demonstrated. Ammonia is transformed into ammonium and exchanged for sodium ions. The ionic concentration of the water directly affects the excretion process; low concentrations of sodium, calcium and chloride ions increase ammonia toxicity. Any factor affecting the osmotic balance influences fish tolerance to ammonia.

Various factors affect ammonia toxicity, including urea, amine and amine oxide derivatives, creatine, creatinine, uric acid, carbon dioxide, and dissolved oxygen concentrations. Carbon dioxide is excreted by the gills and causes some alkalinization of the gill lamellae, when carbon dioxide forms bicarbonate on contact with water. These lamellae are thus exposed to high, though locally restricted, concentrations of ammonia, resulting from high pH.

Fish have specific tolerances, changing with the period in their life history, and with past history. A fish, previously exposed to low concentrations of ammonia, better supports high concentrations after-

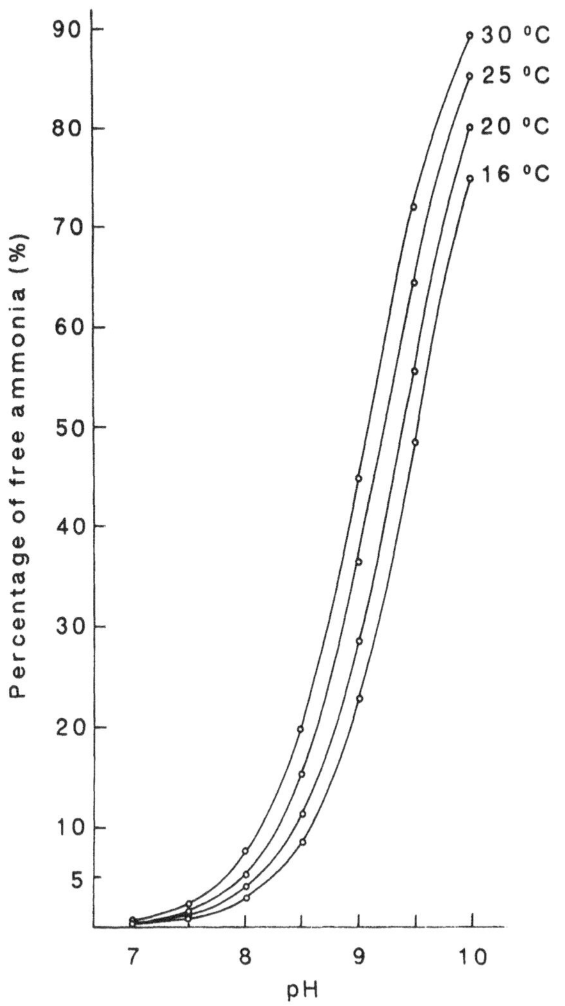

Fig. 30. Proportion of gaseous ammonia, expressed as a percentage of free ammonia, resulting from the equilibrium reaction between the ammonium ion and ammonia at different temperatures and pH's (based on data of Emerson *et al.*, 1975).

wards. Varying concentrations are more harmful than steady ones. Stress reduces resistance of the fish.

Gaseous ammonia is seldom measured, although suitable electrodes exist. Usually, the concentration of total ammonium, gaseous and soluble, is titrated and, with the help of Table 12, the concentrations of gaseous ammonia and ionized ammonium are established, in compliance with pH and water temperature.

Amounts of ammonia excreted by fish

The main ammonia sources in ponds are fish excretion and decomposition of excess unconsumed feed. Attempts to assess the amounts produced per pond have been made using different methods: consumption of feed containing nitrogen, and respiration (Meade, 1985):

- ammonia production (in kg d^{-1}): $W_t \times R_F \times N_D \times N_M \times N_E$
  where W_t = fish weight
  R_F = feeding rate as percent of body weight per day
  N_D = percent protein-nitrogen in the feed
  N_M = percent protein metabolized
  N_E = percent excreted nitrogen excreted as ammonia

- the ammonia production : 0.053 x the oxygen consumption of the fish (in kg d^{-1} of O_2).

In experiments on nitrogen loading and ammonification rates in cement tanks (at the facilities of University of Kalyani, West Bengal) stocked with carnivorous and forage fish, Barat & Jana (1987) observed that daily nitrogen excretion rates of carnivorous fish, *Heteropneustes fossilis* and *Clarias batrachus*, were more important than those of carp, (common and Indian carp, *Labeo rohita*, *Cirrhinus mrigala*, *Catla catla*).

A good approximation is 20 to 80 g N-NH_3(tot.) kg^{-1} d^{-1} of feed. Nitrogen excretion by fish varies in the course of the day, according to the nature of feed consumed, and the period of feeding and duration of digestion. Some 72 to 79 % of the nitrogen of the consumed feed is contained in excreta (Edwards, 1980).

Ammonia is more of a problem when fish are cultured in artificial environments, such as tanks. In African ponds, during periods of active photosynthesis, high alkaline pH values (readings close to 11) are not uncommon, and ammonia toxicity could be a menace.

Nitrogen and nitrogen compounds in water: salient points

1. Nitrogen is present in large amounts in organic matter, and, therefore, the needs for nitrogen are high. Requirements are also linked to those of carbon and phosphorus, in a relatively fixed proportion.

2. Five inorganic nitrogen compounds, i.e., ammonium, nitrate, nitrite, molecular nitrogen and ammonia, and a multitude of organic compounds are present in water. Their respective concentrations change continuously.

3. Nitrogen enters water by nitrogen fixation, mainly by blue-green algae; this process can contribute significantly to the nitrogen budget and primary productivity, even in eutrophic waters. Oxygen inhibits nitrogen fixation, but the process proceeds, nevertheless, in anoxic microzones surrounding the cells.

4. The major nitrogen source in a pond is the ammonification process, which removes ammonium from organic molecules. This bacterial process requires oxygen and light. There is some speculation concerning a coupling between ammonification and photosynthesis. Ammonium released to the water by decaying organic matter is eventually sorbed onto the sediment.

5. Ammonium is subject to nitrification, i.e. transformation into nitrate. This is also a bacterially mediated process, yielding energy, of which light is an inhibitor.

6. Nitrate is used by some bacteria as an electron acceptor, in place of oxygen, being transformed into molecular nitrogen, nitrogen oxide or ammonium. The process is termed denitrification, dissimilation or nitrate respiration.

7. Ammonium is the nitrogen species preferentially absorbed by autotrophs, but, in the absence of ammonium, nitrate or nitrite are taken up as well. The assimilation of nitrate or nitrite requires the intervention of nitrate or nitrite reductase, the activities of which seem to be highest under conditions in which the presence of nitrate and nitrite compounds is most likely. Algal nitrogen uptake competes with that of bacteria.

8. The nitrogen budget of a pond is difficult to establish, as nitrogen compound cycling is rapid and the many nitrogen sources are difficult to measure, e.g. the contribution of zooplankton excretion. A nitrogen budget in a fish pond has shown that primary productivity daily consumed almost all nitrogen dissolved in the water, but half of this nitrogen was returned daily through respiration. The addition of fertilizers compensates for the phytoplankton lost due to settling to the sediments.

9. Ammonium transforms into ammonia at alkaline pH, and is toxic to fish.

10. The amounts of ammonium excreted by fish can be estimated from their oxygen consumption or from calculations based on nitrogen content in feed.

4.2.2. Nitrogen fertilizers

Nitrogen fertilizers available

Different inorganic fertilizers are available (Table 13).

Acidity released by fertilizers

Fertilizers containing ammonium and urea have an acid reaction in water: the transformation of the ammonium ions into nitrate releases protons (as discussed earlier p. 82). Hunt & Boyd (1981) have calculated the decrease in alkalinity produced by the principal nitrogen fertilizers and the amount of lime necessary to neutralize the acidity produced. Each milliequivalent of ammonium liberates 2 meq of hydrogen ions, which react with 2 meq of calcium carbonate; knowing the molecular weight of the different fertilizers, the theoretical amounts of lime necessary can be calculated (Table 14).

Table 13. Chemical composition and nitrogen content (in percent of weight) of the most common nitrogen fertilizers (Brady, 1984)

Name of the fertilizer	Chemical composition	Percentage of nitrogen
Sodium nitrate	$NaNO_3$	16 %
Ammonium sulphate	$(NH_4)_2SO_4$	21 %
Ammonium nitrate	NH_4NO_3 or NH_4NO_3 + dolomite	33 - 34 %
Urea	$CO(NH_2)$	42 - 45 %
Calcium cyanamide	$Ca(CN)_2$	22 %
Liquid anhydrous ammonia	NH_3	82 %
Ammonia	NH_4OH dil. or NH_4NO_3 in NH_4OH or urea in NH_4OH	20 - 25 % 27 - 53 %
Ammonia phosphate	$NH_4H_2PO_4$	11 % (48 % in P_2O_5)
Diammonium phosphate	$(NH_4)_2HPO_4$	21 % (53 % in P_2O_5)

Table 14. Amounts of agricultural lime (in kg of calcium carbonate) per 100 kg of nitrogen fertilizer, to neutralize the fertilizer acidifying effect

Type of fertilizer	Amount of agricultural lime (kg 100 kg^{-1} of fertilizer)
NH_4NO_3	125
$(NH_4)_2SO_4$	152
$(NH_4)_2PO_4$	114
$NH_4H_2PO_4$	87
$(NH_4)_2CO$	167

Acid release by the fertilizers appears only after some time. This is explained as follows. The population of nitrifying bacteria is directly dependent on the amount of substrate available, responding to the supply of fertilizer by a slow transformation rate at first, but increasing after some time.

Destiny of fertilizers in water

The addition of nitrogen fertilizer to a pond is followed by short-lived peaks in ammonium and nitrate. The disappearance of inorganic nitrogen added to water could result from biological immobilization (nitrogen uptake and incorporation into organic compounds), denitrification of nitrate, transformation of ammonium into ammonia and subsequent gaseous escape of the latter, or nitrification of ammonia, followed by denitrification (Bouldin *et al.*, 1974). The relative importance of each process should be assessed (see p. 89 for results of Schroeder, 1987). Amounts of nitrogen sorbed onto and accumulating in the sediment are relatively small. Gaseous escape of ammonia is restricted to periods of high pH and does not appear to be significant.

Relation between nitrogen needs and needs for other nutrients

Nitrogen addition is required when the amounts of available inorganic nitrogen are limiting. After addition, nitrogen is no longer the limiting factor, but some other nutrient or environmental factor takes over. Nutrient needs must always be considered interconnected (see p. 138 & p. 140). Nitrogen and phosphorus are generally more limiting than carbon, but light can also become limiting, for instance in cases of self-shading, when production is very high (see p. 132 & p.135).

Nitrogen and phosphorus needs in a pond are assessed on the basis of concentrations of dissolved nitrogen species and orthophosphate in water. Ideally, the stocks accumulated in sediments should also be taken into consideration. The optimum concentrations of dissolved nitrogen species and phosphate, above which no fertilization is required, have not been clearly outlined. Boyd (1979) considered that 2.0 mg l^{-1} of total inorganic nitrogen and 0.5 mg l^{-1} of dissolved orthophosphate are satisfactory. He (Boyd, 1981b) measured average concentrations of inorganic nitrogen of 200 $\mu g\ l^{-1}$ in productive ponds (sunfish ponds in Auburn, Alabama, U.S.A.), where Secchi disk visibility was 30 cm, and chlorophyll a concentrations reached up to 200 $\mu g\ l^{-1}$, though, on average, values were around 33 to 63 $\mu g\ l^{-1}$. In concrete ponds (in Bangkok, Thailand) where phytoplankton abundance was around 120 mg l^{-1}, Edwards *et al.* (1981) recorded values of 11 to 13 mg l^{-1} of total nitrogen, of which 33 to 68 % was organic, between 0.08 and 0.56 mg l^{-1} as $N\text{-}NO_3^-$, and 2.62 to 8.82 mg l^{-1} as $N\text{-}NH_4^+$. A concentration of 0.1 mg l^{-1} of inorganic nitrogen in ponds (in Auburn, Alabama) is considered low (Boyd & Sowles, 1978). In African lakes, records on nitrogen and phosphate concentrations indicate that nitrate concentrations are usually very low, below 30 $\mu g\ l^{-1}$ of $N\text{-}NO_3^-$ in surface waters, and often are not readily detected. The distribution of ammonia is practically unknown. Concentrations of phosphate and total phosphorus are, on the other hand, generally high, between 150 and 1000 $\mu g\ l^{-1}$ (Talling & Talling, 1965).

Fertilization frequencies

Fertilizers should preferably be applied at low doses and frequently, for example, every two weeks. Fertilization should also take into consideration the feeding regime of the fish. According to the present state of our knowledge, it is impossible to forecast the specific plankton assemblage, and the relative density of algae, which will appear sebsequent to fertilization. Thus, polyculture is more suitable than monoculture, because a variety of species allow a more effective use of the natural feed produced.

Controversy about the need for nitrogen fertilizers

Some authors (Boyd, 1979; Miller, 1976; Murat & Boyd, 1987; Yusoff & McNabb, 1987) consider that nitrogen fertilizers are not necessary, as nitrogen needs can be met by recycling and regeneration of organic matter, or by nitrogen fixation by blue-green algae and bacterioplankton. Other authors (Talling & Talling, 1965; Horne & Viner, 1971; Wurtsbaugh *et al.*, 1985) state that tropical waters are chronically deficient in nitrogen, resulting in severe limitation of primary productivity, the presence of blue-green algae confirming this fact, as they provide missing nitrogen to the water. Observations from African waters confirm low nitrogen concentrations. However, concentrations do not indicate actual fluxes. Also, it is difficult to make general statements on all water types; proper experimentation on the effect of nitrogen fertilizers on production will give a more reliable answer than a long debate. Yusoff & McNabb (1989) discussed this controversy in the light of experiments with varying nitrogen-phosphorus ratios (see p. 154).

Doses of nitrogen fertilizers applied to ponds

Data on dose sizes of inorganic nitrogen fertilizers in the literature vary from 0 to 15 kg ha^{-1} (2 $weeks)^{-1}$ of N. In his classical experiments in fish ponds at Malacca, Indonesia, Hickling (1962) showed that nitrogen fertilizers, in combination with phosphate fertilizers, did not improve the production of the ponds.

Boyd (1976a) tested doses of 0, 1.125 and 4.5 kg ha^{-1} (2 $weeks)^{-1}$ of N, combined with phosphate fertilizers, in ponds in Alabama, U.S.A. Phosphate fertilizers, applied at a rate of 4.5 kg ha^{-1} (2 $weeks)^{-1}$ of P_2O_5 and of 1.125 kg ha^{-1} (2 $weeks)^{-1}$ of K_2O, did not give a significant increase in production. However, the dose of 1.125 kg ha^{-1} (2 $weeks)^{-1}$ of N gave a yield of 947 kg ha^{-1} $yield^{-1}$ of *O. aureus*, as compared to 651 kg ha^{-1} $yield^{-1}$ in the control. This represents an increase of 17.5 kg of fish per kg of nitrogen.

In other experiments at the same location, Boyd (1981b) tested five combinations of nitrogen and phosphate fertilizers, using application rates varying between 0 to 9 kg ha^{-1} (2 $weeks)^{-1}$ of N, i.e. 1, 1.2, 1.6 and 9 kg ha^{-1} (2 $weeks)^{-1}$ of N. He observed that the concentrations of dissolved inorganic nitrogen in the water remained virtually unchanged, at around 155 $\mu g\ l^{-1}$ of total N. The effects of nitrogen fertilizers on fish production are indeed difficult to assess.

Msiska (1983) in ponds at Chikwawa, southern Malawi, tried various combinations of fertilizers at doses of 1.23 kg ha^{-1} (2 $weeks)^{-1}$ of N and/or 2.6 kg ha^{-1} (2 $weeks)^{-1}$ of P_2O_5, but did not obtain a significant increase in fish production. Hepher (1962) recommended the use of 14 kg ha^{-1} (2 $weeks)^{-1}$ of N for fish ponds in Israel, while Miller (1976) reports that Vincke recommends 8.5 to 17 kg ha^{-1} (2 $weeks)^{-1}$ of N, for ponds in Madagascar.

Recommendations for use of nitrogen fertilizers

Fish production depending on natural production necessitates a primary production of about 4 to 6 g m^{-2} d^{-1} of carbon, which requires about 800 mg m^{-2} d^{-1} of nitrogen. About half of the nitrogen is returned to the environment by phytoplankton respiration (Schroeder, 1987). Fish and phytoplankton excretion supply some more nitrogen, which is recycled in the water column. Depending on the nature of the sediment and previous fertilizations, the sediment supplies, by recycling and regeneration, the rest of the required nitrogen to the water column. Losses by denitrification and diffusion are low and a supply of 3 to 7 % of the daily needs may be satisfactory to maintain the natural production of phytoplankton. This would correspond to some 3.4 to 8.4 kg ha^{-1} (2 $weeks)^{-1}$ of N. This also grossly corresponds to the amount eventually reaching the fish in Schroeder's experiments (see p. 89).

To conclude, in the light of actual knowledge, it is suggested to use doses of 5 to 10 kg ha^{-1} (2 $weeks)^{-1}$ of N, along with phosphate fertilizers (see p. 114 for the rates), and to monitor, on a regular basis, the inorganic nitrogen level in solution and the density of phytoplankton, or, in the absence of quantitative measurements, to assess the abundance of phytoplankton. Fertilizer application should increase proportionally to fish load, the amount applied increasing in the course of the production period. A regular adjustment of doses with time is justified by the absence of a storage system for nitrogen in the environment.

Nitrogen fertilizer application

1. An assortment of nitrogen fertilizers is available on the market; nitrogen fertilizers containing ammonium or urea have an acidifying effect on the water.

2. The partitioning of nitrogen added by nitrogen fertilizers to water between the different 'N-users' is difficult to quantify: dissolved nitrogen concentrations are not related to amounts added.

3. Fertilizers should preferably be applied frequently at low doses rather than less frequently at high doses. The rates should take into consideration the increase of fish biomass and be increased accordingly.

4. Nitrogen fertilization is more necessary in African waters than in temperate waters, as nitrogen appears to be more limiting in the former region.

5. The application rates recommended are 5 to 10 kg N ha^{-1} (2 weeks)$^{-1}$.

4.3. Phosphorus and phosphorus compounds

Of the three major elements, phosphorus is considered most frequently limiting, because its concentrations in water are often low. At the cellular level, phosphorus is contained in ATP and ADP, in RNA and DNA, and in numerous enzymes. In animals, phosphorus combined with calcium forms bone, teeth, and scales. The phosphorus needs of animals are met through their food.

For primary productivity, phosphorus is an essential nutrient, its importance having already been appraised by Justus von Liebig in 1855. The intracellular concentrations of phosphorus condition the photosynthetic rate (Senft, 1978). The amounts of available phosphorus are often limiting primary productivity (Hepher, 1958 & 1966, based on his experiments on fish ponds in Israel), and the abundance, the species assemblage and the succession of phytoplankton are a function of phosphorus availability. Over long periods, the amounts of dissolved and available phosphorus determine phytoplankton biomass present in an environment (Straškraba, 1980). According to Schindler (1987), a large portion of the variability in the overall production of fresh waters is explained by external phosphorus loading. In excess, phosphorus causes eutrophication.

4.3.1. Phosphorus cycling

Phosphorus species present in water

As for this element in the sediments, distinction is made between particulate and dissolved phosphorus. Dissolved phosphorus is either reactive (SRP) or non-reactive. Dissolved phosphorus is separated from particulate phosphorus by filtration through a filter of 0.45 μm pore size. Total phosphorus is the sum of all phosphorus forms in water, dissolved and particulate, available and occluded.

The various phosphorus forms in water and the phosphorus stocks of the sediments are (at least partly) in steady equilibrium. Particulate phosphorus makes up the major part, and its relations with soluble phosphorus are quite important (Nalewajko & Lean, 1980).

Sources of particulate phosphorus

Particulate phosphorus is supplied from four sources: organic matter produced by living organisms in water (organic phosphorus), sediments (geochemical phosphorus), from chemical reaction, precipitation or sorption within the water column, and, finally, from feeds and fertilizers. Such phosphorus can be separated into inorganic or organic, and colloidal or non-colloidal particles (Broberg & Persson, 1988):

- plants, bacteria and animals cells, either alive or

dead, and constituting coarse organic particles

- organic detritus; excreta of animals, which constitute fine organic particles
- organic aggregates generated by the flocculation of organic macromolecules, in the form of true organic colloids
- inorganic/organic coprecipitates, complexes or chelates, produced by the inclusion of phosphorus in organic aggregates by metal-phosphorus binding, forming mixed inorganic-organic colloids
- weathering products, such as primary and secondary minerals, and resuspended fine non-colloidal sediment particles
- Inorganic phosphorus precipitated or sorbed onto mineral particles, such as clay, forming true mineral colloids.

Composition of dissolved phosphorus

Dissolved phosphorus mainly consists of orthophosphate and inorganic polyphosphate (DIP, dissolved inorganic phosphorus), and the dissolved organic pool (DOP), part of which probably are nucleotides or polynucleotides (Broberg & Persson, 1988). Polyphosphates and metaphosphates are condensed compounds, with chain and ring structures, produced by biological activity. The soluble organic phosphorus fraction contains phosphate esters and organic acids. Part of these compounds are hydrolyzed by chemical or biological reactions.

The various inorganic phosphorus species are not well soluble and their solubility is pH-dependent. Their successive dissociations are as follows:

$$H_3PO_4 \rightleftharpoons H^+ + H_2PO_4^- \qquad K_1 = 10^{-2.13}$$

$$H_2PO_4^- \rightleftharpoons H^+ + HPO_4^- \qquad K_2 = 10^{-7.21}$$

$$HPO_4^- \rightleftharpoons H^+ + PO_4^- \qquad K_3 = 10^{-12.36}$$

H_2PO_4- and HPO_4^- concentrations are maximum at pH 4.67 and 9.78, respectively.

Composition of dissolved organic phosphorus

A relatively large part of dissolved phosphorus is organic, especially in waters with high biotic loading. The major sources of the DOP are known, but not how much they contribute to the pool. It is also unknown whether DOP is released during active cell metabolism or after cell death. Part of the DOP pool is readily available to biota and rapidly turns over, while part is refractory, probably consisting of nucleotides or polynucleotides (Broberg & Persson, 1988).

The dissolved organic phosphorus pool is comprised of a great number of unidentified compounds, which have been grouped into high (HMWP), intermediate (IMWP or XP), and low molecular weight (LMWP) fractions. This grouping is based on the results of gel chromatography analysis (Broberg & Persson, 1988).

The HMWP fraction consists mainly of colloids, of great ecological importance as carriers and binding surfaces for dissolved organic and inorganic phosphorus. Colloidal phosphorus is biologically available, at least to some extent, though part of it is resistant to enzymatic hydrolysis. It appears to come from percolation through the soil profile. The compounds of this fraction have not been identified, but seem to be made up of a great number of molecules of varying composition and nature, including humic acid-metal-phosphate complexes, organic matter-iron(or aluminium)-phosphate complexes, and ferri-hydroxide-phosphate peptized by organic matter.

Evidence for inorganic phosphorus saturation of high molecular weight humic acids (organic material-Fe-P-Fe-organic material) has been found. This HMWP fraction sorbs orthophosphate only in the presence of microorganisms, which suggests some reversible isotopic exchange with sorbed orthophosphate on iron and/or aluminium oxide surfaces. The nature of these high molecular weight organic compounds probably varies considerably between water bodies. Part of these compounds decompose when exposed to low dosages of UV radiation, which confirms their organic nature. But decomposition by UV radiation does not exclude the presence of metal(calcium or iron)-phosphate polymers, associated or not with organic material. The HMWP fraction is more important in oligotrophic waters, while in eutrophic waters, IMWP and LMWP are more important. Most of the organic phosphorus from the HMWP fraction is retained in the fulvic acid fraction, as opposed to the humic acid fraction.

Phosphorus is, however, not necessarily associated

with organic material. Colloidal iron and phosphorus occur in a ferri-hydroxide-phosphorus complex, which may be peptized by organic carbon. The inorganic nature of this colloidal fraction is further suggested by its acid lability. HMWP compounds may be important regulators of dissolved and total phosphorus contents in the water, being involved in sorption/desorption, precipitation/dissolution and flocculation/dispersion processes (Broberg & Persson, 1988).

The IMWP fraction has not yet been characterized and is probably formed by incorporation of phosphate into seston (Jansson, 1988), by biological activity, leading to HMWP formation. It apparently holds a central position in phosphorus cycling, as it is excreted from seston before being bound to the HMWP fraction, involving a release of orthophosphate. Exposed to UV, IMWP releases phosphate, but enzymes cannot cause such a release. Thus, it reacts as an organic anion complex, of low molecular weight, protected from enzymatic degradation. The possibility of it consisting of phosphomonoesters seems to be ruled out because of its resistance to enzymatic attack. It is more likely that this fraction consists of polyphosphates, as they resist enzymatic decomposition, but not UV radiation. Some have speculated that it could be made up of inositol hexaphosphates (Boström *et al.*, 1988b), which are components of cell walls. These products are protected from hydrolysis by associated lipids and proteins (Broberg & Persson, 1988).

The LMWP fraction is composed of orthophosphates and organic phosphorus compounds of low molecular weight. Several IMWP and LMWP compounds are rapidly degraded by natural enzymes and are, therefore, only temporarily present in water (Broberg & Persson, 1988). The presence of organic phosphorus compounds is not easily detected, as organic compounds release phosphate in acid phase, during the characterization process of inorganic phosphate and the released phosphates artificially increase the dissolved phosphate concentration of the inorganic phosphorus pool.

Among the more easily identifiable organic compounds are excretion and decomposition products, such as phospholipids, phosphomononucleosides, sugar phosphates, and aminophosphoric acids (Chrost *et al.*, 1984).

Phosphorus concentration in water

Phosphorus concentration in natural waters vary seasonally, but, on average, levels fluctuate around 5 to 30 $\mu g\ l^{-1}$ of soluble phosphorus, after filtration through a 0.45 μm filter. Orthophosphate is always present in water, though often in low concentration; the presence of organic compounds is more erratic and probably depends on the biotic load of the pond. Auer *et al.* (1986) have classified waters according to their phosphate concentrations (Table 15).

Oligotrophy is defined by concentrations below a critical level, while at or above the eutrophic level, plankton growth is insensible to any further increase of phosphorus concentration in the water.

In fertilized fish ponds (Auburn, Alabama), concentrations of 20 $\mu g\ l^{-1}$ of $P\text{-}PO_4^{3-}$ and 0.17 mg l^{-1} of total phosphorus are normal (Boyd, 1979). In studies of water eutrophication, maximum phosphorus concentrations and loadings have been assessed and criteria have been established. In temperate waters, phytoplankton production can only be reduced, in the long term, if phosphorus loading is reduced to less than 0.5 to 2.0 g $m^{-2}\ yr^{-1}$ and phosphorus concentrations to less than 80 to 150 $\mu g\ l^{-1}$ (Jeppesen *et al.*, 1990a), or 100 to 200 $\mu g\ l^{-1}$ of total phosphorus or 10 $\mu g\ l^{-1}$ of dissolved phosphorus (Marsden, 1989). These values indicate concentrations in eutrophic conditions, which coincide with those of maximum productivity.

Transformations of phosphorus in water

The phosphorus cycle in the natural environment is complex (Fig. 31), mainly due to the high number of organic and inorganic phosphorus compounds in water (Psenner *et al.*, 1984), and to the difficulty of quantifying phosphorus fluxes between compartments. In this respect, it is more exact to speak in terms of kinetics, than in terms of average concentrations (Shapiro, 1988). The turnover time of some compounds is often short. The reincorporation of phosphorus into sedimentary rocks is calculated in millions of years; this is worrying in view of the rate at which phosphorus compounds are disposed of in sewage, leading to enrichment of natural waters and sediments (Holtan *et al.*, 1988).

The two processes responsible for the decrease of total phosphorus concentrations in water, are the

Table 15. Total phosphorus and dissolved phosphorus concentrations in waters of different trophic levels (according to Auer et al., 1986)

Trophic level	P_{diss} (in $\mu g\ l^{-1}$ of P)	P_{tot} (in $\mu g\ l^{-1}$ of P)
ultraoligotrophic	< 0.5	< 4.3
oligotrophic	0.5 - 1.2	4.3 - 11.5
mesotrophic	1.2 - 8.0	11.5 - 37.5
eutrophic	> 8.0	> 37.5

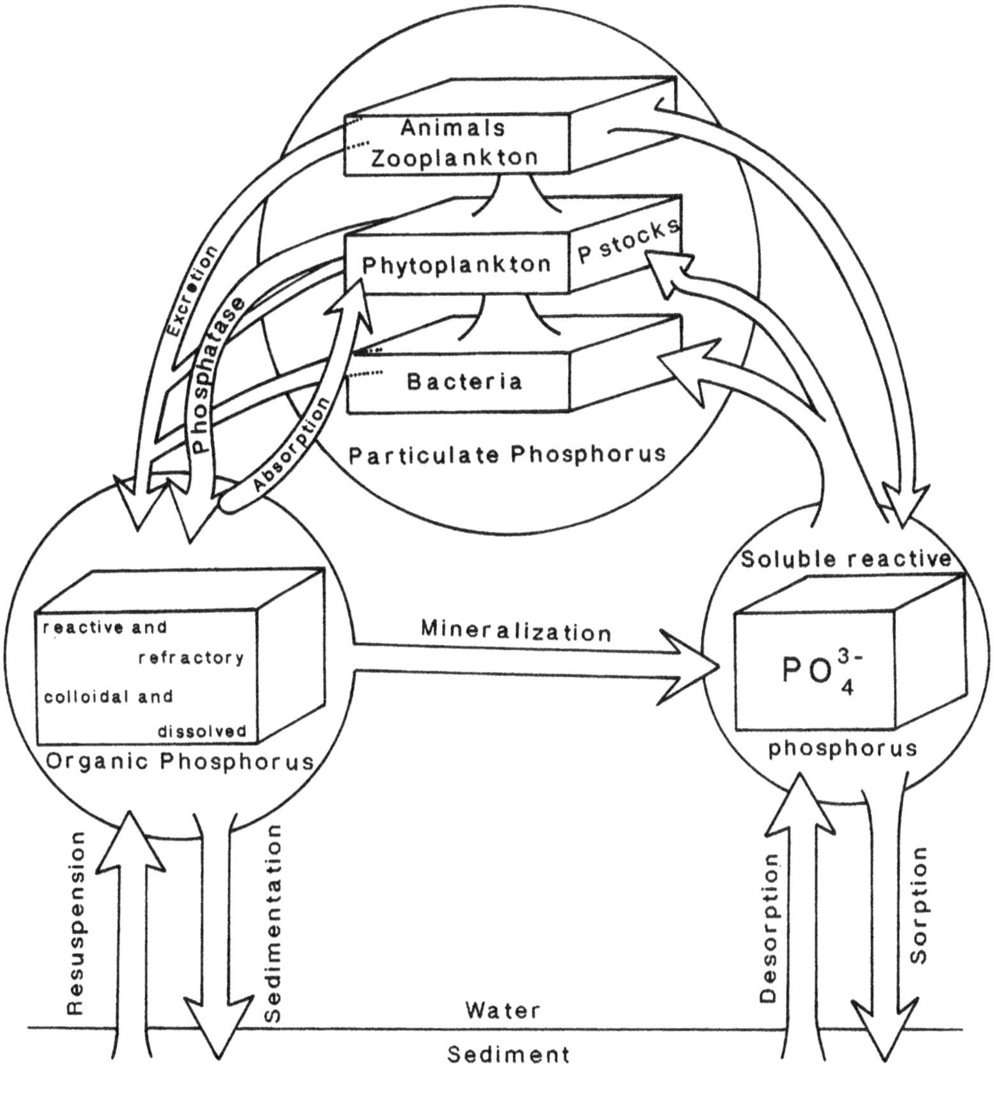

Fig. 31. Phosphorus cycle in water. Transfers of phosphorus between the particulate, the organic and the dissolved phosphorus compartments.

sedimentation rate and the renewal rate of the water (Marsden, 1989).

The parts of the phosphorus cycle which attract most interest are phosphorus assimilation, decomposition of organic phosphorus, phosphorus excretion by organisms, and chemical adsorption.

Time needed for various transformations

In studying the cycling of phosphorus in ponds, the various transformation and transport processes should be quantified (Lijklema & Hieltjes, 1982). The time scale of each process is important.

- Orthophosphate uptake by algae and its incorporation into the foodweb, its active excretion, and its release from dead organisms by decomposition are processes which proceed rapidly, in a matter of hours or days. The incorporation kinetics of phosphorus into biomass are quick. The concentration of dissolved reactive phosphorus and organic phosphorus is directly proportional to the photosynthetic rate and to the excretion rate of zooplankton in water (Golachowska, 1986)
- Sedimentation of organic particulate phosphorus and phosphorus saturation of sediments are slow processes. Algal cells sink and disintegrate at a pace which depends on the volume, the surface to volume ratio, and the density of the cells. Phosphate diffuses out of dead cells during lysis. Part of the phosphorus is quickly immobilized by bacteria, while the rest is taken up again by algae. The sedimentation rate of algae and bacteria is enhanced in the presence of concentrations of dissolved calcium in the water. Self-flocculating aggregates of calcium carbonate, along with algae and bacterial cells and dissolved phosphate, form and sink. In the presence of blue-green algal blooms, all dissolved phosphorus was observed to be removed due to coprecipitation of algae with phosphate and calcium carbonate, while in the absence of the calcium salt, the concentration of dissolved reactive phosphorus remained unchanged (Boström *et al.*, 1988a)
- Supplies and losses to the environment due to fertilization and drainage of the pond, respectively
- Exchanges across the sediment-water interface and release of phosphorus by the sediment. This involves the dynamics of phosphorus release by the sediments, extensively discussed in the first chapter
- The sediment mineralization rate includes the production of dissolved phosphate and leads to anaerobic conditions at the sediment surface. This rate varies greatly
- Adsorption/desorption processes in the water column govern the balance between the different forms of phosphorus, according to adsorption isotherms
- Sediment dilution is the accumulation and burial of sediments in the course of time (Lijklema & Hieltjes, 1982).

Importance of phosphorus mineralization

Organic phosphorus mineralization recycles large amounts, estimated to be 55 to 85 % of the phosphorus required by primary productivity. Microbial decomposition is the main process, accounting for some 70 % of phosphorus recycling. The activity and population size of the microorganisms increase in proportion to the availability of substrate. The remaining part of the decomposition involves physical and chemical processes, such as phosphorus release mediated by UV radiation. Some organic phosphorus is refractory to phosphate release. But, with changing conditions, release of phosphorus becomes possible. The ferri-humate complexes release phosphate under the influence of UV radiation, owing to the photo-reduction of ferric ions, but not under enzymatic attack.

The organic phosphorus pool includes the biotic particulate phosphorus, i.e. phosphorus bound to living organisms in water, which is of course not available. The particulate phosphorus most quickly recycled is that contained in zooplankton and fish excreta, rich in phosphorus and constituting the most important phosphorus source for the water column (Andersson *et al.*, 1988). Actively feeding zooplankton generates plumes with inorganic phosphorus concentrations several orders of magnitude higher than ambient levels, and epilimnetic zooplankton can supply ten times more inorganic phosphate to the surface layer of a natural water body than all outside sources (Tarapchak & Nalewajko, 1986a). Zooplankton therefore holds a primary role in phosphorus regeneration. According to Bartell & Kitchell

(1978), the supply of organic soluble phosphorus by zooplankton in eutrophic waters reaches maximum values of 35 to 125 mg m^{-3} d^{-1} of total phosphorus. Zooplankton faecal material varies in form and density according to taxonomic group, calanoid copepods producing faecal pellets 30 to 300 μm in equivalent spherical radius, which sink rapidly, while cladocerans produce flocculent material or a slurry, more easily recycled (Kitchell & Carpenter, 1987; Andersson *et al.*, 1988). Part of these excreta is soluble in water, while part is refractory. Zooplankton is also responsible for vertical transport of phosphorus, due to their vertical migration behaviour (see p. 162). Healthy phytoplankton cells release soluble and refractory excretion products, while lysing phytoplankton cells promptly release their phosphate content (De Pinto *et al.*, 1986).

Without phosphorus mineralization, the phosphorus stocks in the water column would be rapidly exhausted, i.e. in between 0.2 to 20 days, photosynthesis would stop. Golterman & Kouwe (1980) calculated the recycling rate using the life expectancy of algal populations. The maximum oxygen production of algae is 25 mg (mg chl a)$^{-1}$ h^{-1}, which is equivalent to 9.4 mg (mg chl a)$^{-1}$ h^{-1} of carbon. The chlorophyll a to cellular carbon ratio is 1:35. This means that carbon fixation is 3.22 mg (mg of C_{cell})$^{-1}$ d^{-1}. Algal respiration is assessed at 1 mg (mg chl a)$^{-1}$ h^{-1} of oxygen, or 4 % of the photosynthesis, or 0.38 mg (mg chl a)$^{-1}$ h^{-1} of carbon. The cellular carbon to produced oxygen quotient, expressed in mg, is 9.4:25. The loss of cellular carbon on the death of algae is,

$$\frac{0.34 \times 24}{35} \times \frac{\text{respiration depth}}{\text{euphotic zone depth}}$$

A depth quotient equal to 2 means that there is a loss of 0.52 mg (mg of C_{cell})$^{-1}$ d^{-1} of carbon. The net budget of photosynthesis is 3.22 - 0.52 = 2.7 mg (mg of C_{cell})$^{-1}$ d^{-1} of carbon. This coincides with the maximum oxygen production. In the whole water column, production is calculated to be about 2.6 times that of maximum production. For the euphotic zone, the net photosynthetic budget is therefore 1 mg (mg of C_{cell})$^{-1}$ d^{-1} of carbon. This is also the average growth rate of the algae. Considering a maximum production of 25 mg (mg chl a)$^{-1}$ h^{-1} and a respiration rate of 4 % of the oxygen production, average life expectancy of algae calculated for a period of 1 year is 1 day, which implies that, on average, daily decomposition rate is equal to mean algal biomass. If oxygen production is less, life expectancy of algae increases. This continuous mineralization or metabolic phosphorus cycling is therefore an important contribution to primary productivity, (De Pinto *et al.*, 1986). Phosphorus losses to the sediments have been discussed on p. 32.

Phosphate uptake and influences on phosphorus availability

Phosphorus loading to the water is regulated by calcium concentration and pH, which set the limits to phosphate solubility in water. In hard waters, calcium controls phosphate solubility, while, in acid waters, phosphate solubility is under the control of iron. The range of the solubility product of apatite (10^{-50} to 10^{-56}) sets the limits within which soluble phosphate is likely to be found (Golterman, 1982).

Dissolved organic matter modifies the concentration of dissolved phosphate in water. Inorganic phosphate added to water becomes attached to compounds of high molecular weight (above 100,000 and between 10,000 and 20,000). The degree of this 'attachment' is greater in waters of high humic or iron concentrations. A balance between the fractions of high molecular weight and soluble phosphate exists and the equilibrium is reached in two phases: one quick and one slow. Using Langmuir adsorption isotherms on water samples from the Neusiedlersee, Austria (Gunatilaka, 1982), and from lakes of different trophies (eutrophic, acid bog lake and freshwater estuarine marsh) of Ohio, Georgia and South Carolina (Heath, 1986), two distinct phases could be shown: a first phase which lasts around 1 minute, but is almost completed within 10 seconds, probably corresponding to a rapid adsorption on calcium carbonate, and a second phase lasting for 1 to 10 minutes, corresponding to fixation onto low energy sites on seston. In studies carried out in three acidic lakes of southern Finland (Evo district), Jones *et al.* (1988) noticed that the high molecular weight phosphorus was not reactive (or only partly) to orthophosphate reagents, and exposure to solar light did not release phosphate (Jones *et al.*, 1988). The influence of light on phosphorus release from the humic lake Tjeukemeer, Netherlands, was only significant at acid pH (De Haan & de Boer, 1986).

In water from acidic lakes of southern Finland, humic acids, unlike iron, were not able to directly sorb phosphate to form colloids. However, ferri-humic complexes sorbed orthophosphate and organophosphate ions. Adding iron (III), at a concentration of 1 mg l^{-1}, to water stimulated the movement of soluble phosphate, provided humic matter was not present at high concentrations. In the absence of humic matter, even a concentration as low as 0.1 mg l^{-1}, precipitated all added phosphate (Jones *et al.*, 1988).

In acid waters enriched with humic acids as that of the Tjeukemeer, ferric iron did not show cationic properties and the formation of colloidal iron oxyhydroxide (FeOOH) was postponed, allowing most dissolved phosphate to remain in solution (De Haan & de Boer, 1986). In aerobic water from an acidic lakes in northcentral Oklahoma, the interaction between colloidal organic matter and iron was found to decrease the amount of phosphate adsorption onto iron-organic complexes, thus freeing inorganic phosphorus for phytoplanktonic utilization. Binding of phosphorus occurred when iron was present in the ferric state; ferrous ion-DHM complexes could not bind iron (Franko, 1986).

In alkaline waters, from non-humic lakes, with low concentrations of humate, iron oxyhydroxide dominates. Most of the particulate phosphorus is sorbed onto humic matter, to form humic-iron-phosphate complexes (DHM-Fe-PO_4^{3-}); this reduces the total amount of available phosphate (De Haan & de Boer, 1986).

Fulvic acids are aromatic polycondensates with a molecular weight around 1000. They carry phenolic and carboxylic radicals, which provide them with chelating properties. They are polyelectrolytes, on which some sites, for steric reasons, are not available for ion sorption. They compete for the sorption of metal ions. Inorganic phosphate is sorbed onto inorganic colloids, ionic and colloidal iron, and fulvic acids (De Haan & de Boer, 1986).

Iron sequestration on phosphate by humic matter (in combination with iron) reduces the availability of phosphate. These adsorbed ions remain chemically active, but biologically non-available (Franko, 1986). In acid lakes of southern Finland, as equilibrium between the dissolved phosphorus and these fractions was established, addition to or removal from water of phosphate resulted in additional sorption or release according to the equilibrium. The total amounts dissolved and available remained unchanged (Jones *et al.*, 1988). It is a form of short-term storage, which stimulates the biological incorporation of phosphorus, but on which the addition of nitrogen and iron does not have much effect. The dissolved humic matter complexes formed a chemical regulation system for available phosphates, analogous to a phosphate buffer (Jones, 1990).

Reduction in the size of the dissolved phosphate pool affected the cycling and uptake of phosphorus, phosphatase activities, and carbon uptake. Organisms with low uptake constants, as well as those with alternative strategies for meeting phosphorus requirements, were favoured (Jones *et al.*, 1988).

Bacterial mineralization and role of phosphatase

Phosphomonoester and organic phosphorus compounds release phosphate due to the hydrolytic activity of phosphatases. These enzymes contribute to the decomposition of organic matter, by removing orthophosphate from the organic moiety. They also participate in algal phosphorus uptake. The reaction is a hydrolysis of esters and anhydrides of phosphoric acid, and the enzymes are phosphomonoesterases and phosphomonoesterhydrolysases. These enzymes are also involved in transport across cell membranes and synthesis of new organic phosphate, i.e. transphosphorylation, not discussed here. Phosphatases carry a metal ion as a co-factor, and it has been observed that the addition of divalent cations enhances their activity (Jansson *et al.*, 1988).

Phosphatase sources

Two types of phosphatases are present in water: particulate and dissolved. The particulate phosphatases are associated with phytoplankton and bacterial cells, while the dissolved ones are found in the filtrate after filtration (through a 0.45 μm filter). The latter represent 31 to 87 % of the total amounts in water. Their occurrence is as yet unexplained. They probably originate from bacteria, phytoplankton, or zooplankton (Currie *et al.*, 1986).

Enzymes produced by bacteria have been little studied. Of interest are the amount of bacterial excretions and whether they are passive or active. These dissolved enzymes are able to hydrolyse external substrata, but there is little information on the importance and role of such extracellular excretions.

Algae of all major groups produce intracellular and extracellular phosphatases. The role of these extracellular enzymes in phosphate regeneration, to the benefit of the algae, and secretion during periods of dissolved phosphate deficiency are mainly based on indirect evidence (Jansson *et al.*, 1988).

Zooplankton also produces phosphatases, but as digestive enzymes, and not as components of their food. Apparently, the portion attributed to zooplankton secretion makes up a large part (up to 50 % in Lake Hymenjaure, Sweden (Jansson, 1976)), of the dissolved amounts present in the water. In most cases, the fraction due to zooplankton secretion or excretion cannot singled out.

Dissolved phosphatases do not appear to differ from other enzymes and could originate from the lysis of dying or dead cells. The process leading to their presence in the water, outside of cells, is not yet fully understood, with the importance of active secretion not yet established. Dissolved enzymes are active over periods of up to several days, with a progressive reduction of activity. The inactivation mechanisms are not fully understood either (Jansson *et al.*, 1988).

Differences between alkaline and acid phosphatases

Alkaline and acid phosphatases are separated on the basis of their pH optimum, which is between 9 and 10, and 4 and 6, respectively. Their activity extends, however, beyond these pH zones. The two types are present in algae and bacteria, as internal as well as external enzymes. The alkaline enzymes have been most studied, because they are more commonly found. Differences between the phosphatase types include their location in the cell and their mode of synthesis. Acid phosphatase is more abundant within the cell than outside; the synthesis of the alkaline enzyme is repressed by orthophosphate, while this is not the case for the acid enzyme. This would suggest that acid phosphatases are constitutive, while alkaline enzymes are induced, and dependent on, the cellular supply of phosphate. Several studies have shown that the internal cellular pool of phosphate regulates the synthesis of repressible phosphatases and that the synthesis is derepressed when the internal pool gets exhausted or falls below a certain critical level (Jansson *et al.*, 1988). Dissolved acid phosphatases have been discovered in acid lakes, nearby Stockholm (Sweden), where they appear to substitute the role of alkaline phosphatases present in neutral or alkaline water. They seemed inhibited or repressed by the presence of inorganic extracellular phosphate (Olsson, 1990).

Aluminium and iron have affinities for phosphomonoesters and thereby reduce the action of alkaline phosphatase, via competitive inhibition between the metals and the enzyme for the substrate. In acid waters, this results in a higher phosphatase activity, attributed to the production of greater levels of phosphatase (sometimes approximately ten times higher) by the biota, thus enabling the enzymes to outcompete aluminium for the substrate (Mulholland *et al.*, 1986; Jansson *et al.*, 1988).

This information makes the distinction between the roles of alkaline and acid phosphatases still more unclear and raises the question whether the differences between the two groups of enzymes are not methodological rather than ecological (Olsson, 1990).

Measurement of enzyme activity

Enzyme activity depends upon the concentration of the substrate (Boavida & Heath, 1988) and is expressed by the Michaelis-Menten constant, which establishes the concentration at which the reaction proceeds at half its maximum speed. Low Michaelis-Menten constants indicate high affinities or reactions which proceed at low substrate concentrations, while high constants indicate the reverse. Phosphate uptake has been considered either to follow Michaelis-Menten kinetics (Nalewajko & Lean, 1980), but also to deviate from them (Brown *et al.*, 1978).

According to uptake characteristics, some algae survive in conditions of phosphate concentrations insufficient for others. Phytoplankton is capable of adapting to low concentrations of substrate. Phosphorus starved algae are able to produce phosphatases with higher affinity than those produced by algae in phosphorus-abundant conditions.

Relation between phosphorus deficiency and phosphatase concentrations

Enzyme activity has be used as an indicator of the degree of possible phosphorus deficiency in water (Chrost *et al.*, 1984: 16, Fig. 8). This is only possible if a relationship between the potential activity of alkaline phosphatase (APA) and the algal biomass can be established; this is expressed by the specific APA, which is the APA per particulate phosphorus content (nmol PO_4^{3-} hr^{-1} μg P_{part}^{-1}). The derepression of alkaline phosphatase is associated with low cellular total phosphorus or polyphosphate concentrations, in the range of 6 to 7 μg (mg of C)$^{-1}$ of phosphorus or 3 to 3.5 μg (mg of particulate organic matter)$^{-1}$ of phosphorus for *Chlamydomonas reinhardtii* in chemostat culture (Olsen *et al.*, 1988) or 1.4 μg P mg POC^{-1} for an assemblage of green algae and diatoms from Green Bay, Lake Michigan, U.S.A. (Auer *et al.*, 1986). The characteristic deficiency threshold is encountered when the stocks in reserve fall below 4 times the minimum cellular value (Fig. 32).

Gage & Gorham (1985), studying nine Minnesota lakes in Clearwater and Hubbard Counties, of varying (meso-eutrophy to eutrophy) trophic levels,

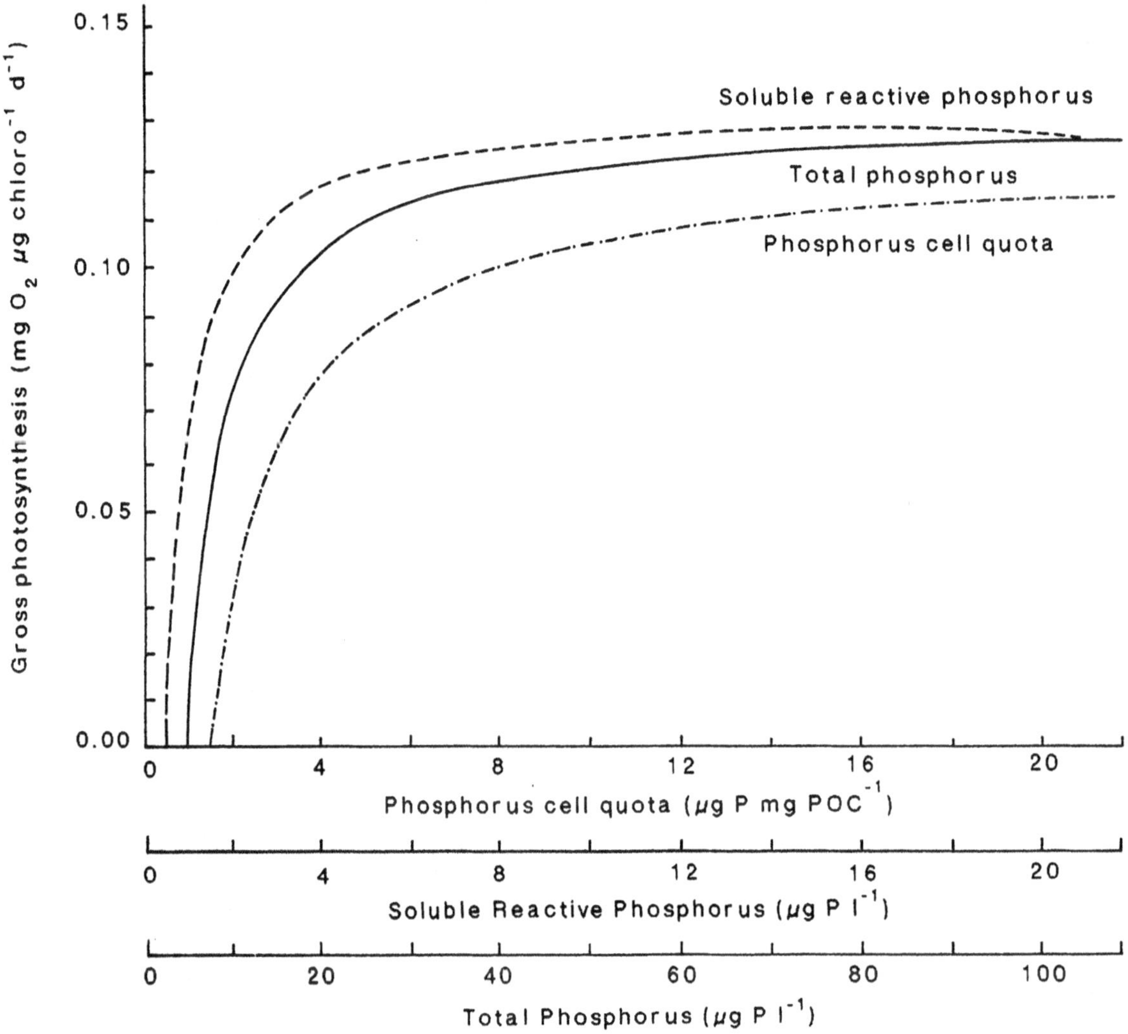

Fig. 32. Gross photosynthesis per unit chlorophyll as a function of (a) soluble reactive phosphorus concentration, (b) total phosphorus concentration and (c) phosphorus cell quota, calculated as phosphorus per unit of particulate organic phosphorus (POC) (modified from Auer *et al.*, 1986).

established phases of alkaline phosphatase activity related to cellular phosphorus concentrations, considering that situations of severe starvation, of rationing, and of excess or surplus phosphorus correspond to equivalent situations of phosphatase activity. A deficiency situation exists when the phosphorus concentration is less than 5 μg (mg of particulate organic matter)$^{-1}$, rationing between 8 to 10 μg (mg of POM)$^{-1}$ of P, and excess above 10 μg (mg of POM)$^{-1}$ of P. The three phases of enzyme activity are: high enzyme activity, medium level activity, and constitutive activity. These activities are in a ratio of 1:2:5, the high level activity being 5 times the activity of the base level (24 nmol PO_4^{3-} (mg POM)$^{-1}$ h^{-1}).

Heath (1986) found that release rates of phosphate from DOP accounted for less than 1 % of the phosphate uptake by seston in five different fresh-water systems (a eutrophic kettle lake, an acidic dystrophic bog lake, an estuarine marsh, in Ohio, and two reservoirs in Georgia and South Carolina) during summer. Abundance of phosphatase during summer phytoplankton blooms suggested that enzyme presence was related with P limitation, but comparisons of calculated phosphate release from DOP (based on Michaelis-Menten kinetics) and measurements of phosphate uptake lead the conclusion that DOP was not a major source of phosphate in summer. Previous observations had shown an increase in phosphatase activity in relation with P limitation and a depletion of internal P stores, specifically in the presence of the blue-green alga, *Aphanizomenon flos-aquae* in eutrophic East Twin Lake, Ohio. Heath (1986) concluded that DOP compounds are of negligible nutritional benefit to phytoplankton.

Although Lake Memphremagog, Quebec is considered highly P-limited, phytoplankton grows at relatively high rates, but does not synthetize large quantities of alkaline phosphatase. Algae obtain their phosphate from other sources than the low dissolved concentrations (Smith & Kalff, 1981).

In four trophically distinct lakes of north-central Oklahoma, Franko (1986) showed high plankton-associated alkaline phosphatase activity related to low ionic orthophosphate concentrations (suggesting P-deficiency) in late summer. It was not, however, the only DOP regenerative mechanism. Photodependent release from UV-sensitive complexes was also operating.

Alkaline phosphatase activity was related to the severity of phosphorus deficiency in 13 lakes in the region of Quebec, Vermont and New York, and correlated positively with the rate constant of orthophosphate uptake (a parameter close to phosphate turnover rate) of phytoplankton (particles above 3.0 μm) (Currie *et al.*, 1986). Other studies on relationships between APA and phosphorus in the ecosystem are listed by Jansson *et al.* (1988).

The relation between low inorganic phosphate concentrations and high APA has lead to the idea that the synthesis of alkaline phosphatase is induced by environmental factors, though the exact mechanism of control of induction and inhibition is obscure (Jansson *et al.*, 1988).

From observations on APA and orthophosphate uptake in several lakes of different trophy, Currie *et al.* (1986) concluded that algae are responsible for more of the APA than for orthophosphate uptake. Bacterioplankton has a higher uptake rate of inorganic phosphate than algae and, therefore, can outcompete them for inorganic phosphate. How exactly algae meet their requirements is not understood. One explanation is that they use organic phosphorus (Currie & Kalff, 1984b), or they might have adapted to low bulk inorganic P concentrations in nature by synthesizing exceptionally high-affinity transport systems (Tarapchak & Herche, 1986). High-affinity systems do not preclude that algae may rely on other sources, however (Tarapchak & Nalewajko, 1986).

In the relationship between dissolved inorganic phosphate deficiencies and extracellular phosphatase production, it is necessary to establish the origin of the phosphatase in solution, and to separate the portion which is not actively secreted by algae from that which is passively released. As long as there is no certainty concerning the amounts and the induction mechanisms of enzyme actively secreted by phytoplankton, the hypothesis of phosphatase secretion by phytoplankton to hydrolyse DOP in view of alleviating the deficiency of ambient inorganic phosphorus, cannot be contented.

The dissolved organic phosphorus concentrations in water are often very low and short-lived, so that the activity of phosphatase on its substrate is not easy to show (Jansson *et al.*, 1988).

Phosphorus uptake by autotrophs

The phosphorus species taken up by bacteria and algae is orthophosphate, PO_4^{3-} (Nalewajko & Lean, 1980). This ion does not change its valence, being neither reduced nor oxidized inside the cell, and is directly dependant on the available phosphorus concentration. Phosphorus is mainly excreted at the same oxidation stage, i.e. orthophosphate, but some is also in organic form, as exudations of living or dead, lysed algal cells, or contained in animal excrements.

Biologically available phosphorus consists of orthophosphate in solution, dissolved reactive phosphorus (SRP), some dissolved non-reactive phosphorus, part of the organic phosphorus pool which, after enzymatic hydrolysis yields inorganic phosphorus, and labile phosphorus, which is likely to yield also some inorganic phosphorus by desorption.

Phosphate uptake by macrophytes, phytoplankton, periphytic and benthic algae, and bacteria is a process of active transport. This transport is mediated by electronic or non-electronic ion pumps, which may be sensitive to iron and/or humates (Franko, 1986; Tarapchak & Nalewajko, 1986a). The energy for such transport comes from photosynthesis or respiration. Phosphate is essential for cellular metabolism and photosynthesis (Fig. 32) and, below a critical cellular phosphorus level, growth (Auer *et al.*, 1986) and photosynthesis stop (Senft, 1978).

Phosphate uptake by phytoplankton is stimulated by light, but high phosphorus concentration is the prime prerequisite for such stimulation (Nalewajko *et al.*, 1986). Uptake is modified by the presence of iron, and humic and fulvic matter in water, which have inhibitory or stimulating influences on the activity of dissolved and particulate alkaline phosphatases, and which form complexes and chelates with phosphates (Franko, 1986).

Phosphate assimilation by phytoplankton depends upon cell size. Smaller cells remove proportionally more phosphate from water than larger cells, owing to the higher surface:volume ratio of the former. (Organism size is considered by Smith & Kalff (1985) as a major determinant of community structure, more than taxon-specific specialization.) In taking up large amounts of phosphate, small cells constitute a pool of stocked phosphate, partly due to the rapid turnover time between the different absorption and excretion phases (Broberg & Persson, 1988).

Periphytic and benthic algae (*Scenedesmus sp.*, *Chlorococcum sp. Oedegonium sp.*, *Tribonema sp.*, *Anabaena sp.*, *Ankistrodesmus sp.*, *Oscillatoria sp.*) collected from a small eutrophic lake (Sövdeborgssjön) in southern Sweden were able, in laboratory experiments, to take up phosphorus from water, but also from sediment or from substrata on which they live (see p. 42) (Hansson, 1989). This gives them an advantage over planktonic algae. The inherent disadvantage of fixation onto a substratum is that light requirements cannot be met to the same extent, so that this factor often becomes limiting. In conditions of limiting dissolved nutrient concentrations, phytoplankton biomass is low and light penetrates deep into the water and, thus, periphyton and benthic algae are favoured, while, in conditions of high nutrient concentrations (eutrophic conditions), plankton algae have the advantage (Hansson, 1988).

Processes of phosphate uptake by algae and bacteria

Phosphate absorption by bacteria is the result of two kinetically distinct systems, one of high and one of low affinity. The high affinity system is activated when the internal reserves of phosphorus are low. The two systems function simultaneously and replenish the inorganic phosphorus pool of the cell. When the pool is full, only the low affinity system is active. The low affinity system supplies the cell with phosphorus at a rate equal to the rate at which phosphorus is metabolized in the cell.

The internal inorganic phosphorus pool has a limited capacity and all the phosphorus which enters the cell is stored in the pool, before being used by the cellular metabolism. The high affinity system is repressible, while the low affinity system is constitutive. Their respective energy supplies are different, the low affinity system being fuelled by protons, while the high affinity system functions with ATP.

Bacteria excrete phosphorus at a high rate. The absorption/excretion cycle is rapid, even when the stocks are full, but excretion is reduced when the cell is deprived of phosphorus (Jansson, 1988).

Knowledge on absorption of phosphate by algae is

less complete. The two-phase absorption scheme also exists, but is not so common. The uptake rate of a phosphorus starved cell is 10 to 100 times greater than that of a non-starved cell. The uptake system is controlled by the internal stocks of phosphate, especially polyphosphates. The internal polyphosphate pool is supplied at times when the phosphorus concentrations in the water are high; this pool can support several generations of growth in the complete absence of an external phosphorus supply. It reflects past phosphorus concentrations in the water. A depleted pool indicates a disruption in supply and low ambient phosphorus concentrations in the past. The energy required for phosphate uptake by algae comes from ATP through photophosphorylation or oxidative phosphorylation. The electronic transport chain is the first energy source.

The positive influence of light on phosphorus assimilation shows the importance of photophosphorylation (Jansson, 1988). Phosphate uptake by algae is 1.8 times higher in light than in the dark. It is maximum at times of highest illumination and inhibited at dawn and dusk (Nalewajko *et al.*, 1986); nevertheless, phosphate uptake also occurs in the dark.

Uptake kinetics of phosphate by algae, and the changes over time in phosphate concentrations of water, influence the specific composition of the planktonic community. Concentrations of phosphate in water can be steady or fluctuating, varying with pulses. This also applies for other nutrients. Competition for dissolved phosphate appears to favour smaller cells, either for reasons of interspecific differences in phosphate uptake or by the constitution of a variable assemblage of species along a resource and nutrient gradient (Kilham, 1986). Picoplankton (0.5 to 2 μm in size) contributes an important part of the total number of phytoplankton species, and, though low in biomass, participates notably in primary productivity. Carbon fixation by picoplankton has been estimated to be 20 % of total carbon fixation (Tarapchak & Herche, 1986).

Phosphate competition between phytoplankton and bacteria

Bacteria and bacterioplankton appear to be the largest consumers of dissolved inorganic phosphorus in water. Changes in dissolved phosphate concentrations are directly related to glucose assimilation in the water and not to bicarbonate absorption or chlorophyll concentrations. Hayes & Phillips (1958) observed that inactivation of bacteria by antibiotics caused the fixation of phosphates by sediments, while, normally, utilization by bacteria kept the phosphate in solution, available to the plankton.

Phytoplankton competes with bacterioplankton for absorption of inorganic phosphate (Currie & Kalff, 1984a; Tarapchak & Nalewajko, 1986b; Smith & Kalff, 1985). Investigations in six different lakes of the North German Baltic area showed that the greatest portion of orthophosphate is absorbed by particles less than 1 μm in size (Rai & Jakobsen, 1990). Competition for phosphorus acts as a phosphorus regulator (Currie & Kalff, 1984a & b).

Phosphorus consumption by bacteria in Lake Nesjøvatn, central Norway, was about four times more important than that by phytoplankton. The P:C ratios of phytoplankton and bacteria were different, being 2.5 to 55 mg P (mg of carbon)$^{-1}$ and 34 to 360 mg P (mg C)$^{-1}$, respectively (Vadstein *et al.*, 1988). Bacteria appear to have an important advantage over phytoplankton (Currie & Kalff, 1984a; Currie *et al.*, 1986), except during conditions of high phosphate concentration. The phosphate uptake system of bacteria is saturated at concentrations lower than those required to saturate the algal uptake system. The proportion used by each of the groups is difficult to quantify, as the use of labelled ^{32}P tracer measures gross absorption of phosphates, and the inorganic phosphorus pool in water cannot be measured precisely (Tarapchak & Nalewajko, 1986). Soluble regulating molecules secreted by some groups of algae affect phosphate absorption by phytoplankton.

Bacterial and algal synergy

The existence of an algae-bacteria synergy has been suggested: bacterial growth is restricted by the supply of energy, in particular, by dissolved, reduced organic carbon. Organic carbon released by growing phytoplankton is therefore a major carbon source for bacteria. It can provide up to 50 % of bacterial carbon requirements. Bacteria, in return, regulate phosphorus supply to phytoplankton (Jansson, 1988).

As phosphate uptake by bacteria is faster and

characterized by a higher affinity, competition for phosphate uptake should be at the advantage of bacteria. However, when bacteria have no direct metabolic use for the phosphorus absorbed, their absorption rate is slowed down and accompanied by a steady excretion. These bacterial excretion products are the phosphorus source for algae, either as organic phosphorus, from which algae remove the phosphate, or, more likely, as directly available inorganic phosphate. Orthophosphate excretion is approximately 100 times faster than excretion of organic compounds (Jansson, 1988). The orthophosphate flux across the bacterial cell membrane is quick, while the absorption/excretion cycle is much slower. The same transport mechanisms are used for the intake and the outflow of phosphorus from the cell. This process is, however, not restricted to bacteria, and has been observed in some algae (Jansson, 1988). It is called phosphate exchange and explains the following phenomena:

- bacteria regulate the net phosphorus uptake. Even when saturated with phosphorus, bacteria steadily import phosphate into the cell, and thereafter reject it. The relationship between absorption and excretion is adjusted to the rate at which intracellular phosphorus is metabolized
- bacteria remove phosphate from the water, but release it rapidly, which contributes to keeping it available for algae; the bacteria thus form some kind of transitory compartment (Jansson, 1988). Phytoplankton develops at high growth rates, in spite of low phosphate concentrations in water. This is explained by one or more of these assumptions
- algae possess a high affinity absorption system (Tarapchak & Nalewajko, 1986a) or
- their requirements for phosphate are met by sources other than the dissolved inorganic phosphorus pool (Currie & Kalff, 1984b; Tarapchak & Nalewajko, 1986a) or
- algae build up stocks during periods when phosphate concentrations in water are high and live on these when supplies become scarce.

Intracellular phosphate stocks in algae

The formation of phosphate stocks is an adaptation of algae to provide for their needs in times of shortage, as pointed out by Ketchum & Redfield in 1949. Intracellular stocks consist of phosphates and polyphosphates, in vacuoles and in the cytoplasm. They reach maximum and minimum values of 30 to 100 mg g^{-1} and 1 mg g^{-1} of dry weight, respectively (Broberg & Persson, 1988). According to Hepher (1958), stocks can reach up to 25 times the minimum cellular phosphate concentration. This has been called luxury consumption of phosphate. At temporary deficiency of phosphate in water, the organisms resort to the use of these stocks. Cellular multiplication is proportional to cellular phosphorus content. In nature, excess intracellular phosphorus seldom exists, as the biomass increases to adjust to increasing phosphorus availabilities, reaching a balance between internal and external phosphorus concentrations. Normal phosphorus amounts in algae are about 2 to 10 mg P g^{-1} of dry weight; in comparison, bacterial phosphorus reaches amounts 5 to 10 times that of algae (Broberg & Persson, 1988).

Importance of phosphate effluxes

In phosphorus cycling, phosphorus effluxes and kinetics of cellular phosphorus also have to be taken into account. The amounts of phosphate rejected are significant, so that the net assimilation of phosphorus appears quite low. Few quantified data exist on these effluxes. The phosphorus metabolism in the cell is important; in certain cases, absorbed orthophosphate is excreted after a lapse of only 1 to 8 minutes, although it is usually longer (Hayes & Phillips, 1958; Nalewajko & Lean, 1980).

Amounts of phosphorus excreted by fish

With their excrements, fish release sizeable amounts of phosphate to the water. Excretion is dependent upon the quantity and quality of ingested food, and on environmental and metabolic factors, such as stress or blood pH. Phosphate absorption in the gut, after the ingested feed has been submitted to enzymatic catalysis, is effected in a passive, but restricted, way. The major part is egested with the excrement, and some more is contained in the urine. The rejected phosphorus content sometimes represents between 61 to 87 % of the feed phosphorus (Edward, 1980; Sternik, 1983, for common carp).

Fish feed contains around 1 g P (100 g of feed)$^{-1}$,

on a dry weight basis, for trout pellets and zooplankton, and some 3 g P (100 g feed)$^{-1}$ for forage fish. When the fish is healthy and feeding conditions are good, a large proportion of the phosphorus in the faeces is orthophosphate, but when fish is starving, most of the phosphorus is organic. For a carp of 100 g, Sternik (1983), found an average daily excretion of around 78 mg P_{tot} kg^{-1} d^{-1} and 4.4 mg P-PO_4^{3-} kg^{-1} d^{-1}, at 25° C in laboratory conditions. This is an impressive amount for a normal stocked fish population, i.e. for a fish loading of 50 kg m^{-3}, it represents a phosphorus supply of 200 μg l^{-1} d^{-1} of P-PO_4^{3-}.

Measurement of phosphorus in water

Phosphorus is measured after transformation into orthophosphate (Boyd & Tucker, 1980). Orthophosphate is analyzed by a method involving the reaction of orthophosphate ions with molybdate ions to form phosphomolybdate, subsequently reduced to a soluble blue complex, detected spectrophotometrically. This analytical method has a tendency to overestimate phosphate concentrations and does not achieve the required precision of 0.01 μg l^{-1} (Shapiro, 1988; Broberg & Pettersson, 1988).

Phosphorus and phosphorus compounds in water: salient points

1. Phosphorus is the most frequent limiting nutrient to primary production in water. Concentrations of dissolved phosphorus are usually low. Phosphorus deficiency affects algal metabolism, growth and photosynthetic activity. In excess, phosphorus causes eutrophication.

2. Phosphorus is present in water as dissolved orthophosphate, organic phosphorus, and particulate phosphorus. Some degree of equilibrium exists between the various phosphorus species. Sorption of inorganic dissolved phosphate onto organic colloidal particles or onto mixed inorganic-organic colloidal aggregates is an important process. Phosphorus is supplied to ponds by organic matter, as a byproduct of primary productivity, animal excreta (zooplankton and fish), and uningested fish feed and fertilizer. Phosphorus is removed from water by sedimentation of seston and drainage of ponds (water renewal).

3. The dissolved organic phosphorus pool contains a group of organic compounds, which has an important influence on dissolved phosphorus availability. Three broad groups have been distinguished, namely, high (HMWP), intermediate (IMWP), and low molecular phosphorus (LMWP) fractions. The HMWP fraction seems to be a decomposition-refractory complex of organic (humate or fulvate) and iron (metal) oxides, having high sorption capacities for inorganic and organic phosphate, sometimes saturated with phosphates, and acting as a regulator of dissolved phosphate. The IMWP or XP fraction results from decaying processes (perhaps, algal cell walls fragments). This fraction releases phosphate under UV radiations, but is refractory to enzymatic action. The LMWP fraction is made of organic compounds of low molecular weight. The IMWP and LMWP fractions are probably decomposition products, only temporarily present in water.

4. Phosphorus cycling in a pond is difficult to quantify, as the turn-over time of many compounds is short and hard to detect, especially phosphate uptake, phosphate excretion by zooplankton and fish, and sorption processes. Other processes, such as sedimentation of phytoplankton to the pond bottom, are much slower, and greatly modified by external and water quality factors. The most important processes responsible for phosphorus changes in water are organic matter decomposition, and phosphate uptake by bacteria and algae.

5. Organic phosphorus mineralization is mostly a microbially mediated process. It provides almost three-quarters of the phosphorus supply. Some physical and chemical decomposition also occurs: UV decomposition or photoreduction. Phosphate sorption and desorption processes by dissolved organic compounds are numerous and as yet unquantified. Most phosphorus release from organic compounds is due to the activity of phosphatases, but some release from humic-iron-phosphate complexes is induced by UV irradiation. Dissolved humic substances act as a kind of buffer for dissolved phosphate.

6. Particulate and dissolved phosphatases are present in water; particulate phosphatase is contained in algae and bacteria, while dissolved phosphatase is released into water by algae, bacteria and zooplankton. The most likely role of these phosphatases is hydrolysis of phosphate from organic compounds, at times of dissolved inorganic phosphate deficiency.

7. Phosphate is taken up by algae and bacteria and competition can occur between these two groups for limited amounts of dissolved phosphate. The advantage in this competition is with the bacteria, which have an uptake system with higher affinity. However, bacteria excrete large amounts of phosphate, so that bacterial phosphate uptake seems to act more as a temporary storage. Bacteria and algae may live in synergy, involving phosphate and organic carbon exchange. Algae form intracellular stocks of phosphate, the process being called luxury consumption; beyond these stocks, any surplus of phosphate in water induces algal multiplication. Stocks of intracellular phosphate can be passed through several generations of algae.

4.3.2. Phosphorus fertilizers

Phosphorus fertilizers available

Phosphate deposits are of three broad apatite groups: deposits of igneous and metamorphic origin, sedimentary phosphorites, and guano and related deposits. Guano is formed on land by sea birds and was the earliest source of phosphate fertilizers. Phosphate rocks are mined for fertilizers and industrial chemicals, such as detergents, toothpaste and pharmaceuticals (Holtan *et al.*, 1988).

When apatite is treated by sulphuric acid, the following reaction occurs:

$$[Ca_3(PO_4)_2]_3.CaF_2 + 7H_2SO_4 \longrightarrow 3Ca(H_2PO_4)_2 + 7CaSO_4 + 2HF$$

A list of phosphorus fertilizers is given in Table 16.

Fertilizers containing a mixture of nitrogen, phosphorus and potassium are characterized by their respective concentrations of each of these elements. A value of N:P:K of 8:16:16 indicates 8 % total nitrogen, 16 % P_2O_5 (or 7 % phosphorus) and 16 % K_2O (or 13 % soluble K).

Destiny of phosphorus fertilizers in ponds

Phosphorus fertilizers added to ponds in Israel caused a rapid increase in dissolved orthophosphate concentration in the water, which disappeared after a short time (Hepher, 1958). Disappearance of phosphorus is the result of absorption by bacteria, macrophytes and phytoplankton, which occurs in a short space of time, sometimes in few minutes, and adsorption by organic colloids in solution or in suspension in water, as well as adsorption by sediment or interaction (chelation or complexation) with ions in the water, as observed by Hayes & Phillips (1958) in eight lakes of different trophy in Nova Scotia, Canada. The main part of added phosphorus to the water of Lake Gunillajaure in northern Sweden was used by the seston, and a small portion sorbed onto sediments (Björk-Ramberg, 1984).

In experiments conducted in ponds at the Auburn University Fisheries Research Unit, Alabama, U.S.A., some portion of the phosphorus supplied by the fertilizers accumulated in pond sediments, which, although acting primarily as a sink for dissolved phosphorus, were also a phosphorus source. Draining the fish pond at harvest had generally a negative effect on pond fertility (Hollerman & Boyd, 1985). In fish ponds in production and undrained for three years, total phosphorus concentration in water was significantly higher than that of ponds drained annually. The phytoplankton biomass of the former was comparatively higher than that of annually-drained ponds, owing to phosphorus mobilization from the sediments (Hollerman & Boyd, 1986). In an undrained pond compared to normally drained ponds of a commercial farm in Upper Galilee, Israel, sediments lost their capacity to remove phosphorus from water (Eren *et al.*, 1977), which resulted in higher average phosphorus concentrations in water to the benefit of primary production.

Table 16. Chemical composition and phosphorus content of the principal phosphorus fertilizers (according to Brady, 1974).

Fertilizer name	Chemical formula	Percentage in P_2O_5	Percentage in P
Superphosphate(*)	$Ca(H_2PO_4)_2$ $CaHPO_4$ $CaSO_4$	16 - 50	7 - 22
Triple superphosphate (*)	$Ca(H_2PO_4)_2$	44 - 47	17 - 21
Ammonium superphosphate	$NH_4H_2PO_4$ $CaHPO_4$ $(NH_4)_2SO_4$	16 - 18	7 - 8
Ammonium phosphate	$NH_4H_2PO_4$	48	21
Ammonium polyphosphate	$(NH_4)_4P_2O_7$ and others	56 - 60	
Diammonium phosphate	$(NH_4)_2HPO_4$	46 - 53	20 - 23
Basic slag	$(CaO)_5P_2O_7.SiO_2$	15 - 25	7 - 11
Steamed bones	$Ca_3(PO_4)_2$	23 - 30	10 - 13
Calcium metaphosphate	$Ca(PO_4)_2$	62 - 63	27 - 28
Phosphoric acid	H_2PO_4	54	24
Superphosphoric acid	$H_3PO_4 + H_4P_2O_7$	76	33

(*) 85 % soluble in water

Interaction of fertilizer with other water characteristics

A high calcium ion concentration induces precipitation of phosphate, thereby decreasing the concentration of dissolved orthophosphate. An alkaline pH also promotes precipitation. If calcium concentration in water is 20 mg l^{-1}, the maximum concentrations of dissolved orthophosphate are 10 mg l^{-1} at pH 8 and 0.01 mg l^{-1} at pH 10 (Boyd, 1979). Alkalinity and pH have a direct effect on the orthophosphate concentration in water. A decrease of carbon dioxide, resulting from photosynthesis or gaseous diffusion, increases pH and causes phosphate precipitation (Hepher, 1966).

The use of phosphorus fertilizers reduces water turbidity caused by clay particles (mineral turbidity). This reduction in mineral turbidity is due to coprecipitation of clay particles, calcium carbonate and algae. An algal polysaccharide inducing flocculation is released during photosynthesis. This algal sensitivity to particles in suspension, shown in limnocorral experiments in a small lake in North Carolina, U.S.A., is specific to certain algal species. *Anabaena* is more sensitive than *Chlamydomonas* or *Chlorella*. *Euglena* does not flocculate at all. Flagellates do not flocculate, and escape by swimming away (Cuker, 1987).

Phosphorus fertilizer doses applied

The use of phosphorus fertilizers is commonly advocated, but doses recommended vary between 2.5 and 20 kg P_2O_5 (eq.) ha^{-1} (2 wks)$^{-1}$. In Indonesia, Hickling (1962) tested effects of three different doses, namely 22.4 kg, 44.8 kg and 67.2 kg P_2O_5 (eq.) ha^{-1} yield^{-1}, on the production of hybrid tilapia during a period of 6 months (6 replications), in monoculture, and in polyculture with grass carps and *Puntius*. Though results were variable, he obtained an increase in production, in comparison with the control pond, of 11.2 kg, 9.67 kg and 10.03 kg of fish per kg of P_2O_5 (eq.) added, for the three doses, respectively. There was a direct proportionality between increase in production and increase in phosphorus fertilizer dose, in a proportion of about 10 kg per kg of equivalent P_2O_5 added.

Boyd (1976a & 1981b) at the experimental fish farm of Auburn University, Alabama, U.S.A., used phosphorus fertilizer applications at rates varying between 2.2 and 9 kg P_2O_5 (eq.) ha^{-1} (2 wks)$^{-1}$, in fish ponds stocked with tilapia, *O. aureus*, or with *Lepomis*. At the same time, doses of 1.2 kg, 16 kg, and 9 kg ha^{-1} (2 wks)$^{-1}$ of N were applied. The average concentration of dissolved orthophosphate in the water increased with fertilizer dose, being 27 μg, 43 μg, and 82 μg l^{-1} of P-PO_4^{3-} for application rates of 2.2 kg, 4.5 kg, and 9 kg P_2O_5 (eq.) ha^{-1} (2 wks)$^{-1}$, respectively. The differences in production between various fertilizer treatments were not large, being relatively good with 9 kg P_2O_5 (eq.) ha^{-1} (2 wks)$^{-1}$ alone or in association with nitrogen fertilizer (considered by Boyd as conditions of superfertilization), and with 4.5 kg P_2O_5 (eq.) ha^{-1} (2 wks)$^{-1}$ with 1.6 kg ha^{-1} (2 weeks)$^{-1}$ of N.

In Israel, Hepher (1962) recommended application of 12 to 16 kg P_2O_5 (eq.) ha^{-1} (2 wks)$^{-1}$, along with 14 kg ha^{-1} (2 wks)$^{-1}$ of N. These doses generated an abundance of algae equivalent to 109 mg of chlorophyll a m^{-3}, a carbon fixation rate of 1.45 to 4.47 g C m^{-2} d^{-1}, and a net algal production of 19 to 32 g m^{-2} d^{-1}. Daily carp production amounted to 0.43 g m^{-2} d^{-1}. A doubling of fertilizer doses, maintaining the application doses, but increasing the application frequency (every week instead of every other week), increased algal biomass to 134 mg of chlorophyll a m^{-3}. The great abundance of plankton caused self-shading of the 20 uppermost centimetres of water and a sizeable reduction of primary productivity. Eren *et al.* (1977) recommended, for Israeli fish ponds, 9 kg P_2O_5 (eq.) ha^{-1} (2 wks)$^{-1}$, three times during the first weeks of production, followed by the same dose once a month thereafter.

In Malawi, Msiska (1983) tested, with success, an application of 2.6 kg P_2O_5 (eq.) ha^{-1} (2 wks)$^{-1}$, while Miller (1976) reported trials by Vincke in Madagascar using 10 to 20 kg P_2O_5 (eq.) ha^{-1} (2 wks)$^{-1}$, along with 8.5 to 17 kg ha^{-1} (2 wks)$^{-1}$ of N. In Côte d'Ivoire, trials with triple superphosphate, at a dose of 25 to 30 kg P_2O_5 (eq.) ha^{-1} mth^{-1}, gave good results, and the dose was recommended.

The Fisheries Department in Malaysia recommended to fish farmers doses of 12 to 15 kg P_2O_5 (eq.) ha^{-1} mth^{-1} (Yusoff & McNabb, 1989). Experiments made in ponds at the Agriculture University of Malaysia at Serdang, to test the efficiency of these doses, alone or with urea at a rate of 16.6 kg ha^{-1} wk^{-1} of N, in comparison with the use of organic fertilizers, gave net productions of up to 1700 kg ha^{-1}, using a combination of four species of carps, at a low stocking rate (0.5 per m^{-2}). The natural productivity was 0.26 g of C m^{-2} h^{-1} and the chlorophyll concentration in water was 109.2 mg m^{-3}. They concluded, in the light of ratios between dissolved inorganic nitrogen and phosphorus in water, that phosphorus was the limiting factor for production (Yusoff & McNabb, 1989).

Application mode of phosphorus fertilizers

Most phosphorus fertilizers are in pellet form, and are applied as such or after solubilization in water.

Sometimes, they are kept inside the sack in which they were packed; this is perforated and immersed in the pond. The fertilizer slowly dissolves and oozes out of the sack. Pellets dumped into the pond sink and get buried in or dissolve on contact with the sediment. Dissolving phosphate becomes sorbed by the mud and is lost to the water. Dissolution prior to the application of solid fertilizer to the water column or hanging of a container containing the solid fertilizer in the mid-water is recommended, as dissolved fertilizers are directly available to plankton (Boyd, 1981a).

In ponds at Auburn, Alabama, U.S.A., Boyd *et al.* (1981) compared three phosphorus fertilizers, triple superphosphate, diammonium phosphate and liquid ammonium polyphosphate, at doses of 9 kg P_2O_5 (eq.) ha^{-1} (2 wks)$^{-1}$, along with 3.5 kg ha^{-1} (2 wks)$^{-1}$ of N. The assessment involved monitoring of orthophosphate and chlorophyll a concentrations. After 10 to 15 days, dissolved orthophosphate and total phosphorus reached the same levels in all ponds, being 5 to 30 $\mu g\ l^{-1}$ of $P\text{-}PO_4^{3-}$ and 0.05 to 0.17 mg l^{-1} of total P. Chlorophyll a concentrations were higher in ponds treated with liquid polyphosphate and this was confirmed even with doses two thirds or one third of the original dose (Fig. 33).

If dissolved prior to application, the effect of tri-

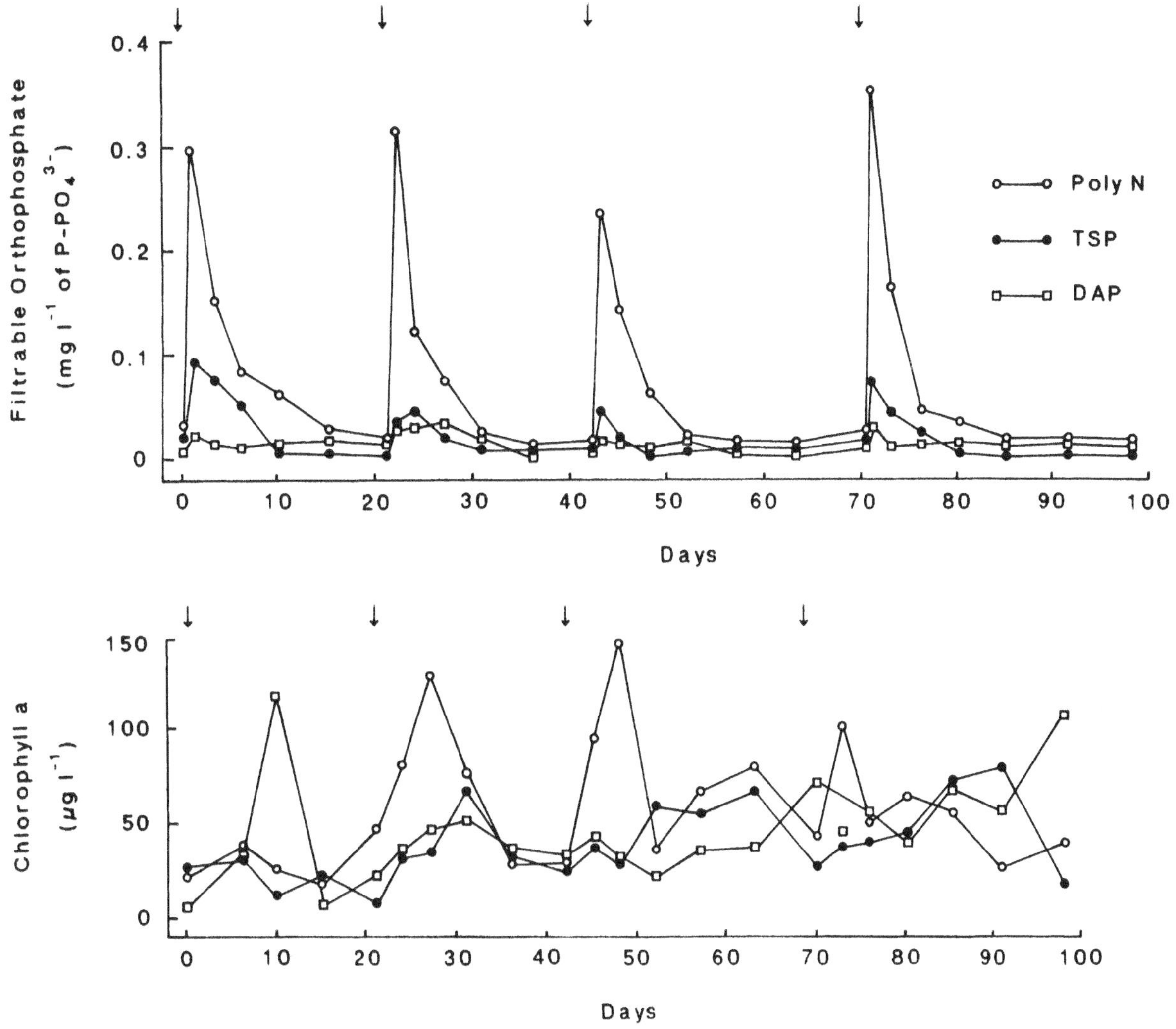

Fig. 33. Comparative effect of three phosphorus fertilizers, i.e. diammonium phosphate (DAP), triple superphosphate (TSP) and liquid ammonium-phosphate fertilizer (Poly N) on filtrable orthophosphate and chlorophyll a concentrations in pond water. The arrows (↓) indicate the date of fertilizer applications (from Boyd *et al.*, 1981) (with kind permission of Prof. Boyd on behalf of the authors & Aquaculture).

ple superphosphate (TSP) was similar to that of liquid polyphosphate. When applying liquid fertilizers, Boyd (1981c) recommends that an atomizer or sprayer be used or that the water is churned to insure a good mixing of fertilizer with water. Fertilizer can also be poured in the eddies of the screw of an outboard engine. Fertilizer has a higher density than water, and, if not thoroughly mixed with the latter, will sink.

Experimentation in ponds to determine the best application rates for nitrogen and phosphorus fertilizers should conform to the following scheme: production trials in stocked ponds in which only fertilizers are applied must run concurrently to trials in which fertilizer is supplemented with the usual feed at the usual feeding rate. The trials need to be repeated at low and high stocking densities, to compare with the growth of fish obtained with natural feed at different densities. The analysis should take into consideration production data, feed conversion rates, and growth rates during different periods (between sampling dates) of production.

Recommended application rates for phosphorus fertilizers

The dynamics of phosphorus transformation within aquatic systems is yet to be fully understood, and the different fluxes and exchanges between the sediment and the water column, between particulate and dissolved compartments, and between the different groups of organisms remain unquantified. It is therefore difficult to predict the amounts of phosphorus, provided by a fertilizer, which really benefit phytoplankton production, and to assess the amounts made available through internal loading of the pond. The amounts of phosphorus required for phytoplankton production cannot be calculated on the basis of an analytical budgeting of the phosphorus cycling in the pond. Some insight can be gained from the data of limnologists studying eutrophication problems and the ways and means of reducing internal phosphorus load. According to this information, the upper limit of external phosphorus loading (in temperate lakes) is considered to be 0.5 to 2 g P m^{-2} yr^{-1} (Jeppesen *et al.*, 1990a). This is equivalent to a fertilizer supply of 7.5 kg P ha^{-1} $(2\ wks)^{-1}$. Considering that frequent renewal of water at each harvest represents an important loss of phosphorus, and that, at each harvest, phosphorus is exported in the form of fish, fertilizer application rates in the range of 5 to 10 kg P ha^{-1} $(2\ wks)^{-1}$ seem reasonable.

These fertilization rates are equivalent to doses of 8.75 to 17.5 kg P_2O_5 (eq.) ha^{-1} $(2\ wks)^{-1}$, and are supplementary to regular feeding of the fish and use of nitrogen fertilizers at 5 to 10 kg ha^{-1} $(2\ wks)^{-1}$ of N. This is equivalent to 12 to 24 kg of urea (or 23 to 47.5 kg of ammonium sulphate) and 17.5 to 35 kg of triple superphosphate per hectare per 2 weeks. In comparison to nitrogen fertilizer, doses of phosphorus fertilizer need to be adjusted less to production. Increasing doses during the production period are less justified, the objective being to maintain surplus phosphorus and saturated intracellular stores.

Phosphorus fertilizer application

1. Many inorganic phosphorus fertilizers are available on the market. Phosphorus fertilizers added to the water interact with other water features, particularly with calcium carbonate in alkaline water, and with iron and aluminium in acid water.

2. Phosphorus fertilizers are preferably applied in liquid form, as solid pellets sink to the pond bottom and are lost from the water column.

3. Recommended application rates for phosphorus fertilizers are 8.75 to 17.5 kg P_2O_5 (eq.) ha^{-1} $(2\ wks)^{-1}$.

4.4. Manure or organic fertilizers

Differences between inorganic fertilizers and manure

A number of organic fertilizers, such as livestock dung, and plant or vegetal detritus, have long been used for fertilizing ponds. Yields obtained with organic fertilizers are usually higher than those obtained with inorganic compounds. In ponds, highest productions are reached in conditions of intense continuous manuring of pond water, in association with animal production. The cost of manure is lower than that of inorganic fertilizers. Their use can, however, be problematic, for several reasons. For example, manure in excess can have adverse effects on production. Also, manure may pass diseases on to humans.

Organic fertilizers contain nitrogen and phosphorus, but also decaying organic matter. Nutrients benefit phytoplankton (and phytobenthos), while the organic matter promotes the development of bacteria in suspension in the water and in the sediment. The heterotrophic production cycle is thus stimulated leading to an increase in zooplankton and zoobenthos. Manure also supplies organic colloids, which increase the sorption capacities of soil and seston.

While the microbial decomposition cycle is boosted with increased supplies of organic matter, the oxygen consumption of the water increases. The supplies of nutrients by manure lead to an increase in primary production, which, in turn, provide more oxygen to the decomposer community. Manure has therefore a dual effect on autotrophic and heterotrophic production cycles. The two cycles are deeply intricated. The phytoplankton population supplies organic matter for decomposition, as most phytoplankton is not consumed by predators. Decomposers recycle and regenerate nutrients to the benefit of phytoplankton and are themselves consumed by zooplankton and fish. Zooplankton is consumed by fish.

Fish fed on plankton and produced in manured ponds have lipid contents (6.2 %) less than those fed pellets of high protein content (15 % lipids) or those fed cereals (20 % lipids) (Moav *et al.*, 1977).

Decomposition processes in water

Manure is a loose, fibrous material, made of particles of various sizes. Dumped into water, it takes time to sediment. Nutrients and colloids ooze out, and organic matter dissolves from the particles in suspension. All these organic flocs are colonized by microbial fauna.

Decomposition processes take place either in water or in sediments. The importance of recycling of nutrients in water is often underestimated, partly because there is controversy concerning methods for measuring bacterial productivity (Moriarty, 1990; Costa-Pierce, 1990). Costa-Pierce & Craven (1987) concluded from studies on microbial production in aquaculture ponds in Kahaku, Hawaii, using the incorporation of adenine into ATP, RNA and DNA, that more than 50 % of heterotrophic production is microbial. In Lake Titicaca, Peru/Bolivia, Carmouze *et al.* (1984) observed that almost 90 % of the organic matter is recycled within the euphotic zone whence heterotrophic activities are particularly intense.

Manure provides a substrate for microbial development, and microbial population size and activities are tightly correlated to the available substrate. The substrate is characterized by its C:N or C:N:P ratios; ratios of 17:1 or 20:1:0.2 ensure a good microbial growth. In an aerobic environment, 20 to 50 % of the available carbon in organic matter is used by bacteria, while in anaerobic water, the efficiency of carbon utilization is reduced. Microbial production is essential for fish production, as the former can contribute to almost half of the latter. Microbial production is consumed by zooplankton, especially ciliates (microzooplankton), and by detritivorous fish.

In investigations on Canadian shield lakes in the Experimental Lake Area (ELA), decomposition of organic carbon in the hypolimnion of three lakes of different trophy (artificially eutrophied and artificially acidified) proceeded aerobically or anaerobically, according to dissolved oxygen concentrations. In the absence of oxygen, concentrations of available electron acceptors and redox potential determined the reduction process, which included, for instance, sulphate and nitrate reduction (Kelly *et al.*, 1988).

The decomposition process on sediments is also determined by oxygen supply. Sediment oxygen demand (almost equivalent to the biological oxygen demand, BOD) increases with microbial activity. Aerobic decomposition in sediment is more limited by oxygen than that in water. It is useful to remember the importance of oxygen demand for the decomposition of organic fertilizer, when estimating the maximum manure load to be used in a pond (Schroeder, 1980). Decomposition in water and on/in sediment is a function of bacterial population densities.

Liquefied dung supplies organic colloids to the water. It remains in suspension and serves as food and substrate for the bacterial population of decomposers, which live in the water column. As such, fertilization benefits the water column more than the sediments. The surfaces exposed to microbial attack are much greater than those of lumps of solid dung. Oxygen availability is greater in water than in sediments. The released nutrients are directly available to the phytoplankton, and do not have to diffuse from sediments to the euphotic zone. With the use of liquid cow dung, good productions have been achieved in Israel (Moav *et al.*, 1977).

In lake Aydat, in Central France, two major types of bacteria are involved in the degradation of suspended organic matter, i.e. free living organisms, of small size (0.2 to 1.2 μm), and fixed organisms, of greater size (1.2 to 63 μm) (Marvalin *et al.*, 1989). A great proportion of the dissolved organic matter is polymeric material, of high molecular weight, which can only be assimilated by bacteria after decomposition by extracellular enzymes.

Exoenzymatic hydrolysis is the first and probably the limiting step of organic matter decomposition. In turbid waters, exoenzymes are closely associated with particles in suspension and remain active even when adsorbed. The activity of exoenzymes increases in the presence of high concentrations of suspended material. Exoenzymes most frequently considered are alkaline phosphatases (see p. 102), aminopeptidases, and endopeptidases. Aminopeptidases decompose proteins; they appear, for instance, quickly after the death of an algal bloom, but their activity does not persist (Boon, 1990).

The microfauna which develops on organic matter is consumed by ciliates, flagellates, and rotifers. The recovery of energy from organic matter to the benefit of the foodweb is called the microbial food loop and is discussed on p. 170.

Exoenzymes release nutrients, which are used by algae, thus fostering primary productivity. An increase of algal biomass is not necessarily accompanied by an increase in grazing. The algal cells disappear after some time and are returned to the decomposer organisms.

Types of fish cultured in manure-fertilized ponds

Fertilization with manure can be applied to ponds where fish with various diets are cultured in polyculture, making use of the different feed resources of the pond. Primary productivity is used by phytoplanktivorous fish, secondary productivity by zooplanktivorous fish, and detritus by detritivorous fish. Some fish, especially tilapias, have wide feeding regimes, capable of using a broad range of food items (see p. 184). For example, *O. mossambicus*, generally considered a phytoplanktivorous fish, can browse on the flocculent layer at the surface of the sediment, as in Lake Valencia, Venezuela (Bowen, 1981). This flocculent layer consists of organic matter undergoing decomposition, colonized with decomposers and microzooplankton. *O. niloticus* from Lake George, L. Albert, and the middle Niger Delta was observed feeding on bottom deposits on some occasions (Trewawas, 1983). Different emphasis on possible food items is reflected in the relative coarseness or fineness of the pharyngeal dentition and, to some extent, of the teeth on the jaws (Trewawas, 1983).

Detritivorous or benthophagous fish ingest organic mud, digesting bacteria which develop within, and egesting, in their excrement, all what cannot be digested. These excrements are immediately recolonized by microorganisms.

About 90 % of the phytoplankton produced in ponds consists of nanoplankton (cell sizes between 3 and 20 μm), unavailable to fish, because of their small size. This assertion is, however, refuted by some researchers, who state that tilapias are able to agglutinate, using mucus secreted by the oral and pharyngal epithelium, the small phytoplankton cells brought into the buccal and branchial cavities. In Lake Kivu, nanoplankton and the planktonic bacterium *Spirillum* are indeed used as food by *O. niloticus* during most of the year (Trewawas, 1983).

Fish feeding on phytoplankton can foster some taxa, such as the blue-green alga *Microcystis*. Fish have direct and indirect effects on phytoplankton assemblages in ponds. This aspect will be discussed below (p. 176).

Nutritional value of manure

Some fish, especially tilapia, feed directly on fresh excrements. They can be observed, for instance, below pigsties or at the entrance of feeder canals bringing sewage from stables, feeding on excreta floating in the water. This lead to the idea of using dung directly in the preparation of artificial feed pellets. Analysis of cow dung and chicken manure indicated that this contains an amount of metabolizable energy, in the order of 2510 to 3347 kJ kg^{-1} and 3766 to 5021 kJ kg^{-1} and 10 to 15 % and up to 25 % gross N-proteins, respectively. Msiska (1981) found chicken manure to consist of 14.2 $\pm$ 4.0 % gross protein.

Trials with pellets containing 30 % protein, made out of dung, as feed has been tested with a variety of fish (common carp, tilapia, channel catfish, goldfish) and has not given satisfactory results (Schroeder, 1980). Other trials in outdoor tanks in northern Israel, where part of the ingredients were replaced by cow dung, gave results which varied with the species cultivated. The replacement of 50 % of the pellet ingredients by cow dung gave better results in tilapia (*O. aureus*) production, but not in *Clarias gariepinus*. These differences appeared only when feed was given in daily amounts corresponding to 5 % of the actual biomass; at 2.5 %, no difference appeared. The most likely explanation for this increase in tilapia production is that the dung contained in the feed had a stimulatory effect on primary productivity, which benefitted the tilapia culture, but was not used by the catfish (Degani *et al.*, 1984).

The organic nitrogen portion of manure, previously considered to be gross protein and measured by Kjeldahl analysis, is actually mainly uric acid and non-protein nitrogen. These substances have no direct nutritional value for fish. More precise analysis has indicated that cow dung contains less than 1 % pepsin-digestible proteins, more than 65 % total fibres, less than 1 % starch or carbohydrates, and a caloric level below 3347 kJ kg^{-1} of metabolizable energy. This makes it a feed of low nutritional value. In comparison, a feed of medium level quality contains about 25 % proteins, 65 % starch and 12552 kJ kg^{-1} of metabolizable energy (Schroeder, 1987).

Detritus and dung directly ingested by tilapia are mainly stripped of their bacterial colonies and digested. This was confirmed by observations in aquaria at the University of Sterling, UK; *O. niloticus* (average weight of about 3 g) fed actively on free-living non-attached bacteria, which were introduced in the water at concentrations of 10^5 to 10^6 bacteria ml^{-1}. Bacteria were found in the fish stomachs (Beveridge *et al.*, 1989). See also the observations of Bowen (1981), already mentioned (p. 52).

Disadvantages of the use of manure

A disadvantage of manure is the presence of excess amounts of organic fertilizer, causing a dissolved oxygen deficit, in particular during the hottest hours of the day and in the early morning (shortly before dawn). This is a consequence of high microbial activity. Another threat in the use of manure is the release of ammonia, especially when pH is alkaline. As discussed earlier, ammonia is toxic. The use of dung in ponds increases the risk of transmission to humans of parasitic diseases with fish as intermediate and humans as final hosts. However, no such disease has yet been detected.

Mode of use of fertilizers - precautions for their use

Precautions to be taken when using organic fertilizers are related to the disadvantages listed above. Manure should only be spread in water in small amounts, several times a week. The spreading should be as homogeneous as possible over the whole surface. In experiments conducted at the Wuxi Municipal Fish Farm, in Jiangsu province, China, ponds of large size (7333 m^2) are more difficult to manage than ponds of smaller size (between 667 and 3800 m^2), and give lower production values. The effects of manure were more drastic in ponds of small size; the fluctuations of dissolved oxygen were wider and were causing problems to fish (a variety of Chinese carp) more quickly (Zhang *et al.*, 1987).

One of the first factors affected by high decomposition rates is DO. The value of dissolved oxygen at or a little before dawn should be regularly checked. A low or critical DO value in the early morning indicates excessive heterotrophic respiration. The presence of fish gasping for oxygen at the surface is typical of oxygen depletion. A decrease of luminosity, owing to an increase of cloud cover or rain, decreases photosynthetic rate (Fig. 20) and, consequently, oxygen release. The spreading of manure should be suspended as soon as indications of a dissolved oxygen deficit are perceived and physical and chemical features of the water should never be allowed to fluctuate widely.

Manures containing much ammonia should be used in extra-small amounts at a time, in combination with close monitoring of pH. The presence of ammonia in water in concentrations around 30 to 100 mg l^{-1} of $N\text{-}NH_4^+$ or the presence of hydrogen sulphide indicate a threatening anaerobic situation.

Manure in an advanced stage of decomposition fertilizes better than less decomposed organic matter, because the nutrients diffuse more easily to the water and because the ultimate decomposition can take place during the sinking of the organic matter towards the pond bottom. This allows enrichment with nutrients of the whole water column and a decrease in oxygen demand of the sediment; the microorganisms digesting the organic matter remove the oxygen needed for their activities from the water and not from the sediment. Dissolved oxygen removed from water is more easily replaced than that taken from sediments.

In laboratory experiments using soil collected from a brackish water fish pond, in West Bengal, India, manure decomposed less rapidly in brackish water than in fresh water, the rate of decomposition decreasing with increasing salinity. The amount of nitrogen released into the water, in the form of nitrate and ammonium, decreased and the amount of organic carbon found in the soil after a given period was higher at increased salinity (Chattopadhyay & Mandal, 1980).

Manures are usually added to the water. Another method is to spread dung or compost on the bottom of dried ponds before impoundment, or even to work it into soil by shallow ploughing. The exchange capacities of the soil are increased by incorporation of organic matter; the water retention capacity of the soil is greater and infiltrations are reduced (Chakroff, 1976).

Organic fertilization requires proper pond management and constant monitoring of production conditions. Ponds best suited for organic fertilization should be shallow, ideally with a mean depth of 1 m. A judicious choice of species and of proper stocking densities allows the best use of different food resources produced and the maintenance of a balance between the various natural production processes within the pond.

Types of organic manure used and their composition

Organic fertilizers used are mainly solid or liquid excreta of livestock animals (cattle, pigs, sheep, ducks, chicken), slaughter house wastes (rumen contents, blood, etc.), compost, and domestic or industrial (from agricultural industries) sewage water. Fish culture may be combined with animal production: livestock is held in stables nearby or even over the pond, so that excreta and feed surplus find their way directly to the pond.

The chemical composition of animal excreta is variable, depending not only upon the nature of feed ingested, but also on other ingredients added to form the final manure product, such as urine, straw litter, and/or excess feed. Moreover, they are characterized by a great variability in water and nutrient content. Their nutrient to dry matter content ratio is almost 20 to 30 times lower than that of inorganic fertilizers. A ton of organic fertilizer is equivalent to about 45 kg of N:P:K of 10:5:10 (Delmendo, 1980). The nitrogen/phosphorus ratio is not balanced, phosphorus being less abundant. Organic matter contributes to the formation of humus. An organic fertilizer not properly stored or mixed with soil quickly looses its quality. Nitrogen is leached or lost by gaseous diffusion.

In Tables 17 and 18, average compositions of excreta of livestock animals and amounts of excrement produced daily are indicated.

Comparative value of manures

At the Ginosar Farm, Israel, comparative trials of dry and liquid cow dung and chicken manure, at a rate of 5 kg of manure (dry weight) per day, with applications five times a day, to carp ponds in which fish feed was in pellet form, demonstrated the superiority of chicken manure over cow dung. As

Table 17. Average composition, in humidity, nitrogen, phosphorus, and potassium, and faeces/urine ratios, of excreta of common livestock animals (modified from Little & Muir, 1987)

Type of animal	Faeces to urine ratio	Humidity	Nitrogen (in %)	Phosphorus (in % of P_2O_5)	Potassium (in % of K^2O)
Dairy beef	75:25	85	4.0	2.7	3.4
Feedlot beef	75:25	85	7.8	2.5	3.6
Chicken	100	65	6.0	7.9	3.7
Pig	53:47	85	4.7	4.3	2.7
Sheep	67:33	70	3.9	2.4	3.5
Horse	75:25	75	2.3	1.3	1.4

Table 18. Amounts of excrements, expressed in fresh weight and dry weight per animal, and per 100 kg of live weight, per day, produced by common livestock animals (modified from Little & Muir, 1987)

Animal	Fresh excreta per animal and per day (kg)	Dry weight per animal and day (kg)	Fresh excreta per 100 kg live weight and day (kg)	Dry weight per 100 kg live weight and day (kg)
Beef	30.0	4.5	6.0	0.9
Chicken	0.7	0.27	35.0	13.4
Ducks	1.0	0.38	33.3	12.7
Pigs	8.0	1.2	14.5	2.17
Sheep	2.1	0.71	7.0	2.4
Horse	24.0	8.2	6.3	2.15

compared to the control, chicken manure increased fish production by 21 %, while this value was 11 % for cow dung. The feed conversion rate of pellets was decreased by 12 and 6 %, respectively, for chicken manure and cow dung, as compared to the control. These manures, in liquid or dry form, at the rates used, were well mineralized, as shown from a comparison between the dissolved organic matter, organic carbon, total nitrogen, and ammonium concentrations in sediment of fertilized ponds versus unfertilized ponds. This indicates that organic fertilizers did not accumulate unduly on the pond sediment (Rappaport & Sarig, 1978).

In ponds in Camayagua, Honduras, other comparative trials between manures and inorganic fertilizers, in production ponds of *O. niloticus* indicated that better productions are achieved with chicken manure than with cow dung or inorganic fertilizers, at equivalent fertilization rates (Green *et al.*, 1989). In Ibadan, Nigeria, trials using chicken sweepings, containing 30 to 40 % manure, and spread 6 days a week on ponds as feed and fertilizer, yielded a production of 6.75 tons ha^{-1} yr^{-1} of *Clarias isheriensis* (Faybenro & Sydenham, 1988).

Effects of manure on water quality

Manures affect water quality in different ways, but mainly with regard to nutrient content. Ammonium and ammonia concentrations vary but sometimes reach high values: 0.16 to 0.78 mg l^{-1} and 0.22 to 1.24 mg l^{-1} of $N\text{-}NH_3$ in ponds near Kinmundy, Illinois, U.S.A., fertilized with the excreta production of 39 to 66 pig ha^{-1} (Buck *et al.*, 1978); and 0.56 to 4.05 mg l^{-1} and 4.25 to 9.28 mg l^{-1} of $N\text{-}NH_4^+$ in concrete tanks in Bangkok, Thailand, fertilized with sewage water (Edwards *et al.*, 1981). Such values do not seem to affect the environment adversely. Nitrate values, however, remained low: less than 0.01 mg l^{-1} $N\text{-}NO_3^-$ (Buck *et al.*, 1978); and 0.15 to 0.83 mg l^{-1} and 0.01 to 0.03 mg l^{-1} $N\text{-}NO_3^-$ (Edward *et al.*, 1981).

Plankton concentration and primary productivity obtained with manure

The levels of phytoplankton production reached in ponds subjected to organic manuring vary widely. Buck *et al.* (1978), using 39 to 66 pigs ha^{-1}, which is equivalent to about 40 to 70 kg of excrements (dry weight) per day, to fertilize ponds in Illinois, recorded a primary productivity of 15,000 organisms ml^{-1}, equivalent to 125 μg l^{-1} of chlorophyll a. In Malawi, Msiska & Cantrell (1985), conducted experiments in cement tanks fertilized with chicken manure. They obtained variable chlorophyll a concentrations reaching up to 80 μg l-1, with an average of 50 μg l^{-1}.

In Israel, at Dor Research Station, Noriega-Curtis (1979), fertilizing with liquid cow dung at a rate of 67 to 70 kg (dry matter) d^{-1} ha^{-1}, recorded a planktonic population, collected on a plankton screen of 37 μm, of 2.1 to 10.6 μg l^{-1}, and a primary production of 3.6 to 5 g C m^{-2} d^{-1}. In Bangkok, Edwards *et al.* (1981), using sewage water for fertilization, reported phytoplankton concentrations between 36.5 and 175 mg l^{-1}, and primary productivity between 1.62 and 6.67 g C m^{-2} d^{-1}. They considered a phytoplankton production of 70 mg l^{-1} sufficient to sustain optimum growth in *O. niloticus*.

Considering the variability in size of the plankton production obtained, it appears that primary production parameters are not easy to forecast. Neither plankton concentrations achieved, nor the plankton assemblage which develops, can even approximately be predicted (see p. 157). The different phytoplankton species present in an assemblage do not have the same nutritional value and are not grazed upon in the same proportions. The problem is to know which species should be promoted and how to promote them. These two questions have not yet been answered, mainly because information on the selectivity of fish feeding upon natural food is scarce and the ecophysiology of tropical plankton (and also epiphytic and epilithic fauna and flora) is still largely unknown. The plankton assemblage changes with time, with species appearing and disappearing. In cement tanks in Bangkok, Edwards *et al.* (1981) observed that euglenoid algae, *Euglena*, and *Cryptomonas*, which concentrate at the surface, induced a decrease of dissolved oxygen in deeper water. A relationship between decreased fish production and dissolved oxygen depletion has not been clearly established, but is surmised. The sedimentation rate of these algae is sizeable, which would suggest that they are not consumed to a great degree. The dynamics of species changes within plankton communities has still to be explored. This whole subject will be discussed in details in various sections of chapter 3.

Yields obtained with manuring

The values of fish production obtained vary widely and depend upon stocking density, which needs to be balanced with the volume of natural food available. In ponds of the Aquaculture Research Centre of Texas A & M University, U.S.A., Burns & Stickney (1980) observed productions of *O. aureus* in the range of 2.5 to 16 kg ha^{-1} d^{-1}, in conjunction with manuring with poultry droppings from 500, 1000, 2000, and 4000 birds per ha. However, as stocking rates were low, at 0.16 g m^{-2}, fish yields were also low. In Israel, Noriega-Curtis (1979) recorded fish productions of 21 kg ha^{-1} d^{-1}: the stocking density was low at 8000 ha^{-1}, consisting of 4000 carp, 3000 tilapia and 1000 silver carp per hectare. Common carp does not readily consume phytoplankton and, thus, does not exploit properly all natural food resources present.

In Israel, at the Dor Research Station, Moav *et al.* (1977) reported fish productions of 32.6 kg ha^{-1} d^{-1}, with 4 species in polyculture: common carp, tilapia (*O. aureus* or hybrids), silver carp, and grass carp. The ponds were manured with liquid cow dung. The conversion rate of fertilizer into fish was calculated to be 2.7, based on the dry matter weight of the liquid fertilizer. Comparison of the results of these trials with results from ponds cultured simultaneously and in analogous conditions, where fish were fed a spectrum of artificial feeds, such as protein-rich pellets, showed that liquid manure fertilization resulted in yields similar to feeding with protein-rich pellets.

In Bangkok, Tilapia (*O. niloticus*) productions, resulting from manuring with sewage water, of about 50 kg ha^{-1} d^{-1} (44 to 51.5 kg ha^{-1} d^{-1}) were found by Edwards *et al.* (1981). The organic matter present in the culture medium was assessed through COD, chemical oxygen demand, measurements. Production in tanks during a period of 12 weeks reached values, which, when recalculated and extrapolated, were equivalent to 20 tons ha^{-1} yr^{-1}. Only proper management, accompanied by strict monitoring of water quality, and an optimum fish density, allow the achievement of such high production rates; sometimes, even such monitoring and control cannot prevent unexpected drops in production (see p. 190).

Unexplained water quality deterioration in the culture medium has been observed in intensive conditions. The reasons for this may be suboptimum dissolved oxygen levels, accumulation of growth inhibiting factors, a decrease in natural food organisms, and a decrease in nutritional value of primary producers (Moav *et al.*, 1977).

As indicated earlier, the rate of diffusion of nutrients from sediments to water is slow. A means to accelerate this diffusion would be to promote convective water movement. Water from deeper layers would bring nutrients to the surface, while surface water would carry dissolved oxygen downward, thus making use of the supersaturation of oxygen in the surface waters during periods of high photosynthesis. Such a device for convective water circulation may considerably improve pond conditions (see p. 199).

Number of livestock heads needed for integrated fish farming

To combine pond fish culture with animal production, one needs to know how many heads of livestock animals should be used to fertilize a pond, as the number of animals determines the amount of droppings which will find their way into the pond. This number varies with intensity of production, and with size and weight of the animals. For pigs, the numbers recommended vary between 30 to 85 pigs ha^{-1}, and, for ducks, between 1000 and 3500 ha^{-1} (Buck *et al.*, 1978; Delmendo, 1980; Stickney *et al.*, 1979; Little & Muir, 1987). Pigsties and duck pens can be built directly over ponds, on stakes, or nearby. In the latter case, the pens need to be cleaned regularly with water and the excreta and excess feed washed into the pond. When the pens are over the water, the floor should be a lattice work, such that detritus falls directly into the pond.

Pigs are raised for 5 to 6 months, until they reach a commercial weight (60 to 100 kg). During this period, they require suitable and balanced feed, and proper health care. Chicken and duck are raised for 1 to 3 months, till they reach their market weight.

Maximum amounts of manure to be used

Organic fertilizers are more efficient in aerobic than in anaerobic pond conditions. It is useful to determine the maximum amount of organic matter that a pond can digest. At ideal temperatures, i.e. above 18° C, maximum sediment oxygen consumption has been estimated to be 3.5 g O_2 m^{-2} d^{-1}. Each gram of oxygen consumed is able to oxidize 0.9 g of organic matter, which is equivalent to the decomposition of about 32 kg ha^{-1} d^{-1} of organic matter. If the bacterial population of the water (of 1 m depth) is equivalent to 1 square meter of bottom sediment, the total amount of digested organic matter is around 70 kg ha^{-1} d^{-1} (dry weight), or about 100 kg ha^{-1} d^{-1} of manure. An equal amount of organic matter is synthesized each day by the bacterial population. In practice, up to 150 kg ha^{-1} d^{-1} (dry weight) of manure have been dumped, without problems, into ponds with high fish biomasses. In stocked ponds, manuring is increased concurrently with fish growth, the daily applied fertilization rate (in dry weight of organic matter) being 2 to 4 % of the standing fish biomass (Schroeder, 1980).

Schroeder (1987) studied the carbon and nitrogen cycling in manured fish ponds in Israel in an effort to understand the reasons why fish production seemed to be limited to a maximum of 30 to 35 kg ha^{-1} d^{-1}. In ponds stocked with silver carp or grass carp, hybrid tilapia and common carp, he observed, using stomach content analysis, that only the silver carp directly consumed live algae, while the other species fed on detritus, generated by algal sedimentation and decomposition by microorganisms, and algal consumption by invertebrates and transformation by microbial processes. These microbial processes, and the passage through the invertebrate pathway, were estimated to represent 3 and 10-fold losses of organic carbon, respectively. The primary production of the pond supplied, directly or indirectly, with the manure, about 50 % of the carbon required for fish production.

Organic fertilizers, role and use

1. A great variety of organic fertilizers are available, their advantages being that they are cheap and efficient. Organic fertilizers or manures supply not only nitrogen and phosphorus to the pond, but also organic carbon and organic colloids, which promote organic matter decomposition within the water column. The heterotrophic production cycle is thus stimulated.

2. Decomposition processes taking place in the water are still largely unexplored and represent a bottleneck to organic fertilizer use in ponds. Microbial productivity has been only grossly assessed and the different production processes affected by organic fertilizer additions to ponds have not been quantitatively differentiated.

3. The quality of fish raised in organically fertilized ponds is higher than that of fish raised on artificial feed.

4. The maximum recommended application rates for organic fertilizers are 100 to 150 kg ha^{-1} d^{-1} (dry weight) in intensive production, or an equivalent amount of livestock excreta.

4.5. Other nutrients involved in productivity

4.5.1. Silicon

Importance of silicon

Silicon is a non-metallic element present in the valves and cell walls of diatoms, representing 10 to 30 % of their dry weight. Diatoms are an important phytoplankton group, also with epiphytic and epilithic forms, with regard to primary productivity and natural food for some fish species. As diatoms are the organisms which make most use of silicon, their biology is closely linked to the cycling of this element.

Species and concentrations of silicon in water

Silicon is present in water as orthosilicic acid, $Si(OH)_4$, and the hydrated form of silica, SiO_2. The transformation of dissolved silicic acid into amorphous silica and back by organisms forms the core of the biological silicon cycle (Paasche, 1980). Silica is also present in colloidal form, which is a highly polymerized form of silica, and biogenic silicon, i.e. amorphous silica, included in diatoms and Chrysophyceae. In interstitial water of Lake Ontario, L. Erie and L. Superior, Canada, aluminium and iron (III) oxyhydroxides and biogenic silicon formed a cryptocrystalline complex of ferro-aluminium silicate, $(Al_{1-x}.Fe_{1-x}\ (SiO_2)OH)_{6(1-x)}$ (Nriagu, 1978).

The concentration of silicon in freshwater lakes of

temperate regions, in different parts of the world, is around 1 to 5 mg l^{-1} (Paasche, 1980: 261, Table 7.1), or less than 10 mg l^{-1} (Lund, 1965), but up to 100 mg l^{-1} in saline African lakes of class III conductivity (Talling & Talling, 1965: 445, Fig. 6). In ponds, values range from 1 to 80 mg l^{-1} (Boyd, 1979). Concentrations in seawater are low, silicon being present almost exclusively in living or dead diatoms, or as biogenic silicon.

Silicate comes mainly from rock and soil erosion and run-off water. In rocks, the average ratio of phosphorus, nitrogen and silicon atoms is 35:2:10 000. The Si:P ratio of soils, which should be around 285:1, is in the range of 110:1, according to field determinations (Golterman & Kouwe, 1980). The solubilization rate of silicate is 2 to 3 times less than that of phosphate. Besides run-off water, sediment-water exchanges, and dissolution of diatom valves supply silicon to the water. Dissolution of diatom shells takes place in the sediments and is accelerated by the presence of salt (Paasche, 1980). High pH and alkalinity promote dissolution of silicon compounds (Talling & Talling, 1965). Lake overturn and vertical mixing of water return significant amounts of silica to the photosynthetically active upper zone, thus boosting diatom growth (Talling, 1969). In Lake Titicaca, Peru/Bolivia, sedimentation of diatom valves proceeds fairly quickly and this silicon is largely lost to production (Carmouze *et al.*, 1984).

Concentrations less than 25 μg l^{-1} of Si-Si$(OH)_4$ are considered silicon depletion; concentrations of 0.5 to 0.8 mg l^{-1} are needed to sustain a diatom population. Decreases in silica concentration in water are linked to increases in diatom biomass, examplified in phytoplankton species succession in temperate lakes described in the PEG-model (see p. 158).

Silicon assimilation by algae

Silicon is assimilated by algae as orthosilicate; the colloidal form is unavailable to them (Lund, 1965). Formation of new valves (silicification) in diatoms is restricted to the period of cell division which lasts around 1 to 2 hours (Paasche, 1980). All silicon is, however, not contained in the valves, part of it being involved in algal cellular functions, such as regulation of protein synthesis, and hydrocarbon and chlorophyll formation. This metabolic silicon does not exceed 10 % of total silicon (Paasche, 1980). Absorption kinetics of silicon are directly dependent on silicification rate, as cells can store only insignificant amounts of silicon beyond that used for cell wall formation (Paasche, 1980). Factors affecting silicon absorption, as well as concentrations limiting to diatom growth, are not fully understood. Silicification rate of cells is dependent on temperature, and light; high illumination reduces absorption (Lund, 1965).

Decrease in size of diatom cells may be related to decreases of silicate concentrations in water. When silicate concentrations are low, the limiting effect of phosphate on growth is more pronounced.

Influence of silicon concentrations on plankton composition

Varying proportions of silicium, phosphorus and nitrogen in water promote different planktonic groups. According to Kilham *et al.* (1986), dominance of diatom species in different East African lakes is the result of specific requirements for silicium, phosphorus and light: *Synedra*, *Nitzschia*, and *Stephanodiscus* have increasing phosphorus and decreasing silicium requirements; *Melosira distans* has high light but low phosphorus requirements. The opposite is the case for *M. nyassensis*. Paleolimnological studies in lakes extending from Lake Abhé, Ethiopia to L. Cheski, Zambia, have shown the dominance of *Stephanodiscus astraea*, the eutrophic character of their water, and the importance of internal nutrient regeneration processes (Kilham & Kilham, 1990).

In Lake Michigan, U.S.A., in the presence of Si:P ratios of over 100 and high N:P ratios, diatoms are abundant; when the Si:P ratio is below 5, green algae dominate, while for low N:P ratios, blue-green algae are more abundant. Diatoms have high silicium requirements, but lower phosphorus requirements than green algae (Kilham, 1986).

Measurement of silicon

Silicon concentrations are determined colorimetrically by transformation of silicon ions to molybdosilicic acid, reduced to a blue heteropolyacid, and measured spectrophotometrically. Phosphate in the sample reacts to form molybdophosphoric acid, which is destroyed by oxalic acid. The separation of biogenic silica from quartz and silicate minerals is important, but not always easy. Several methods exist, mainly aiming at making use of the greater solubility of diatom silica as compared to mineral particles, or attempting to convert diatoms cell counts into cell surface areas and thus into silica amounts (Paasche, 1980).

4.5.2. Sulphur

Sulphur is one of the six major elements of living matter. It appears in the amino acids cystine, cysteine, and methionine, and is, therefore, present in all proteins. It is also part of some co-enzymes.

Sulphur is abundant in water and soil and makes up some 7.8 % of the solutes in sea water. The sulphate ion, SO_4^{2-}, is soluble and therefore the most common form. Sulphate is absorbed by plants, thus entering the cellular metabolism. In the cell, sulphate is reduced to sulphide and is present as the -SH radical in molecules. This radical is chemically active and forms disulphide bonds between chains of amino acids and participates in the quaternary structure and active sites of macromolecules. Incorporated in organic molecules, sulphur is transferred to animals. Excess sulphur is rejected in urine, in the form of sulphate.

Another sulphur source is decomposition of organic matter of plant or animal origin, including excreta. The transformation of these products by bacterial action, in particular by the thiobacilles, the chlorothiobacilles and the rhodothiobacilles, or by chemical action produces sulphide, hydrogen sulphide and elementary sulphur in reducing conditions, and sulphate in oxidizing conditions:

$$2H_2S + O_2 \longrightarrow 2H_2O + 2S + \text{Energy}$$

$$2S + 2H_2O + O_2 \longrightarrow 2\,SO_4^{2-} + 4H^+ + \text{Energy}$$

In coastal environments, such as mangroves, the presence of sulphur drastically modifies the chemical and biological characters of the water. The accumulation of organic matter in mangrove soil causes an enrichment in sulphide, which profoundly affects sediment acidity once sulphides are transformed into sulphate.

Sulphur deficiency causes a sudden increase in the synthesis of polyphosphates in bacteria. Their role is related to the regulation, at the intracellular level, of ATP, ADP, phosphonucleotides and orthophosphate; a deficiency in sulphur halts synthesis of nucleic acids.

4.5.3. Iron

Importance of iron

Iron is an oligoelement, abundant in rocks and sediment. It plays an important role in living matter, is involved in a number of redox reactions, in the synthesis of chlorophyll, and occurs in proteins and enzymes, such as myoglobin, hemoglobin, and cytochromes. Iron, in dissolved or available forms, is needed at concentrations between toxicity and deficiency limits (Huntsman & Sunda, 1980).

Iron species in water

Iron is usually present in oxidized form, iron (III) hydroxide, in natural waters. Iron oxide is sometimes combined to organic matter, such as humic substances, and to phosphorus, silicium, calcium, and trace metals. In natural waters, particles containing ferric oxide are of different origins. In some organic complexes, iron atoms are coordinated to functional organic groups and may be stabilized in ferrous form (Tipping *et al.*, 1982).

In Eastwhite Water, a small productive lake in Cumbria, U.K., iron concentrations in water samples taken at different periods showed an important increase of iron (III) in water at overturn, which can be explained by the ascendant movement of sedimentary iron (II) or iron from anoxic water layers to be oxidized in aerobic water.

Part of the iron in suspension in water, not sedimented by centrifugation, was in the form of small particles stabilized against flocculation by adsorption onto organic matter. Adsorbed humic substances impart negative charges. Iron (III) oxide produced by ferrous iron oxidation at or nearby the oxic-anoxic border during stratification was associated with humic substances, which were also negatively charged and underwent slow flocculation.

The degree of flocculation increased with increasing ion concentration, especially of multivalent ions, of opposite charge, such as magnesium or calcium. Flocculation depended also on total ionic charge and the amount of material in suspension. It involved associations with particles similar to those containing iron. Flocculated iron was mainly amorphous, consisting exclusively of iron or iron associated with clay minerals or particulate organic matter (peptized form). Calcium, in large amounts (equimolar amounts), silicium, phosphorus and sulphur were also consistently present. Iron complexes with organic compounds also occurred; they consisted of monomeric iron (III) atoms coordinated to humic functional groups, sufficiently stable not to undergo hydrolysis. A group of largely undefined iron compounds was found in the supernatant after centrifugation (Tipping *et al.*, 1982).

The importance of chelators

Most ferrous compounds are poorly soluble. However, owing to the presence of chelators and various complexing compounds, considerable amounts of iron are kept available in the water. EDTA is one of the artificial chelators most used in laboratories, while citric acid is used in bio-assays (Huntsman & Sunda, 1980).

In natural waters, humic and fulvic acids, and phenolic, quinone, enolic, carbonyl, amino and carboxylic groups have chelating and complexing properties. They are derived from soil and decaying plant material. Humic matter sometimes reaches concentrations of up to 70 mg l^{-1}, binding metals on the carboxylic and phenolic radicals, which are primary binding sites. Fulvic acids are less effective than humics in stimulating algal growth. As a matter of fact, humic acids are characterized by their capacity to bind iron and there are strong correlations between humate concentrations and amounts of dissolved iron. Iron may exist in peptized form rather than in true solution. Adsorption of other ions occurs, with a preference for divalent cations, such as copper, cobalt, and nickel, which are three to four times more abundant than the trivalent cations such as aluminium and iron (Huntsman & Sunda, 1980).

Humic substances stimulate algal growth, probably because they hold not only essential oligo-elements, but also common nutrients, such as phosphate and silicate. Iron is mainly linked to hydroxy and carbonate groups, which predominate in fresh water and are pH and alkalinity dependent. Chelate formation is a common binding mode, with low kinetics, but is not fully understood. It allows the temporary fixation of ions, which are transported through water, the pH of which would normally cause precipitation, and released when the environment permits. The rate at which they are released corresponds to the uptake rate of the organisms present. This system maintains suitable concentrations of these ions for organisms, preventing toxic concentrations from being reached. Chelators thus act as buffers for metals. The total amount of chelating agent should, however, not be excessive, leading to over-chelation (Huntsman & Sunda, 1980).

Iron concentration in water

Ferrous ions should be absent from aerobic water. They are, however, present, because a certain number of organic compounds, such as humic and tannic acids, and high chlorinity reduce the oxidation rate of ferrous iron. The chelating capacity decreases with salinity. Dissolved iron concentrations in natural waters are higher than those predicted by the solubility constant of iron hydroxide. In freshwater, Fe^{2+} and $FeOH^{+}$ are predominant, while, in sea water, $FeCl^{+}$ ions are more common. In freshwater, Fe^{2+} can compete successfully with calcium and magnesium ions for organic chelating agents, forming moderately stable Fe (II) complexes. In sea water, high concentrations of ions of the alkaline earth group outcompete ferrous ions for available chelators. Mercury, silver and cadmium ions are preferably complexed to chloride ions (Huntsman & Sunda, 1980).

Generally iron concentrations in natural waters are variable, depending upon rainfall, the amount of material in suspension, erosion, and human contamination (pollution) (Huntsman & Sunda, 1980). Normal average concentrations in temperate oxygenated water are in the order of 10^{-7} mol l^{-1}. Considering the solubility constants of iron compounds, the concentrations should be no more than 10^{-15} mol l^{-1}. The difference is due to the presence of chelators (Huntsman & Sunda, 1980).

Iron is always more abundant in anoxic sediment, in reduced form, and is recycled during mixing or overturn events, when sediments are resuspended.

Effect of iron on fish and algae

Excessive concentrations of ferric hydroxide in water cause suffocation of fish. Iron hydroxide precipitates in alkaline water, at a pH of 7.8. Ammonia excretion alkalinizes water, on and near the gills, which results in precipitation of iron hydroxide on the gill, leading to suffocation.

Phytoplankton is able to live in water containing variable amounts of iron and trace metals. Some species have evolved mechanisms enabling them to make use of varying metal availabilities. Some are able to bind metals outside the cell, by producing extracellular chelating agents, in particular hydroxamic acids. These hydroxamate compounds virtually outcompete all other ligands. They are produced by many species of bacteria and fungi, but the only phytoplankton group to do so seems to be the blue-green algae. To be effective, these hydroxamates have to outcompete other ligands, such as humates and fulvates. Some eucaryotic algae have developed a different strategy; insoluble iron (III) is reduced to the soluble ferrous form, by the mediation of extracellular products at the external cell surface. The general importance of this reduction mechanism in iron uptake has, so far, not been determined (Huntsman & Sunda, 1980). The high affinity process of blue-greens for impounding available iron in water influences competition with other algal groups. During blooms of blue-green algae, other plankton algae cannot thrive. The growth of blue-greens results in a high nitrogen fixation rate and a high iron demand. This induces production and excretion of hydroxamate siderophores or siderochromes by *Anabaena flos-aquae*, which has chelating properties for iron, outcompeting other species in culture (Murphy *et al.*, 1976; McKnight *et al.*, 1980). By controlling water quality, these algae insure their own dominance (Hart, 1982; Smayda, 1980).

Iron determination in water

Iron in water is determined spectrophotometrically. The amounts of dissolved iron measured are often higher than predicted, due to the occurrence of particulate iron. Some dissolved iron hydroxide is sometimes pooled with the particulate fraction. The separation of particulate and dissolved iron is thus not easy to perform. The rapid development of bacteria and phytoplankton on the walls of aquaria is due to the adsorption of iron onto surfaces (Huntsman & Sunda, 1980).

Importance of silicon, sulphur and iron in water

1. Silicon is supplied to the water by the sediments and run-off water. The silicon concentration in water is usually not limiting for algae, but may determine the structure of the plankton assemblage and its succession, as some algae, like the diatoms, have strict and high silicon requirements.

2. Sulphur is supplied to water by organic matter decomposition; microbial sulphur cycling is related to organic matter decomposition.

3. Iron is supplied by the sediments and is an important oligo-element in many organic molecules. Iron in oxidized form is insoluble in water, being present as iron oxyhydroxide. Iron ions are also present in organic complexes, such as humates. The Fe-DHM complexes are binding sites for phosphate and other nutrients. These iron complexes maintain iron (and other trace metals) in water at concentrations which range between deficiency and toxicity levels. Iron oxide in water harms fish, as it may precipitate on the gills causing suffocation. Algae have developed mechanisms to guarantee their iron supply: blue-green algae excrete hydroxamates, which have a high chelating affinity; other algae are able to reduce iron at the cell surface, before absorption.

Chapter 3.

Productivity and fish production

1. Primary productivity of water

1.1. Natural productivity and fish production

Types of aquatic organisms present in water

The aquatic environment supports various communities of living organisms. They constitute the biotic load of a pond. Natural productivity is the capacity to increase this biotic load over time. Assessment of this productivity is required in order to determine how much an environment can generate or produce using its natural resources.

Living communities are made up of different groups: benthos, plankton, periphyton, and fish. The names of the first three designate the place where they occur in the environment: in or on the sediment, within the water, and fixed on a substratum. Organisms can also be grouped according to their role in the ecosystem: producers, consumers, decomposers. Primary production is the rate of formation of organic matter due to photosynthesis, by all photosynthetically active organisms. Secondary production refers to increases in biomass of zooplankton, zoobenthos and fish, while decomposers recycle dead organic matter, using the energy and nutrients contained therein. These separations are somewhat arbitrary, as will be seen later.

Plankton designates the community of pelagic organisms, composed of various groups, which are in suspension in water and hence restricted mobility, often less than that of the water which carries them. Plankton is grossly divided in phytoplankton or photosynthetic organisms, and zooplankton or heterotrophic organisms. The smallest algae composing the phytoplankton belong to the group of blue-green algae and are also called bacterioplankton. Plankton may also be classified according to the size group to which they belong.

Seston is the particulate matter in suspension in the water column, comprised of approximately two thirds detritus and one third living organisms.

Periphyton consists of different types of communities, firmly attached to or moving freely on substrata of various kinds, such as leaves of aquatic plants or stones, forming dense felts of brown and green scums and composed of living cells thriving on dead cells underneath. It can be divided in phytoperiphytonand zooperiphyton. Phytoperiphyton is made of groups of algae (diatoms and green algae), thick mats of algal filaments or gelatinous lumps of algae (greens, such as desmids and *Spirogyra*, and blue-greens); algae which move freely on substrata, especially sediments, such as green algae and diatoms (Westlake, 1980). Zooperiphyton is composed of ciliates, rotifers, crustaceans, Bryozoa, Hydrozoa, sponges, Oligochaeta (Naididae). The periphyton attached to aquatic plants is sometimes called aufwuchs. The species composition of periphyton of tropical water is still largely undocumented.

Benthos designates the organisms on and in sediments and can, again, be divided in phytobenthos and zoobenthos. Algae of the phytobenthos are found on the surface of sediments, as, for instance, epipelic diatoms and blue-green algae. Organisms belonging to the zoobenthos have been discussed earlier (see p. 50). Zoobenthos is grouped according to size range, in microbenthos (bacteria, ciliates), meiobenthos (rotifers, nematods, cladocerans, ostracods, copepods) and macrobenthos (Turbellaria, molluscs, Oligochaeta, insect larvae, decapods).

Little information exists to date on the relative importance of periphyton and benthos and their roles in fish ponds. Periphyton develops in littoral zones of lakes, at shallow depths, where nutrient availability is important, and they are readily available food for invertebrates and fish. The response of phytoperiphyton to nutrient enrichment is usually poor. Larger individuals increase in proportion with increasing trophy, but the response to eutrophication is variable. Periphyton communities are dependent upon substratum quality, slope and fetch (Cattaneo, 1987). The development of benthic algae in shallow brackish water ponds in the Philippines is promoted to serve as food for the milkfish (lab-lab). In the following discussion, it should be remembered that, though the information is centered on pelagic algae, the contribution of periphyton and benthic algae must be included in primary production. However, their relative contributions remain to be quantified.

The last group of the living community focusing our interest is the fish. Fish species consist of primary herbivores or phytoplanktivores, and secondary consumers, zooplanktivores or predators, and their growth and production is dependent on natural food available to them (but see p. 179).

According to Noriega-Curtis (1979), fish production cannot be predicted solely on the basis of primary productivity of ponds fertilized with manure, because autotrophic production is not the only energy source for the fish. Protein biosynthesis through the microbial decomposition cycle (or the heterotrophic production cycle) is certainly as important as that of phytoplankton production. Apportionment of each of the production cycles to fish production will need to be assessed by future work.

Spectrum of algae found in ponds

The phytoplankton community consists of a large number of taxa. According to cell size, the following groups can be distinguished: picoplankton (from 0.02 to 0.2 μm), nanoplankton (from 0.2 to 2.0 μm), and phytoplankton (from 2.0 to 200 μm) (Sieburth *et al.*, 1978). Organisms over 20 μm in size are called netplankton. Picoplankton and nanoplankton, often ignored in plankton studies, can contribute significantly to primary productivity, often in excess of what their biomass would suggest (Hardy *et al.*, 1986; Stockner & Antia, 1986; Stockner, 1988). In experiments with bags suspended in lake Brooker, Florida, U.S.A., and which were filled with lake water ten times concentrated, Scheda & Cowell (1988) showed that they were efficiently consumed by filtrating zooplankton and played a key role in energy transfer towards higher trophic levels. Increases of herbivorous (phytoplanktivorous) rotifers affected nanoplankton biomass.

The most diverse algal group in tropical African water bodies are the green algae or Chlorophyceae, which make up between 50 to 85 % of the total number of species found. Amongst the green algae, the desmids are the most important algal order. Ponds and pools recently filled with rainwater are invaded by green algae, of the order Chlorococcales, belonging to the genera *Scenedesmus*, *Ankistrodesmus*, *Crucigenia*, and *Chlorella*, and also brown algae or Cryptophyceae, Chrysophyceae (of the genus *Ochromonas*) and Euglenophyceae. Green algae form dense population blooms in aquaria left in open air. If left undisturbed, such tanks will eventually fill with blue-green algae or Cyanophyceae. Zygnematalean algae, another order of the green algae, are filamentous, and can almost completely fill temporary pools where they form loosely entangled, gelatinous masses. *Spirogyra*, along with the cyanophyte *Lyngbya borgertii*, is reported to form entangled masses in pools in the Ivory Coast. In fertilized ponds, the phytoplankton is almost entirely made up of euglenoids. Green, flocculent, or felty growths of algae covering submerged stems and leaves are commonly slimy and consist of various zygnematalean algae, belonging to the genera *Spirogyra*, *Zygnema*, *Mougeotia*, and occasionally entangled with *Oedogonium*, which is usually a truly benthic form. Diatoms are opportunistic, and rapidly colonize newly submerged surfaces. They are the dominant algal class colonizing wooden surfaces. They become associated with surfaces in large numbers when forming blooms. They are a major component of epiphytic and epilithic colonies, along with blue-greens (John,

1986).

Species numbers of blue-green algae or Cyanophyceae (also termed Cyanobacteria) and of Euglenophyceae represent each ca. 5 to 15 % of the total number of algae species. Bacillariophyceae or diatoms are quite diversified and make up 15 to 25 % of species diversity. Chrysophyceae, Xanthophyceae, and Dinoflagellates represent only 1 to 2 % of the total number of species (John, 1986). Diatoms require lower light intensities than green algae (Chlorophyceae) for their development. Blue-green algae can withstand higher light intensities and higher temperatures. Almost half of the species are widely distributed, not specific to Africa (Iltis, 1980).

Algae associated with sediments are called epipelic and consist of assemblages of diatoms and filamentous blue-green algae. These algae possess a vertical migration rhythm similar to that reported in temperate regions. Upward migration occurs just before dawn (John, 1986).

Effects of phytoplankton on the environment

Phytoplankton development benefits ponds for the following reasons (Erez *et al.*, 1990):

- a dense phytoplankton bloom can produce ten times more oxygen than it consumes. This oxygen production is important in compensating for respiration and is supplied to the pond without added energy expenditure
- phytoplankton represents an efficient means of removing excess nutrients or toxic micronutrients, especially ammonium, from the water
- phytoplankton acts as a screen to light penetration in the water and prevents development of benthic algae. The development of scums of blue-green algae on the bottom promotes production within the sediment of a reductant surficial layer, rich in hydrogen sulphide, which is highly toxic. These algal scums remove nutrients and postpone the return to normal growth for algae after an algal crash
- phytoplankton is good food for planktivorous fish. Exploiting primary productivity is a cheap way to produce fish
- the development of phytoplankton prevents the development of macrophytes, which is unfavourable for fish culture.

Relations between phytoplankton and planktivorous fish

In ponds relying on primary productivity for fish nutrition, daily phytoplankton production, as well as ways to increase the latter, should be known. The increase in supply of available phytoplankton is used to calculate the amount by which the stocking rate of fish can be increased, or the amount by which the fish load in a pond has increased, due to fish growth. Phytoplankton availability has to be adjusted to fish consumption to achieve high production. Consumption of phytoplankton by fish and effects on pond primary production will be discussed later (p. 184 & p. 179).

Common features for lakes and ponds

An abundant limnological literature on nutrient cycles and on the dynamics of living communities exists. Application of information collected in natural water bodies to fish ponds must be done cautiously, taking into consideration the common features, but also the differences between the two types of environments.

Morphometric differences are linked to the respective sizes of the water bodies; natural lakes are always some order of magnitude larger than ponds, the area of which never exceeds few hectares. The water mass of lakes is relatively stable against changes in physical and chemical variables. Ponds are usually not more than 3 m deep, while lakes with maximum depths of 3 m are considered shallow. This entails large differences in the exchanges between sediment and water. Lakes are permanent water bodies, with an external water supply, from a river or at least run-off water, which also supplies nutrients. In a fish pond, the water supply is limited to pond impoundment and does not supply a significant amount of nutrients.

Despite such differences of scale, and the greater stability of lakes, lacustrine information is important in understanding the functioning of ponds.

Production in tropical and temperate regions

According to Nilssen (1984), production processes in tropical and temperate regions are fundamentally similar. Kilham & Kilham (1990) consider that, in tropical aquatic environments, the biological control of elementary cycles is predominant over physical processes and morphometric parameters, unlike in temperate lakes. The consequence is that nutrient regeneration is more important and nutrient recycling in the tropics faster. Endogenous processes of a lake dominate the external loading. Lakes show less long-term stratification and the frequent overturn of water allows resuspension of sediments, which promotes mobilization of sequestered nutrients and a better oxygenation of the water. The reduction of water density makes mixing easier, and to a greater depth. In a study of 39 natural lakes in Cameroon, Kling (1988) observed that light penetration is deeper and water temperature fluctuations are more narrow than those of temperate lakes. The regeneration rate of phosphorus is, in the long term, more important than its immobilization. In a great number of lakes, more than 85 % of phosphorus is recycled from zooplankton excretion, and phytoplankton and detritus decomposition, the external supply not exceeding 5 %. Temperature also acts to increase enzymatic reactions. Silica is supplied internally from the sediments by solubilization (Kilham & Kilham, 1990). More profound differences exists in the taxonomic composition of tropical and temperate zooplankton (p. 165).

Natural productivity: what the concept covers

The natural productivity is the capacity of the living community to increase its biotic load over time. The community of living organisms can be separated according to their position in the food web as producers, consumers and decomposers. This separation is based on their mode of harnessing the energy needed for their metabolic activities. Three main processes emerge: photosynthesis, the use of solar energy to elaborate organic compounds; grazing and predation, or consumption of living organisms containing high-energy compounds; decomposition or consumption of dead organic matter, again to make use of high-energy compounds. These three processes are complementary and interdependent.

1.2. Determinants of primary productivity

Primary productivity characteristics

Primary productivity is the production of organic matter or carbon fixation by photosynthesis, using inorganic nutrients and light as an energy source.

To measure primary productivity is to assess the rate of photosynthesis, i.e. the rate at which chemical energy and organic matter is added to the system. The rate of production of organic matter is governed by the rate of input of light into the water column and by the capacity of the organisms present to use it for gross photosynthesis. Gross photosynthesis is the transformation rate of radiant energy to chemical energy, accompanied by equivalent conversions of carbon dioxide and water into organic compounds, with release of oxygen. However, while performing photosynthetic activities, the plant uses energy, converting some of the organic compounds back into carbon dioxide. Respiration performed in the light is called photorespiration, while respiration which occurs at night is called dark respiration. Photorespiration and dark respiration are different; thus, net photosynthesis cannot be correctly and precisely measured. This is why the term photo-assimilation was coined. This term designates the difference between the rate at which organic carbon becomes assimilated into the plant and the dark respiration rate, stressing that this is not the true net photosynthetic rate (Westlake, 1980) (see p. 132). For blue-green algae, it has been shown that photorespiration may be twenty times greater than dark respiration (Westlake, 1980). Net photosynthesis is a function of the availability of nutrients, and, eventually, of the increase in phytoplankton biomass. In the presence of high biomass, photosynthesis decreases owing to self-shading.

Nutrients needed for photosynthesis are supplied by the environment or from outside. Carbon is supplied by carbon dioxide and is primarily used to produce glucose, the basic energy molecule. Nitrogen and phosphorus are mainly supplied by decomposition of organic compounds, partly also by the atmosphere, through nitrogen-fixing organisms, and by geochemical phosphorus released by the sediment. Supplies of nitrogen and phosphorus contained in the feed and fertilizers should also be included. The major share of nutrients available to primary production comes from recycling and regenerative processes. The nutrient concentration in water may limit the specific rate of gross photosynthesis directly, but except in situations where carbon is limiting, this is not common (Westlake, 1980) (see also p. 155). Very rarely, situation arise (in conditions of extremely dense algae blooms, and hard water with high pH (Tilzer, 1989)) where the supply of CO_2 severely restricts the extent of phytoplankton photosynthesis (Wetzel, 1983). In an oligotrophic lake in Pennsylvania, U.S.A., in conditions of intense production, the growth of periphyton (86.1 % of which were Chlorophytes) was limited by the availability of dissolved inorganic carbon and was enhanced by a supplement of organic and inorganic carbon and nitrogen (Fairchild & Everett, 1988).

The photosynthetic capacity

Photosynthetic capacity is defined as the maximum specific rate of photoassimilation reached at high irradiances or in light-saturating conditions, expressed per unit of chlorophyll a, in mg O_2 (mg chl a)$^{-1}$ h^{-1}, or sometimes in dry or wet weight (Westlake, 1980). This capacity is dependent on temperature, on nutrient availability and on the amount of photosynthetic organisms present in the water, i.e. standing crop (Fig. 34). For phytoplankton, photosynthetic capacity varies in the range of 0.5 to 38 mg O_2 (mg chl a)$^{-1}$ h^{-1}, which corresponds to a population renewal (or a life expectancy for the algae) of 180 to 3 hours. For macrophytes, it is estimated to be ca. 2 mg O_2 (mg chl a)$^{-1}$ h^{-1} (Westlake, 1980).

Primary productivity is always a function of light; production versus light intensity reveals a production optimum. This production optimum can be divided by a measure of biomass (chlorophyll, biovolume or particulate carbon) to give a light-independent measure of the carbon turnover in phytoplankton (Heyman & Lundgren, 1988).

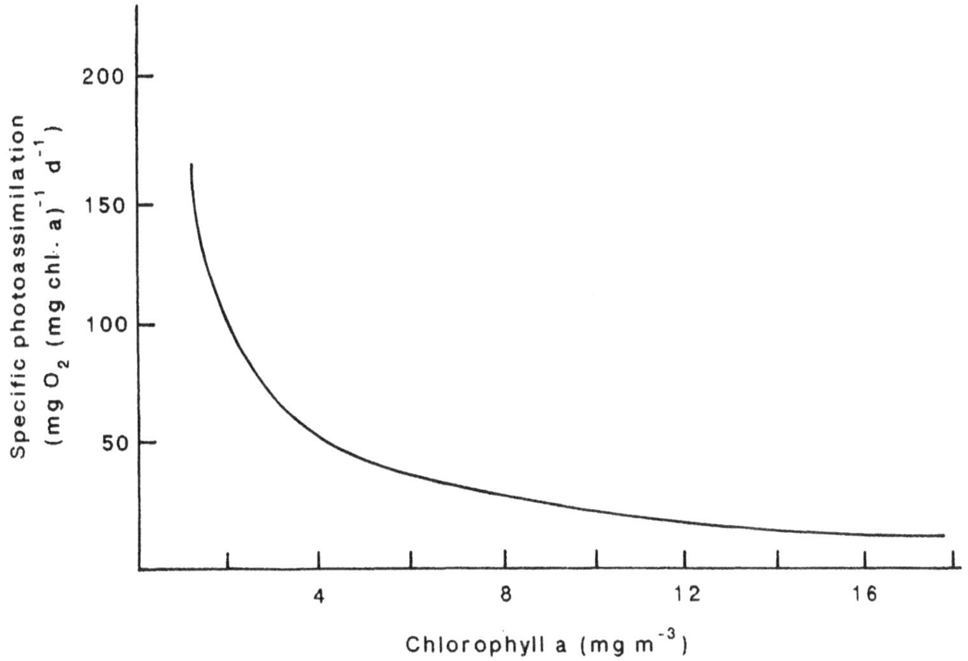

Fig. 34. Relationship between mean daily specific photoassimilation and mean algal biomass concentration (euphotic zone), expressed as chlorophyll a concentration, in a Czech reservoir (after Javornický, 1980).

Methods to measure photosynthetic rate

To measure photosynthesis, one of two conventional methods is usually adopted: oxygen production in dark and light bottles, the oxygen method of Gaardner & Gran (1927), or carbon-14 fixation (Steemann Nielsen, 1952; Harris, 1980). Both methods have serious limitations. With the spiked carbon method, no reliable estimate of respiration is possible (Westlake, 1980). Some of the fixed carbon is directly lost to respiration and part of this released carbon dioxide is used anew in photosynthesis.

The increase of oxygen in a bottle is not a true reflection of the oxygen released by photosynthesis, as some amount is consumed by respiration (community respiration) of seston (the heterotrophic and autotrophic organisms present in water) (Stack, 1973). The respiration rate in the light, the photorespiration, and in the dark, the dark respiration, are obviously not equal. Nighttime carbon losses usually comprise 30 to 50 % of the previous day's uptake and are far in excess of daytime respiration (Tilzer, 1990). Therefore, the assumption that the respiration rate in the dark bottle provides a basis for the calculation of net photosynthesis in the light bottle is challenged (Westlake, 1980).

Carbon fixation calculated by the dark and light bottle method gives a value for carbon fixation about twice that found by the carbon-14 method. The following rule is therefore adopted: results obtained using the carbon-14 method with phytoplankton are multiplied by three to obtain the oxygen output of photoassimilation. With higher plants, result obtained by the carbon-14 method are multiplied by 2.5 (Westlake, 1980).

In a given environment, oxygen consumption increases in direct proportion to the size of the bacterial population. The respiratory activities of bacteria can constitute up to 30 % of the total oxygen consumption. The daily bacterial consumption rate is estimated to be 0.1 10^{-9} mg O_2 per cell. The average rate of specific community respiration is assessed to be 1 mg O_2 (mg chl a)$^{-1}$ h^{-1}, with a normal range between 0.1 to 4.5; up to 10 mg O_2 (mg chl a)$^{-1}$ h^{-1} have been recorded, during periods when high bacterial activity was suspected (Westlake, 1980).

Factors determining photosynthetic rate

Factors determining photosynthetic rate are besides light, nutrient concentrations and volume of producers present, i.e. the biomass of phytoplankton, periphyton and macrophytes, including their species composition. In addition, the presence of inhibitors of photosynthesis, and the density of herbivores should be taken into account. Producer biomass is assessed on the basis of chlorophyll measurements. Factors affecting cell growth, such as temperature, will be discussed on p. 137.

Maximum densities of chlorophyll a

The maximum chlorophyll content is around 2 g of chlorophyll a m^{-2} for macrophytes and 500 mg (range: 180-620 mg) of chl a m^{-2} for phytoplankton, which is equivalent to 200 and 50 mg m^{2} of dry matter for the macrophytes and phytoplankton, respectively. In Lake George, Uganda, phytoplankton density has been found to reach 400 mg of chl a m^{-2} (Horne & Viner, 1971).

In chemostat culture, Agustí *et al.* (1987) showed that the maximum density of algae in a population is a function of the -1.27 power of cellular volume. In a study involving 11 species of freshwater algae, followed in batch culture, the maximum density varied as the -0.79 power of cellular volume, independently of environmental conditions. Physiological constraints and self-shading are probably responsible for this value (Agustí *et al.*, 1987; Agustí & Kalff, 1989).

Phytoplankton density and photosynthetic rate

With increasing phytoplankton density, i.e. increasing biomass, photosynthetic rate decreases, owing to alteration of pigment content or modifications of photosynthetic enzymes. The daily mean specific photoassimilation, expressed in terms of algal volume, declines rapidly as cell volume increases. This suggests a deterioration of certain factors in the environment of the planktonic algae, leading to a decrease in chlorophyll a content; chlorophyll synthesis lags behind cell growth, which leads to a decrease of specific photoassimilation in terms of carbon volume, while, in terms of chlorophyll, it remains more or less constant. Two main factors are involved: self-shading in shallow waters by an abundance of plankton, and nutrient depletion (Westlake, 1980).

1.2.1. Influence of light

Autotrophic and heterotrophic organisms

The energy required for cellular metabolism is supplied to animal organisms by oxidation of energy-rich organic compounds, while plant-like organisms use solar light, transform it into chemical energy and elaborate organic compounds. Animals are heterotrophic organisms, while macrophytes and algae are autotrophs. Autotrophic organisms introduce energy into the system, which is captured with the help of photosynthetic pigments. They are therefore the base of production. Solar energy is also the cheapest source available.

Among heterotrophs, different subgroups are distinguished: chemoheterotrophs are organisms growing without solar light, while photoheterotrophs use light as an energy source, but organic compounds for their nutrient requirements. Mixotrophic organisms can adjust their metabolism to prevailing conditions. A certain number of phytoplanktonic organisms are mixotrophic, especially amongst the blue-green algae of the genus *Oscillatoria* (Feuillade & Feuillade, 1990). Observations on two lakes in Quebec, Canada revealed that some mixotrophic Chrysophyceae of the genus *Dinobryon* depend more on ingested bacteria than on photosynthesis for their metabolism (Bird & Kalff, 1987).

Light absorption by water and stratification according to light intensity

Part of the solar rays falling on the water surface is reflected, while rays which penetrate the water mass are refracted. The photosynthetically active radiation (PAR) at zero depth, I'_0, represents about 40 % of the incident radiation, I_0, and is composed of the visible spectrum, between 400 and 700 nm (Westlake, 1980). It is progressively absorbed with depth. Absorption depends on wavelength, on turbidity, and on colour.

Photosynthesis is nil when irradiation drops below 1 % of incident radiation. This is the limit of the euphotic zone, or the extinction depth. The optical depth defined by the subaquatic irradiance relative to the surface irradiance (Westlake, 1980), creates a vertical stratification of the water column. The vertical attenuation coefficient varies with wavelength, is different for each wave-band (Westlake, 1980) and is dependent on the biotic and non-living particles in the water.

Photo-assimilation rates as a function of depth present a vertical stratification. Three zones can be distinguished: a lower one where light is limiting and photosynthetic rate is low, corresponding to about 15 to 1 % of the incident light; a middle one where light is saturating and photosynthetic rate is at a maximum, corresponding to 60 to 15 % of the incident light; and an upper or inhibition zone, of low photosynthetic activity, where high irradiances lead to photoautoxidation of the photosynthetic pigments (Fig. 35) (Westlake, 1980).

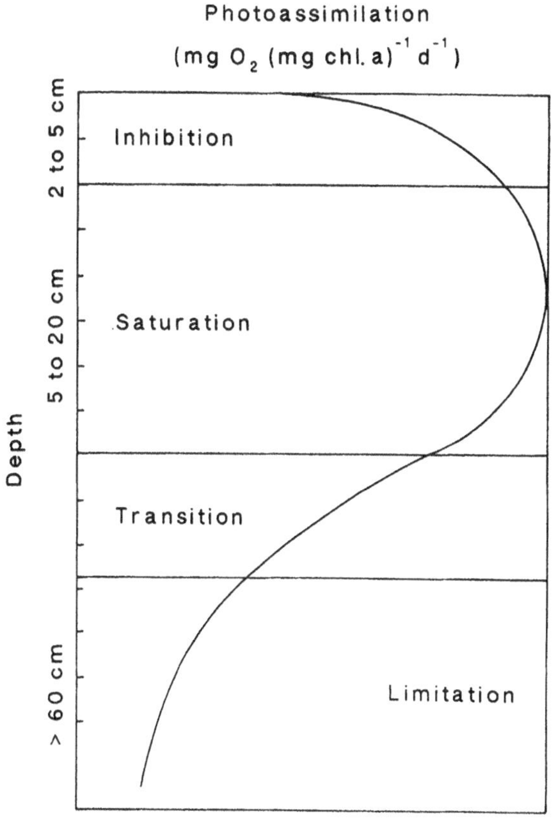

Fig. 35. Generalised diagram of photoassimilation in a homogeneous phytoplankton population according to depth and light penetration within the water mass (from Tilzer, 1980) (with kind permission of Prof. Tilzer, Dr. LeCren & Cambridge Univ. Press).

Oxygen production in a pond is a function of light and depth. The light compensation point is reached at the irradiance value (E_c) where photosynthetic oxygen liberation just equals respiratory oxygen consumption. Above this depth or that irradiance value, net photosynthesis is achieved and photosynthetic rate increases linearly with irradiance, E_d, up to a certain value. After that it curves and eventually levels off. In the range of irradiances where photosynthetic rate does not vary with E_d, photosynthesis is said to be light-saturated. The rate has the value of P_m, the maximum specific photosynthetic rate, sometimes referred to as the photosynthetic capacity. With further increases in irradiance, P begins to decrease again, a phenomenon referred to as photoinhibition.

The light intensity required to saturate photosynthesis, and the compensation point vary markedly from one species to another. These parameters also depend on CO_2 concentration and temperature. The light compensation point changes seasonally, lower in winter or spring than in summer or autumn; whether this is due to differences in temperature or whether other factors are involved is uncertain (Kirk, 1986).

Light absorption or attenuation and trophic loading

The incident irradiance is attenuated underwater, by absorption and scattering (Westlake, 1980). The absorbing and scattering components are the water itself, the plankton, the seston and dissolved substances. Their relative importance in the process and how they compete for spectral energy are largely unknown, partly because of technical difficulties (Yentsch, 1980). The degree of absorption and scattering increases with increasing trophic loading of the water. Irradiance in the green, yellow and even red is least attenuated (Westlake, 1980). Detritus primarily absorbs ultraviolet light (Yentsch, 1980); UV and infrared irradiances are both rapidly attenuated. In ponds with high plankton and organic loads, all light is rapidly attenuated and photosynthetically active radiation is reduced to a low value. Minimum attenuation in water with high phytoplankton abundance occurs in the orange-red, while the phytoplankton has its most marked effects on light in the blue spectral region (Westlake, 1980; Yentsch, 1980).

Use of light in photosynthesis

Autotrophs utilize light as an energy source. Two photochemical reactions, occurring in series in photosynthesis harness this energy. Light reaction 1 is associated with the reduction of NADP (photosystem I); light reaction 2 brings about the liberation of oxygen from water (photosystem II). Each photoreaction takes place in a reaction centre in association with specialized light-harvesting pigment-protein and electron transfer agents. The two systems are referred to as the photosynthetic units. Each photosystem has a chlorophyll a/carotenoid-protein which harvests light and is intimately associated with the reaction centre. The pigment-protein represents a funnel through which all the excitation energy collected by other pigment-protein must pass before it is delivered to the reaction centre. Complex I is made of chlorophyll a/ß-carotene-protein and is present in all plants, photosystem II is not yet so well characterized (Kirk, 1986).

For a plant to be able to make efficient use of light quanta absorbed at a given rate, the activity of the electron transfer components must be high enough to ensure that the excitation energy collected by the light-harvesting pigments is utilized as fast as it arrives at the reaction centres. At low light intensity, the autotroph is achieving its highest conversion efficiency and quantum yield. As the incident light intensity increases, the rate of absorption of quanta reaches a point at which excitation energy arrives at the reaction centres faster than it can be used by the electron transfer components and/or the CO_2 fixation enzymes. Some of the additional absorbed quanta are utilized for photosynthesis, and some are not, the energy of the latter being eventually dissipated, mainly as heat. In the light-saturated state, the electron transfer and/or the CO_2 fixation enzymes are working as fast as they are capable, and so any additional absorbed quanta are not used for photosynthesis at all.

The amount of photosynthesis achieved per unit light absorbed is lower at some wavelengths than at others. For *Chroococcus* (blue-green alga), *Chlorella* (green alga), *Navicula* (diatom), a fall in quantum yield is observed in the red absorption peak (680-710 nm) of the chlorophyll and in the region of carotenoid absorption (440-520 nm, blue portion of the spectrum). This is particularly marked for the blue-green alga, significant for the green alga, slight for the diatom. One explanation is that light absorbed by the carotenoid fucoxanthin in diatoms is efficiently used, while light absorbed by carotenoids in blue-green and green algae is used with much lower efficiency. The more likely explanation is now considered to be the fact that the two light reactions must be energized for photosynthesis to occur. Algae have different light-harvesting pigment arrays. Monochromatic light is strongly absorbed by one photosystem, but weakly by the other. Photosynthetic efficiency is low (Kirk, 1986).

Therefore algae have to be exposed simultaneously to both wavebands (the blue and the red bands), which have an synergistic effect. This is referred to as enhancement. When photosystem II is absorbing light faster than photosystem I, some of the excitation energy is transferred to photosystem I: this is referred to as the spill-over hypothesis. The reverse is not possible: light absorbed in system II can drive both light reactions, but that of system I drives only light reaction I. The fall-off in quantum yield in certain parts of the spectrum indicates that those spectral regions absorb mainly due to photosystem I pigments.

The rate of light collection does not only depend on their pigment complement but also on their position and orientation within the cell. In brown algae, chloroplasts moved to the cell walls facing the light at low irradiance and to the side walls parallel to the light direction at high irradiance. The change of position is completed within 1-2 hours.

In the filamentous green algae *Mougeotia* and *Mesotaenium* which have a single, flat centrally-located rectangular chloroplast in each cell, the characteristic movement of the chloroplast in response to a change in the light regime is to rotate along its longitudinal axis, rather than to move around the wall. In planktonic algae, where the cells contain one or a few chloroplasts, the plasts shrink and aggregate in bright light (Kirk, 1986).

Effects of limiting or excess light on phytoplankton

Light deeply affects algal growth, as will be discussed later (Factors affecting growth rates, p. 138).

Reactions to light intensities vary according to species. Light requirements, the ability to adapt to low or changing light intensities, and resistance to high light irradiations determine the outcome of competition between algae (Nicklisch & Kohl, 1989). Some, such as the blue-greens, thrive at high light intensities, which would kill others by photo-oxidation.

In some conditions, reduced growth is caused by self-shading of algae, which can occur independently of lack of nutrients. In a survey on phytoplankton communities in 165 Florida lakes, this caused a modification of the phytoplankton assemblage, in which small diatoms were replaced by blue-greens of large size (Agustí *et al.*, 1990).

Excess light has an inhibitory effect. This effect increases in the course of time. The higher the temperature at a given light intensity, the more rapidly inhibition ensues. Phytoplankton can recover from the inhibitory effect of intense light if transferred to a lower light intensity.

Mechanisms of photoinhibition have been studied in detail in higher plants. The spectrum of photoinhibition of electron transport showed its main activity in the ultraviolet region, with a peak at 250-260 nm. Photoinhibition also occurred in the visible region but with a much lower quantum efficiency. Between 400-700 nm, the action spectrum followed the absorption spectrum of chloroplast pigments, with a distinct chlorophyll peak at 670-680 nm. Lesions appear primarily to affect the light reactions of photosynthesis, possibly by damaging the reaction centre of photosystem II.

Photoinhibition in the surface layer is not as frequent as was originally thought. Except under very still conditions with no wind or waves, there is always circulation of water in the upper layer. This avoids phytoplankton from being exposed to intense light, thus preventing photoinhibition. The transport downward must, however, not reach too low.

Efficiency of the use of solar energy in ponds

In a fish pond under intensive culture using manure and artificial feed, exploited according to traditional Chinese practise, a total efficiency of solar energy utilization by primary production was 1.2 %. The efficiency of transformation of plankton by planktivorous fish reached 4.2 %, which gave an efficiency of utilization of solar energy by filter-feeding fish of 0.05 %. Compared to natural aquatic ecosystems, in which the efficiency lies around 0.1 to 0.5 %, such fish ponds are extremely efficient (Li, 1987).

Photosynthesis determinants and influence of light

1. It is not possible to measure net photosynthetic rate exactly, as part of the organic compounds elaborated by the photosynthesis are directly used (by respiration) for metabolic needs. Such respiration, occurring simultaneously with photosynthesis, is termed photorespiration as opposed to dark respiration, which occurs in darkness. Photoassimilation is an approximate measure of photosynthetic rate.

2. Photosynthesis is determined by the amount of light available, nutrient availability, the volume of autotrophs, and temperature.

3. Light is progressively absorbed in water and by particles in suspension in the water. The availability of light according to depth leads to a stratification with regard to photosynthesis. Three zones can be distinguished: that of light inhibition causing photoxidation of the pigments, that of light saturation, and that of light limitation.

4. An increase in phytoplankton biomass results in a decrease of photosynthetic rate, owing to self-shading and nutrient depletion.

1.2.2. Influence of phytoplankton biomass

Phytoplankton abundance and measurement of phytoplankton density

In a pelagic system, the amount of phytoplankton determines the rate of photosynthesis. The phytoplankton is composed of an assemblage of algal species. Inventorization of such assemblages requires good systematic skills and has to be done regularly as species assemblages change with time and with modifications of water quality. The phytoplankton is grazed upon by the zooplankton and phytoplanktivorous fish and this grazing pressure also modifies phytoplankton composition.

Phytoplankton density is measured by using plankton netting of adequate mesh size and measuring the quantity of phytoplankton filtered from a known volume of water. It can be expressed as dry weight, as chlorophyll or by numbers of taxa per unit volume or surface area. The biomass of phytoplankton recorded is the standing crop and can be compared to the carrying capacity of the aquatic environment, which is the maximum density attainable.

To express photosynthetic capacity of phytoplankton, it is necessary to have a common conventionally, the amount of chlorophyll a is used as the reference unit, because the ratio of chlorophyll a to cellular organic carbon is fairly constant, in the range of 21.3 to 24.4 mg of chlorophyll a per gram of cellular carbon (Westlake, 1980). Use of chlorophyll as a biomass estimator has, however, been questioned by Heyman & Lundgren (1988), who state that the chlorophyll content of algae may vary up to 5 times with temperature, light, nutrient, and species. Westlake (1980) showed a negative parabolic relationship between phytoplanktonic biomass and chlorophyll a (Fig. 36). A fairly constant ratio between chlorophyll a and carotenoids has also been found (Westlake, 1980).

Phytoplankton biomass changes results from the growth of different species composing it, and from environmental factors modifying it.

Phytoplankton growth

Phytoplankton growth is expressed as a rate, i.e. an increase of biomass with time. This increase is proportional to a specific growth rate, μ, and biomass, B:

$$\frac{dB}{dt} = \mu . B$$

The growth of phytoplankton shows a J-shaped curve in unlimiting conditions, and an S-shaped curve in restricting conditions (Fig. 37). Density rises until a factor becomes limiting. Limiting fac-

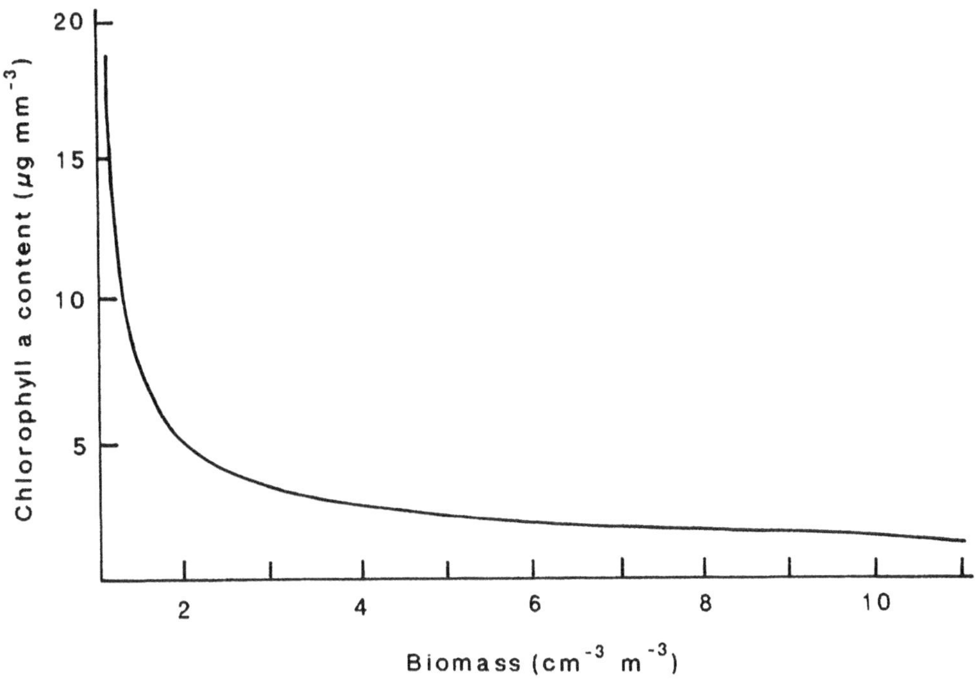

Fig. 36. Chlorophyll a content of phytoplankton (expressed in µg mm^{-3}) against mean biomass concentration (cm m^{-3}) in the euphotic zone of a Czechoslovakian reservoir (modified from Javornický, 1980).

tors are nutrients, or light. The specific growth rate varies and is fastest (the maximum specific growth rate) at the point of inflection of the S-curve (Ahlgren, 1988; Heyman & Lundgren, 1988).

Factors affecting specific growth rates

A common model of restricted growth is the model of Verhulst (1838), where growth rate is plotted against carrying capacity, i.e. maximum attainable biomass. Another group of models considers growth of a population in relation to nutrient concentrations, either extracellular or intracellular. In fact, nutrient concentration determines carrying capacity.

Specific growth rate is also dependent on external nutrients (S) (Monod model (Monod, 1942)) and internal nutrient concentrations (q) (Droop model (Droop, 1974)) (Fig. 38). The relations between the external nutrient concentration and the specific growth rate, and between the internal nutrient content and the specific growth rate are both hyperbolas. In the case of the internal nutrient level, the origin of the curve does not depart from zero but from the minimal cellular nutrient level, q_o. For

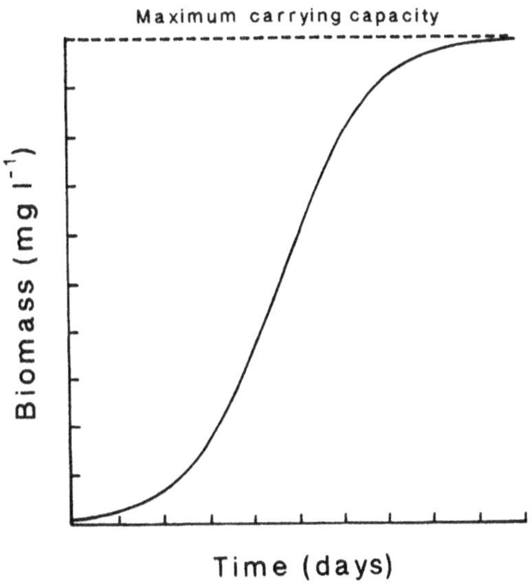

Fig. 37. Generalised growth curve of algal biomass as a function of time (modified from Ahlgren, 1988).

each nutrient, such a relation can be established. Such models consider the role of each nutrient independently, i.e. without interaction between nutrients. It is a widespread opinion that the single nutrient

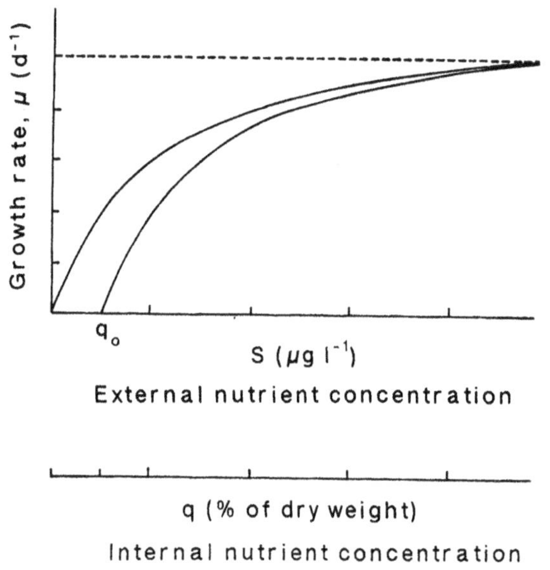

Fig. 38. Generalised specific growth rate curve for algae as a function of (1) external nutrient concentration, S, and (2) internal nutrient concentration, q. q_o is the minimal internal nutrient concentration or minimal cell quota. The dotted line shows the maximum specific growth rate associated with S and q, respectively (modified from Ahlgren, 1988).

which is in short supply, determines the algal growth rate, independently of other nutrients. This is called the threshold or Liebig-Blackman model. Opposed to this is the interactive or Mitscherlich-Baule model, which considers suboptimal nutrient levels to act simultaneously, leading to multiplicative effects (Ahlgren, 1988) (see also p. 140). Clear relations between phytoplankton growth and nutrient concentrations are difficult to show in the field; for example, biomasses for a given level of phosphorus might vary by a factor of 5 to 10. In the case of phosphorus being the limiting nutrient, any supply of phosphorus will quickly be removed, and recorded concentrations will appear low, though production will be high. If a population is heavily grazed upon, biomass can be kept low in comparison with carrying capacity, but specific production will be high (Heyman & Lundgren, 1988). This is common in tropical waters.

The specific growth rate depends also on light and temperature. Interactions between light and nutrient limitation are complex and the respective effect of each factor cannot be singled out. But, light limitation is more severe than nutrient limitation, because nutrients cannot be utilized if sufficient energy is not provided. This in indirectly confirmed by plankton adaptations to changing light regimes (Tilzer, 1990).

Nutrient content increases at low light intensities, as has been shown for silicon and nitrogen (Fig. 39).

In the case of nitrogen, this is an adaptative mechanism, associated with increases in the size of the photosynthetic units. The combined effects of light and nitrate limitation are greater than the sum of the individual effects.

The effect of light on phosphorus metabolism is more complex. Light stimulates phosphorus uptake by algae (Jansson, 1988). In cultures of *Skelotonema costatum* and *Dunaliella tertiolecta*, it was shown that under P limitation the value for light saturation decreased. In both cultures, the respiration rate under P limitation was reduced to a lower value (25 %) than under N limitation (40 %) (McAllister *et al.*, 1964). Phosphorus limitation reduces the availability of phosphate for high-energy phosphate compounds, and affects cellular respiration. Excretion of dissolved organic P increased with light, but N and P limitation reduced excretion. Growth rates were also reduced under such P and N limitations (Kuenzler, 1970).

A three-way interaction between light, P and carbon dioxide has been also shown. To sustain a

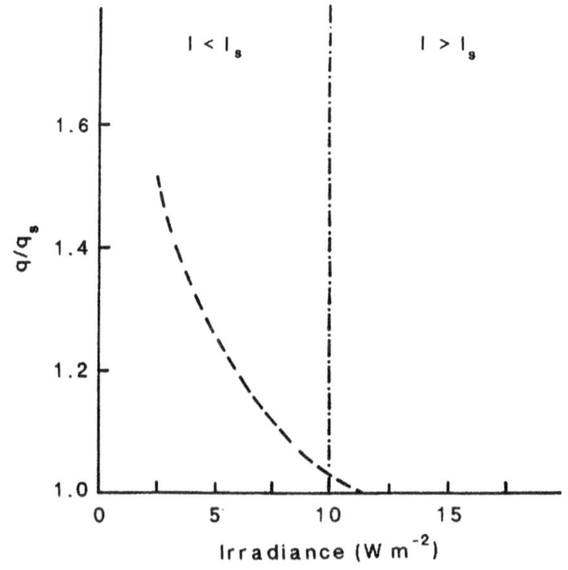

Fig. 39. Ratio between the actual cell quota, q, and the cell quota, q_s, at light-saturation, I_s, as a function of light irradiance, I (modified from Ahlgren, 1988).

given growth rate, the alga needed 10 times higher CO_2 concentrations at both P and light limitation than when the factors were limiting separately. At higher growth rates, light became the primary limiting factor (Young & King, 1980).

The effect of light and nutrient limitation on protein and polysaccharide synthesis was studied in several experiments. Some experiments followed the fate of fixed CO_2 in algal cells (*Merismopedia tenuissima*, a blue-green) under light and nutrient limitation. The percentage of total fixed carbon increased with μ and the increase was stronger under P than under N limitation and is explained because protein synthesis is obstructed by N limitation (Konopka, 1983). The amount of C incorporated into protein decreased with increasing light, but in P-limited conditions, incorporated C was constant under different light conditions. The percentage of total fixed C in polysaccharides decreased with μ in P-limited conditions and was independent of μ in N-limited conditions. In nutrient-sufficient conditions, the percentage of polysaccharides increased with light (Konopka & Schnur, 1980 & 1981). Polysaccharides and protein synthesis should be mirror images of each other, when they are measured as percentage of total fixed carbon (see Hecky & Kilham (1988): 799, Fig. 1) (Ahlgren,1988).

Comparing the percentage of C incorporated into protein against a light gradient for P-limited and nutrient-sufficient algae, (preconditioned at two different light intensities) it was shown that below 20 μE m^{-2} s^{-1} (light limitation) the percentage of protein increased with increasing light. Between 20-100μE m^{-2} s^{-1} (both light and P-limitation) it decreased, and above 100 μE m^{-2} s^{-1} (only P-limitation), it was independent of light. P-limited algae incorporated only half of the C into protein than nutrient-sufficient algae. Algae preconditioned to low light showed smaller (especially at low light intensities, 20 μE m^{-2} s^{-1} under light limitation) differences between non-limited and P-limited conditions, than algae preconditioned in strong light. There may be interaction between light and nutrient limitation, which affects the physiological state of algae (Konopka, 1983). One of the consequences of this observation is that nutrient-sufficient populations cannot be distinguished from nutrient-deficient populations without due consideration of light conditions (Ahlgren,1988).

Interacting effects of light, N, and P was investigated in chemostats on the blue-green *Synechococcus linearis*. In a narrow range of irradiance, both N and light or P and light simultaneously influenced the growth rate of the alga. With decreasing irradiance, the internal content of the limiting nutrient increased. The q_o for N clearly increased with decreasing light, but not the q_o for P. The critical N/P ratio (R_c, where the algae change from N to P limitation) increased with decreasing light (Healey, 1985). Since μ is strongly correlated with light, μ should also be correlated with R_c. With decreasing light or μ, algae need relatively more N than P.

As conditions become adverse, an increasingly higher internal content of the limiting nutrient is required. Three explanations are suggested (1) that in depleted conditions, a physical factor, light, or a major nutrient may be able to substitute each other (Healey, 1985); (2) that in non-optimum conditions algae adapt by increasing the concentrations of enzymes per cell, and as enzymes consist of essential nutrients, their requirements increase (Steemann Nielsen & Hansen, 1959); or (3) that uptake mechanisms of N and P are less influenced by low light than growth metabolism. The metabolic utilization of such nutrients is therefore reduced, hence their accumulation (Ahlgren, 1988).

Ammonium ions are smaller in size than phosphate and should be taken up with less energy expenditure than phosphate. The second explanation is considered the more likely.

Temperature affects growth rate, because it affects the kinetics of enzymatic reactions (Fig. 40) and determines maximum growth rate. Contents of phosphorus, nitrogen and, to a lesser extent, carbon are increased at lower temperature. The minimum cell quota varies with temperature, being minimum at the optimum temperature and increasing at low and high temperatures (Fig. 41). Average cell size decreases with increasing temperature to reach a minimum, before increasing again (U-shaped relation) while biomass remains fairly stable (Fig. 42) (Ahlgren, 1988).

Specific growth rate and biomass

The specific growth rate controls the amount of biomass produced. Growth rate decreases from a maximum at low biomass to an insignificant value

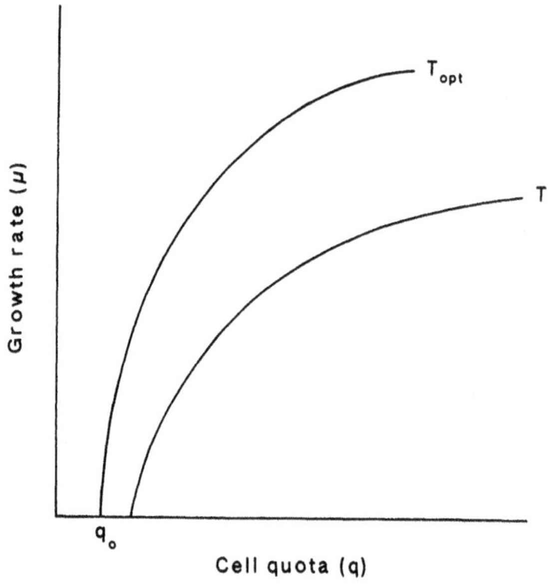

Fig. 40. Generalized specific growth rate curve for algae as a function of the internal nutrient concentration or cell quota, q, at optimum temperature, T_{opt}, and a different temperature, T. The minimum cell content, q_o is minimum at the optimum temperature (absolute minimum cell quota) and the maximum specific growth is maximum at optimum temperature (absolute specific growth rate) (from Ahlgren, 1988) (with kind permission of Prof. Ahlgren).

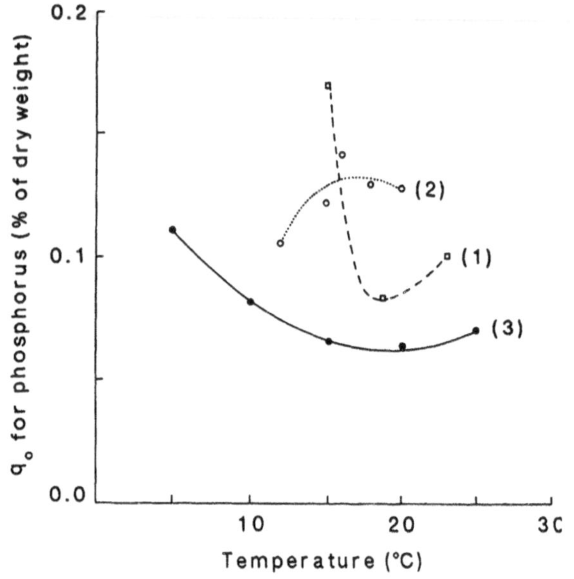

Fig. 41. Minimum cell quota, q_o, for phosphorus in percent of dry weight against temperature. (1) *Monochrysis lutheri*, (2) *Scenedesmus sp.* and (3) *Scenedesmus quadricauda* (from Ahlgren, 1988) (with kind permission of Prof. Ahlgren).

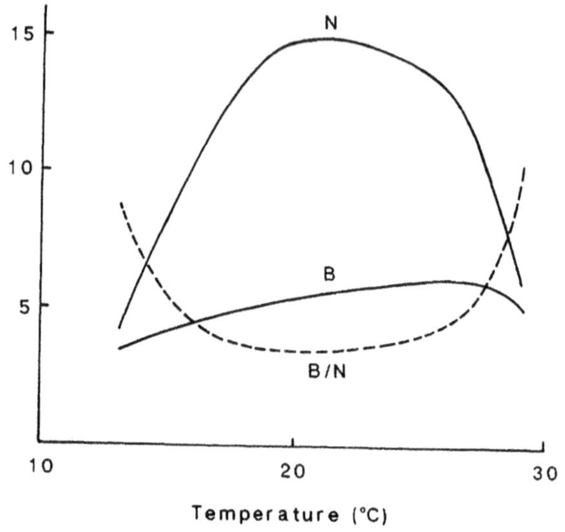

Fig. 42. Schematic representation of the relation between algal cell number, N, biomass, B, and average cell size, B/N as a function of temperature in a chemostat culture The Y-axis is graded in relative units (from Ahlgren, 1988) (with kind permission of Prof. Ahlgren).

close to carrying capacity (Fig. 43). At low growth rates, the same amount of phosphorus will be used more efficiently and produce a higher biomass than at higher growth rates. This applies also for nitrogen, but not in the same proportion. The amount of nutrient consumed to produce a given biomass is defined as the growth yield (Ahlgren, 1988). The relation between yield and growth rate at different temperatures for phosphorus, for instance, shows that yield and growth rate increase with temperature. The higher the temperature, the higher the possible growth rate and yield (Fig. 44) (Ahlgren, 1988).

A shift from nitrogen-limitation to phosphorus-limitation does not appear when algae are exposed to different N/P ratios in culture media. In the field, production of algal biomass in eutrophic lakes was reduced only when phosphorus and nitrogen inputs were both decreased, not when one or the other nutrient was singly decreased (Ahlgren, 1988). This tends to confirm interactive effects of nutrients on phytoplankton growth. Similarly, Yusoff & McNabb (1987) showed in experiments performed in ponds at Serdang, Indonesia, that nitrogen and phosphorus supplies to waters of low nutrient content had an

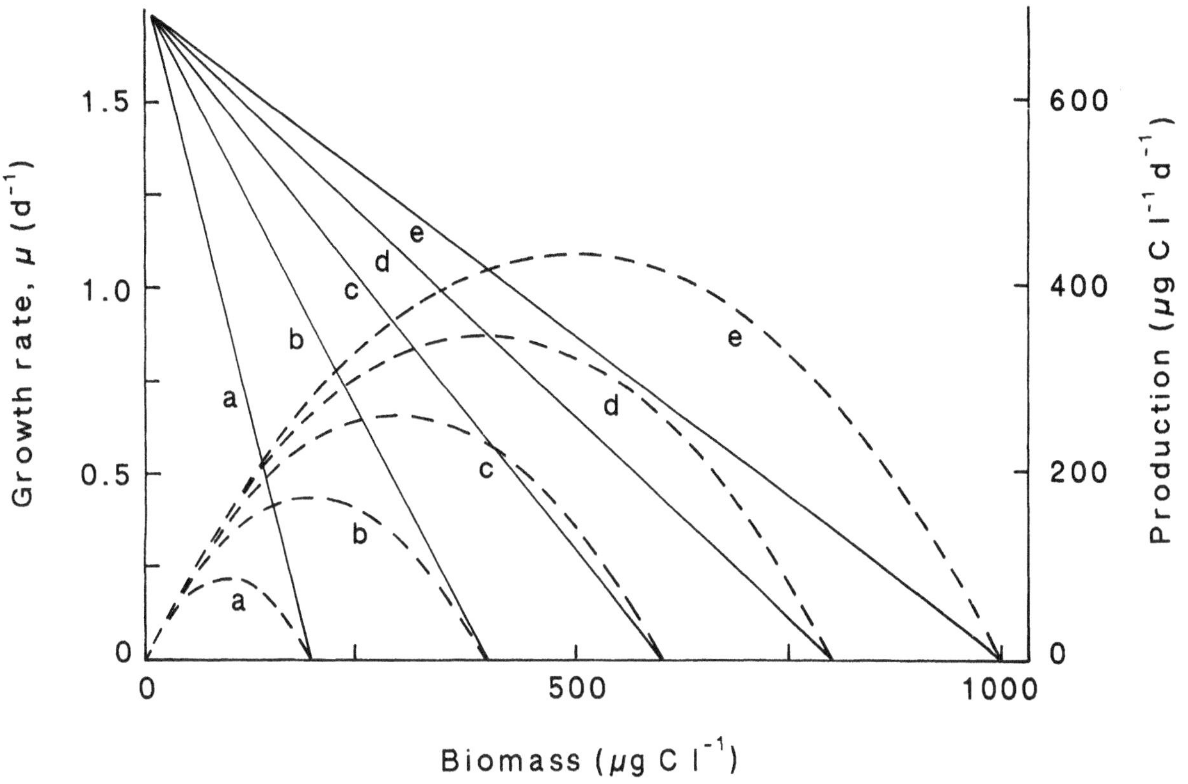

Fig. 43. Schematic relation between specific growth rate, production and biomass. Straight lines indicate specific growth rates and broken lines, production. The letters denote relations at different carrying capacities increasing from (a) 200 μg C l^{-1} to (b) 400 μg C l^{-1}, upto (e) 1000 μg C l^{-1} (after Heyman & Lundgren, 1988).

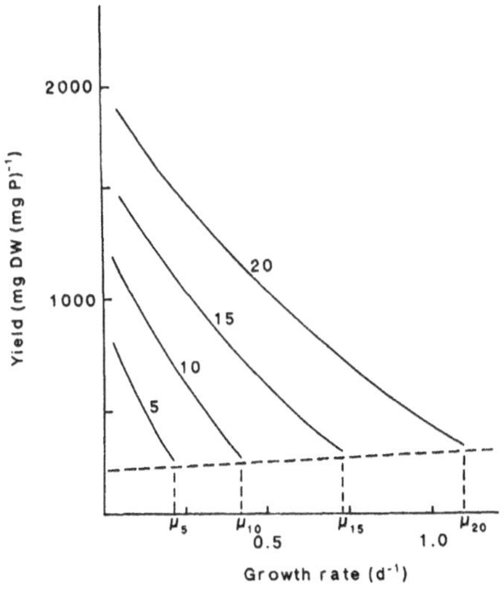

Fig. 44. Generalized diagram of phosphorus yield, i.e. the biomass produced per unit of phosphorus consumed, against specific growth rate, μ, at four different temperatures, 5, 10, 15, and 20 °C. μ_5, μ_{10}, μ_{15}, and μ_{20} indicate maximum specific growth rates at the different temperatures (from Ahlgren, 1988) (with kind permission of Prof. Ahlgren).

interactive effect; applications of one nutrient alone were not effective, as the other nutrient rapidly became limiting.

Community production and biomass

Biomass increase versus biomass (community production divided by biomass, P/B) appears to remain fairly constant within the trophogenic layer, when light is adequate. The main factors determining P/B are biomass level, carrying capacity (Fig. 43) and temperature, while nutrient level does not seriously affect this ratio. This means that all changes to biomass, such as elimination of phytoplankton, modify the P/B ratio. Elimination of phytoplankton biomass is of prime importance in regulating production and biomass, as a low biomass would increase drastically if there was not some kind of elimination (Heyman & Lundgren, 1988).

Three different types of phytoplankton dynamics

can be distinguished: (i) under light-limiting conditions, algae are characterized by a high cellular nutrient content and a low growth rate. The finite source of nutrients determines biomass, and temperature determines maximal growth rate (Fig. 45); (ii) under light-sufficient conditions, temperature determines the maximum growth rate, and nutrients determine carrying capacity. Biomass remains well below carrying capacity. Temperature simultaneously affects grazing, sedimentation, mineralization, and species composition; (iii) under initial conditions of neither nutrient nor light limitation, biomass increases close to carrying capacity, inducing nutrient-deficiency and light limitation due to self-shading, i.e. bloom conditions.

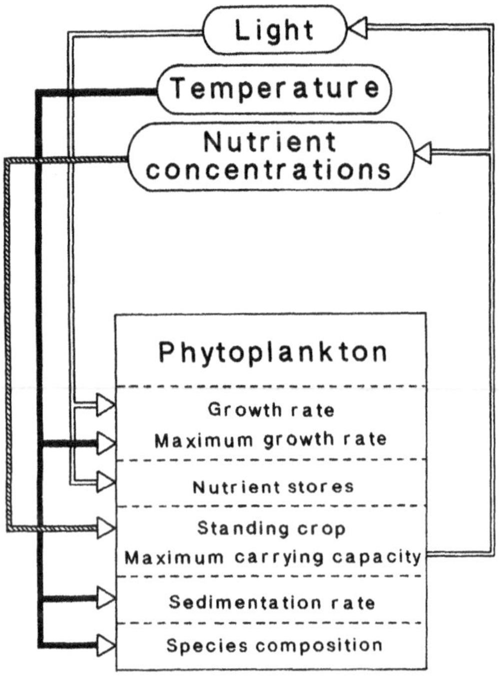

Fig. 45. Effect of different water characteristics on phytoplankton dynamics. The right-hand side arrows indicate a feed-back effect of the algal biomass having reached carrying capacity and causing self-shading and nutrient-limiting conditions. Not included are all the indirect effects, such as of temperature or standing crop on nutrient regeneration, of sedimentation rate on nutrient concentration, etc.

The bloom situation will be discussed later. When biomass is close to carrying capacity, growth rate is slow and hampered by a lack of nutrients and by self-shading (Heyman & Lundgren, 1988).

Steps in the increase of plankton biomass

From the management point of view, it is important to know what is the minimum plankton population density, from which exponential growth of the population will start, as well as the maximum population density which will lead to a reduction in photosynthetic rate by self-shading. The species spectrum which can possibly give rise to an algal bloom is also important.

In the evolution of phytoplankton biomass, four successive periods may be distinguished: a period of low stable biomass, followed by exponential growth, then a high standing biomass (bloom), ending eventually with an algal crash. It is during the period of low biomass that dissolved nutrient concentrations (orthophosphate, silica, nitrate, nitrite) are maximum. Not all species bloom (Krom *et al.*, 1989a); blooms occur in blue-green algae, diatoms, dinoflagellates, and green algae, but not amongst Chrysophyceae (Heyman & Lundgren, 1988). Populations of blue-green algae may reach a maximum of 400 mg DW m^{-3}.

Effects of water circulation on phytoplankton production

Water movements modify phytoplankton production. In stagnant water, nutrients and phytoplanktonic biomass are low at the beginning of production. In the course of production, nutrient concentration in the water and phytoplankton density increase. After some time, biomass reaches a given load, the carrying capacity, determined by light and nutrient availability. At high phytoplankton densities, anaerobic and aerobic patches develop and toxic metabolites accumulate. The conditions for an algal crash and for deterioration of water quality are set. In the case of running water, phytoplankton biomass is determined by nutrient supply. An exchange of water between 1 to 5 times a day allows the daily phytoplankton growth rate to be multiplied up to 5 times (Krom *et al.*, 1989b). According to Erez *et al.* (1990), the dilution rate should not exceed the specific growth rate, and daily replacement of water should be maximum half the volume of the pond. The outflowing water removes metabolites produced in the pond. A periodic crash of algal populations cannot, however, be prevented. If water flow is too fast, phytoplankton is washed away and diluted, which further reduces phytoplankton growth. Only bacteria and algae then grow on the dykes (Krom *et al.*, 1989b).

Water flow has another direct effect on photosynthesis, by fostering carbon dioxide uptake. Immediately next to the surface of any phytoplankton cell, there is an unstirred layer of water across which CO_2 must diffuse. Thickness of this layer is about 5 μm around small cells like *Chlorella*, and around 30-150 μm for larger cells and macrophytes. If the plant is subjected to a water flow, the thickness of this layer decreases and the uptake of carbon dioxide is increased. Morphological adaptations which increase the surface to volume ratio of phytoplankton cells also help to overcome the diffusion problem (Kirk, 1986).

Oxygen release and algal biomass

Algae, besides being food for fish, contribute to the ecosystem by releasing oxygen. Net oxygen production per algal biomass unit decreases, when algal biomass increases, because nutrients and light tend to become limiting and photosynthetis is depressed. Oxygen production is no longer sufficient to satisfy the need of total community production, entailing an oxygen deficit and environmental stress. Paradoxically, excessive phytoplankton biomass may lead to dissolved oxygen depletion. Nutrient concentrations in water are not homogeneous and oxygen depletion is therefore occurring in patches; light is subject to the same heterogeneity, some areas being less illuminated than others (Smith, 1988).

Plankton population composition and trophic state of the water

Phytoplankton composition changes in the course of season and time, adapting to changes in the trophic state of the water body. The relative importance of the different taxonomic groups in a plankton assemblage depends on trophic state, on nutrient supply and on zooplankton grazing. Complex interrelations exist between herbivorous zooplankton, nutrient concentrations, and phytoplankton standing crop (discussed further below) and prevent as yet modelling of these relations (Uehlinger & Bloesch, 1989). Phytoplankton succession (seasonal change) is discussed on p. 158.

Phytoplankton biomass and primary productivity

1. Primary production relies not only on the amount of phytoplankton present in a pond, but also on its rate of increase, i.e. the growth rate of this biomass.

2. Phytoplankton growth depends on the carrying capacity of the environment and on nutrient concentration, both intracellular and external. Light limitation causes an increase in intracellular nutrients and in the size of the photosynthetic units. Light and phosphorus limitation result in a reduction of cell respiration, while light, carbon and nitrogen limitation result in a reduction of organic compound elaboration. Temperature determines maximum growth rate. The factors determining growth under light-saturating conditions are, in order of importance: temperature, nutrient concentrations, and carrying capacity. Growth rate is inversely related to phytoplankton biomass.

3. Nutrient requirements vary with light availability. Under conditions of low light irradiances, and at low growth rates, phosphorus is used more efficiently than at higher growth rates or higher irradiances.

1.2.3. Elimination of phytoplankton biomass

Types of elimination or biomass decreases

Amongst the factors directly involved in determining standing crop, those discussed above are related to biomass increase. Factors related to biomass decrease include predation or grazing by zooplankton and fish, parasitism and diseases caused by viruses and bacteria, sedimentation, and cell lysis, caused by environmental stress.

Elimination of phytoplankton can also result from physical factors: dilution or concentration of nutrients, evaporation and drying up of the pool, and changes in climatic conditions, leading to excessive turbulence capable of damaging cells, excessive temperatures, and harmful light intensities.

If phytoplankton biomass is measured by organic carbon content, algal exudations can also be included among factors contributing to algal elimination.

Plankton sedimentation

Losses by sedimentation of algae from the euphotic zone are a significant process. Sedimentation is not a purely physical phenomenon, but is affected by species composition and by nutrient status of the cells (Heyman & Lundgren, 1988). A number of green algal species, heavy and large-sized, are maintained in the water only by continuous turbulence. During calm periods and when stratification occurs, losses by sedimentation can be sufficiently considerable to cancel out daily production. In an Eastholstein lake, Germany, elimination of algal cells was more important during nutrient deficiency (Meffert, 1989).

Near the surface, sedimentation reduces overall algal biomass, but buoyant blue-green algae accumulate. In Windermere, U.K., densities of colonies of *Anabaena lemmermannii* reached 975 kg m^{-3}, with intact gas vesicle, and 1022 kg m^{-3}, after gas vesicles collapsed (Walsby *et al.*, 1991).

Sedimentation rate increases with a decrease in water density and viscosity, i.e. with increasing temperature, and differs between species. Mean floatation rate of colonies of blue-greens in lake Windermere was 181 $\mu m\ s^{-1}$ and 186 $\mu m\ s^{-1}$, with intact gas and collapsed gas vesicles, respectively (Walsby *et al.*, 1991). Algal biomass is usually greater in the lower part of the epilimnion. The ascendent migration of algae is regulated by changes in buoyancy. In the case of blue-green algae, the density of a cell is modified by the formation and resorption of gas vacuoles (Walsby & Reynolds, 1980). These gas vacuoles ensure vertical movements. When irradiation is too high, the gas-vacuole collapses, and the cell sinks; the vacuole is rebuilt under conditions of low light intensity (Westlake, 1980).

In *Aphanizomenon flos-aquae* from two lakes in Indiana, U.S.A., an irradiation above 100 μmol photons $m^{-2}\ s^{-1}$ reduced buoyancy. In the presence of 2 μmoles of KH_2PO_4, irradiation must be above 200 μmol photons $m^{-2}\ s^{-1}$ to cause cell descent (Konopka, 1989). Carbohydrates are used as a ballast to help cells sinking. Cellular turgidity, linked to carbohydrate synthesis, has been considered responsible for the disappearance of the gas-vacuole, but this seems unlikely. Cell turgor pressure is linked to the presence of a gas-vacuole, modifying the size of the latter, and hence, buoyancy (Walsby & Reynolds, 1980). Gas vacuoles are only present in prokaryotic algae, i.e. blue-greens, while eukaryotic algae posses only fluid vacuoles. In planktonic diatoms, they are well developed, serving as a means of reducing average cell density and in nutrient storage. Other algae have evolved particular shapes, hair-like expansions, or large volume vacuoles, to lessen density (Taylor, 1980). In the case of flagellates, migrations are active.

Algal consumption

A study of consumption of diatoms, filamentous blue-green algae and green algae by cladocerans, copepods and rotifers indicates that herbivores are capable of consuming filamentous algae at the same or higher rates than unicellular algae. Some phytoplankton species are even heterotrophic or phagotrophic and consume other phytoplankton. Parasites, pathogenic bacteria and diseases also affect phytoplankton, as well as growth-inhibiting substances, which are either auto-antagonistic in populations of aged algae or hetero-antagonistic in species which exclude each other, such as *Aphanizonemon* and *Scenedesmus* (Westlake, 1980).

Analysis of grazing of caddis-fly larvae

(*Neophylax automnus*) upon diatoms has shown that almost half of the diatoms are still alive in the larval faeces. Algae of large size are consumed more than those with a small size. Those which adhere better to the substratum escape predation to a greater degree. As a result of lower predatory pressure, their relative number increases (Peterson, 1987) and because of grazing, the structure of the diatom population changes.

The quantitative or relative importance of each of the factors causing elimination of algae is difficult to establish; it appears that grazing cannot be generalized to be the main cause of algal destruction, though it may be important in eutrophic waters (see also the sections hereafter).

Phytoplankton defenses against predation

The importance of zooplankton grazing is controversial, as zooplankton is not always capable, for anatomical reasons, of feeding directly on all algae. A certain number of filamentous or colonial algae reach a volume superior to the maximum particle size that can be ingested by zooplankton. Other algae possess thick gelatinous sheaths (Vanni, 1987a). Some blue-green algae, such as *Anabaena flos-aquae*, are not consumed due to production of toxic or harmful products, and not to the morphology of the filaments. Resistance to digestion as mentioned above is also a defense. Blue-green algae (*Microcystis*, *Aphanizomenon*, *Oscillatoria*, *Anabaena*) Dinoflagellates (*Peridinium*), Chlorococcales, Eugleninae resist well to grazing and digestion by fish, as numerous viable cells are present in the faeces of herbivores, such as silver carp and *Oreochromis aureus* (Lazzaro, 1987; Prowse, 1964; grazing of *O. aureus* and *Sarotherodon galilaeus* in Lake Kinneret on *Peridinium*: Gophen, 1983; Landau, 1983; Ben-Tuvia & Reich, 1983).

Controls of the dynamics of phytoplankton density

Control of phytoplankton biomass is either achieved by grazing (top-down) of herbivorous zooplankton and phytoplanktivorous fish, or bottom-up, by nutrient availability (McQueen *et al.*, 1986). The control of primary productivity by predation is termed the trophic cascade (Carpenter *et al.*, 1987). Determination of the predominant process in overall control of phytoplankton is established from relations between phytoplankton and zooplankton, by comparisons of zooplankton filtration rates to population size, and responses of phytoplankton to nutrient enrichment. Results range from dependency to independency, with high to low grazing pressure, including pressure equal to production, and are seasonally changing. Experiments in limnocorrals, with removal of portions of the plankton, show that zooplankton greatly affect phytoplankton biomass. A negative relation between zooplankton and phytoplankton biomass is indicative of a grazing-controlled system.

Top-down control, when apparent, is always restricted to some period of time. Though no general statement is likely to be true, the type of control appears linked to trophic state of the water body: in eutrophic systems, control of primary production is from below, while in oligotrophic systems, grazing controls phytoplankton.

Importance of elimination of algae

1. Algal elimination is an important process. Phytoplankton is grazed upon by fish and zooplankton. Algae, which are not consumed, supply organic matter to decomposers, which regenerate nutrients for phytoplankton growth.

2. The main fraction of algal elimination is due to sedimentation: grazing is seldom a limiting factor.

3. Phytoplankton growth is controlled by a combination of grazing (top-down control) and nutrient supply (bottom-up control).

1.2.4. Phytoplankton behaviour and life strategies

Importance of life strategies

Phytoplankton species have evolved a variety of life-history strategies. These involve such aspects as nutrient uptake affinities (high versus low), capabilities to withstand grazing by motility or protection, ability to form resting stages, adaptations to low or high irradiations, to different temperature optima, and to living at low or high densities, i.e. close or away from carrying capacity. Production to biomass ratios are fairly constant; this implies that, if production rate is high, natural elimination to maintain a stable biomass will be high and vice versa.

The life-history strategy of phytoplankton can be appraised on the following criteria (Kilham & Kilham, 1980):

- the magnitude of the intrinsic rate of increase
- the amount of resource expenditure used to produce offspring
- the extent of environmental saturation with individuals
- the degree of niche overlap
- the degree to which productivity or efficiency is maximized.

In the context of phytoplankton management, the discussion will focus on species' ability to utilize environmental resources, considered at a patchy, ephemeral and microhabitat scale, species distribution within the water and its modification, and changes induced in the water by algae, such as exudations, blooms and crashes.

r and K strategies of phytoplankton species

The influence of nutrients operates through the nutrient assimilation capacity of the different species. Several strategies can be distinguished: some species specialize in assimilation, i.e. they absorb nutrients quickly in considerable quantities and use them for cell multiplication; others specialize in stocking nutrients in intracellular stocks; still others show high affinities, which allow them to absorb nutrients at very low concentrations.

r-selected species have evolved primarily in temporary habitats, where they cannot maintain their maximum density for any length of time owing to competition from other species, or to rapid changes in the environment. They allocate most resources to reproductive efforts. K-selected species are adapted to operate close to the species' carrying capacity in the environment, being selected for competitive ability and efficient resource utilization. Non-reproductive activities receive more resources, and genetic survival is optimized.

r-selected species are synonymous with fugitive, colonist, weedy or opportunistic species, while K-selected species are equilibrium or autochthonous species (Kilham & Kilham, 1980). r-selected species have rapid growth capacities and can use varying nutrient concentrations, such as pulsed supplies; they have high absorption affinities. When the cell quota are depleted, they show growth rates depending upon nutrient concentrations available in the water; they are usually edible to consumers. K-selected species are unable to react quickly to nutrient enrichment of the water, showing cell quota often close to that of minimal intracellular concentrations, and normal growth rate close to the maximum. They can sustain stable populations for long periods of time; they are usually inedible (Kilham & Hecky, 1988).

The plankton paradox

The paradox of the plankton was formulated by Hutchinson (1961), who asked how a high species diversity in the pelagic of lakes and ponds can be maintained. The water column is a typically homogeneous, unstructured environment; thus, it is puzzling how niche diversification can occur here. Besides a permanent state of non-equilibrium of the communities, the existence of a series of microhabitats, ephemeral and patchy in nature, not revealed by conventional samplings, but providing spatial and temporal heterogeneity (Smayda, 1980) is advocated as an explanation. The ephemeral character of these patches does not allow one species to develop in such abundance that it can outcompete other species.

Vertical distribution and migration of phytoplankton

The vertical distribution of algae in the water is stratified. This stratification is the result of active migration and of passive sinking (Table 19). In Lake

Table 19. Causes prompting phytoplankton migration

Downward migration	• to take up nutrients/vitamins released by the sediments or produced by microorganisms • to avoid deleterious irradiation causing pigment bleaching • to avoid zooplankton grazing
Upward migration	• to escape light limiting conditions

George, Uganda, phytoplankton distribution is homogeneous at dusk. During daytime, the lake changes from isothermic to stratified, followed by mixing during late afternoon. Algae sink during the day, but begin to distribute uniformly when thermal stratification disappears (Ganf & Horne, 1975).

In deep, aphotic water, nutrients are often more concentrated and some chemical gradients occur around the thermocline and in the epilimnion (Westlake, 1980). Algae move to those areas to take up and stock nutrients, which are assimilated later in euphotic conditions. Aggregations of algae (*Cryptomonas*, for example) in mid-column may be linked to the presence of vitamins at that level. These vitamins are produced by bacteria, whose presence is, in turn, connected to an accumulation of detritus in the deoxygenated hypolimnion (Westlake, 1980). Some algae, such as *Oscillatoria agardhii* var. *isothrix* may begin their growth in the deoxygenated hypolimnion and later move upwards (Saunders, 1972; Westlake, 1980).

In Wellington reservoir, near Pearth, New Zealand, diel migrations were promoted by advective currents, produced by physical processes (Viner, 1989). At mid-day, photosynthesis close to the surface decreases (light inhibition), but increases in deeper waters. In the afternoon, the photosynthetic rate is reduced by nutrient depletion, by the increase of respiration, by pigment bleaching, and by zooplankton grazing. Seasonal variations superimposed on this are changes in light regime, temperature, wind-induced turbidity, supplies of run-off water, nutrient availabilities, water stratification, and species composition. However, in tropical waters, seasonal changes are damped or non-existant (Westlake, 1980).

Algal exudations

Primary productivity is the increase in algal carbon per unit area and time. It is measured by the amount of carbon fixed, but some bias originates from the discrepancy between carbon uptake and carbon assimilation, i.e. its incorporation into organic molecules. Some of the carbon absorbed is reconverted into carbon dioxide by respiration. Some is excreted as glycolate and other organic compounds, such as carbohydrates (hexoses: Chrost, 1989), amino acids and larger molecules; such excretions are called algal exudations.

Glycolate excretion occurs in two circumstances, provided that glycolate production exceeds its rate of consumption: (1) when growth rates are low, due to algal senescence or nutrient depletion; (2) when cells are metabolically stressed or subject to photoinhibiting conditions (Harris, 1980).

Exudations are produced by healthy algae (Bjørnsen, 1988; Hino, 1988, for blue-green algae, in lake Barato, central Hokkaido, Japan). A proper explanation of the physiological mechanism behind these exudations is lacking. One interpretation is that the active release or passive permeation (over-flow) of photosynthetic products is the result of carbon uptake in excess of the transformation or assimilation capacities of the cell. Exudations are nitrogen-rich compounds of low molecular weight, such as amino acids. They are quickly utilized by pelagic bacteria and are a carbon supply for them under limiting conditions. They have been qualified as an 'income tax' or as a 'property tax' (continuous loss) contributed by the algae to the bacteria (Bjørnsen, 1988). In a eutrophic lake, in Michigan, U.S.A., the presence of pelagic bacteria could not, however, be solely explained on the basis of these exudations. Other food sources, for instance, organic carbon from decomposition, were needed by pelagic bacteria (Storch & Saunders, 1978).

Laboratory experiments on the effects of light intensity and temperature on excretion of dissolved organic carbon by three different species of algae (*Chlorella vulgaris*, chlorophyte; *Isochrysis galbana*, prymnesiophyte, *Synechoccus sp.*, blue-green alga) revealed a tight coupling between carbon fixation and net excretion of dissolved organic carbon. Excretions increase under extreme irradiation and at temperatures which depress photosynthesis. The dissolved organic carbon flux towards the bacterial food web is important even when conditions are suitable for photosynthetic activities, though the percentage excreted is low (Zlotnik & Dubinsky, 1989).

Phytoplankton is used as food by zooplankton, which also prey on bacteria, which, in turn, live partly on algal exudations. This intricacy of trophic relations will be further discussed later (p. 173).

Some organic substances are released by algal cells undergoing lysis. Unlike exudations of healthy cells, they are macromolecules, like polysaccharides (Chrost, 1989).

Algal blooms

Algae are said to bloom when a population suddenly develops in such a way as to become dominant. In lake Okaro, New Zealand, high algal biomasses were the result of low and steady growth periods, maintained over long periods of time, during relatively stable environmental conditions, rather than quick responses to exceptional conditions lasting for short periods of time (Vincent & Dryden, 1989). Blooms coincide with stratification, high photosynthetically active radiation, and high supplies of organic matter and nutrients (Paerl, 1988).

The dominance of a species is the expression of its competitive superiority for limiting resources. The appearance of a bloom of nitrogen-fixing blue-green algae often indicates nitrogen limitation, combined with other algal characteristics, such as inedibility (Sterner, 1989). Other characteristics promoting blue-green algae dominance are positive buoyancy of the cells, which increase the average exposure of the algae to light. The interactive effects of diffusion by turbulence, photic depth, mixing depth, and daily migrations on growth rate and production of algae differ fundamentally for algae with positive buoyancy and algae with negative buoyancy. Maximum growth is higher for species with positive buoyancy than for species with negative buoyancy, until self-shading densities are reached. The capacity of blue-green algae to migrate vertically allows them to cope with the spatial separation of nutrients and light (Humphries & Lyne, 1988).

The density of algae with negative buoyancy (naturally sinking) increases with depth, while that of algae with positive buoyancy (naturally rising) is greatest at the surface. Among the characteristics ensuring positive buoyancy are low density of the cells, a low surface/volume ratio and coloniality. Usually these algae have a low growth rate, and are restricted to eutrophic waters. This is confirmed by observations that blooms develop during periods of low nutrient concentrations (Humphries & Lyne, 1988).

Nutrient concentrations do not reflect nutrient supply, as a high turn-over of nutrients can satisfy the demand of a large biomass. The standing crop of algae increases until an equilibrium is reached between demand and supply of nutrients; at that stage, prevailing nutrient concentrations in water reach a minimum. The advantage of species with negative buoyancy compared to species with positive buoyancy depends on the nature of the stratification period; species with positive buoyancy perform better when stratification is destroyed daily. The biomass of species with negative buoyancy is continuously diluted by sedimentation and its increase of biomass is lower than that of species with positive buoyancy (Humphries & Lyne, 1988). Species which are subjected to blooms generally possess resting stages, cysts or akinetes (Paerl, 1988).

Algal crashes

An algal crash is the sudden death of a bloom and the release, from this algal biomass, of considerable amounts of organic matter. There result stressed conditions for the pond: the concentrations of carbon dioxide and ammonium increase (see p. 76) and there can be transformation of ammonium to toxic ammonia, which can lead to fish mortalities. The concentrations of nitrogen and phosphorus increase strongly, owing to decay of organic matter settled on the sediments (ammonification processes). These decompositions remove oxygen from the water, leading to anoxy, and release of phosphate from the sediments (Søndergaard *et al.*, 1990) (see Table 7, p. 39, 40 & p.39).

Causes of algal crashes

Though important for the management of phytoplankton populations, the process whereby biomass increases till populations collapse is not fully understood. A detailed discussion on possible causes is presented by Smith (1988) (Table 20). In most cases, algal crashes happen in communities dominated by blue-greens, with one species dominant. Species most commonly affected by blooms and subsequent crashes are those with cells of large size or colonial forms, which are less vulnerable to grazing. Several causes have been considered: photolysis, attack by pathogenic microorganisms, nutrient deficiencies, and grazing by microzooplankton (ciliates and flagellates). Photolysis is the destruction of cells under the effect of intense light. Blue-green algae, which belong to the group of algae most exposed to high light intensities, grow moderately well under conditions of intense irradiation at the water surface. They produce carotenoids for their photoprotection and possess the specific capacity to form scums of dead algae floating at the surface.

Photolysis does not explain algal crashes, as they frequently appear during overcast periods and sudden temperature drops. The hypothesis of pathogenic viruses or bacteria causing algal death does not seem likely either, as they are omnipresent (e.g. in Lake Erie, U.S.A., Fraleigh & Burnham (1988)) and there is therefore no reason why they should suddenly become harmful. A more probable reason for the death of algae is nutrient deficiency. The appearance of nitrogen-fixing blue-green algae is often promoted by nitrogen-limiting concentrations. The presence of a bloom changes the wind mixing depth of the water. In lakes Rotongaio (eutrophic) and Okaro, New Zealand, they remained at the surface because they were able to sustain varying light conditions, from intense radiations to low light intensities. Lastly, the zooplankton did not graze upon them (Vincent, 1989).

When a dominant species comes close to its carrying capacity, the population is stressed by decreasing nutrient supplies, which could lead to mass destruction. The fate of nutrient-deficient algae has never been clarified: do the algae get senescent or do they die and disappear? There is no convincing evidence that algal crashes are caused by nutrient deficiencies. In Lake Sobygård, Denmark, Jeppesen *et al.* (1990c) observed that a crash was preceded by an increase of sedimentation rate, suggesting that some nutrients were limiting. When algal crashes occur during overcast or cloudy weather, an increase of intracellular nutrient demand, as a result of the lower light intensities or temperature drop, could cause their death. Calm weather, reducing the mixing depth of water and preventing a supply of nutrients to the areas with high phytoplankton densities, enhances stress. A dense population reacts in a more extreme way to changing environmental conditions. Severe nutrient depletion halts or reduces protein synthesis, which induces photolysis, in a way analogous to the action of chloramphenicol (an inhibitor of protein synthesis). Jeppesen *et al.* (1990c) noticed a considerable increase of the filtrating zooplankton population, concurrent with the initiation of an algal crash.

Table 20. Possible causes for algal crashes

photolysis
temperature jumps
attacks by pathogenic microorganisms
nutrient deficiencies due to high standing crop or light-limiting conditions
grazing by microzooplankton

As none of the hypotheses discussed can be considered as the sole cause of algal crashes, a combination of causes acting synergistically seems likely.

Effect of zooplankton on algal blooms

Zooplankton grazing on phytoplankton blooms involves a two-way reaction. Algal blooms promote the abundance of algae of large size and of low edibility. The large-sized zooplankton, which does not find adequate food, disappears progressively. This leads to a decrease in the proportion of nutrients recycled by the zooplankton, which, in turn, affects phytoplankton growth (Krom *et al.*, 1989a). Interactions between algae and zooplankton are discussed in more detail later (p. 162).

The presence of algal blooms affects the whole plankton community. Blooms of blue-green algae modify the behaviour, the composition, and the structure of the zooplankton community. In Rotangaio, New Zealand, for example, *Anabaena minutissima* var. *attenuata*, produces toxins (Forsyth *et al.*, 1990) which decrease the filtration and ingestion rates of limnetic zooplankton and their vertical migration. An inhibition of zooplankton feeding by mechanical interference with algal filaments also occurred.

Algal disappearance or physiological death

In the absence of a clear dominance of particular algal species or of characterized blooms, large portions of the phytoplankton, composed of a range of algae, die, but in an inconspicuous, unnoticed way. Each species lives according to its metabolic requirements and when these cannot be met, the alga disappears progressively (Smith, 1988).

Phytoplankton management

Management practices should be aimed at preventing the appearance of algal blooms and subsequent algal crashes. Ideally, the objective would be to maximize plankton production and optimize quality (Tilzer, 1989) (see also p. 157). In the light of the above discussion, it appears that the way to maintain the phytoplankton standing crop at or just below its carrying capacity, is to supply nutrients in sufficient amounts. Optimum fertilization will reduce the probability of an algal crash. When primary production is at a maximum, ammonium uptake is also at a maximum and 0.15 of a mole of nitrogen is assimilated for each molecule of oxygen released (Smith & Piedrahita, 1988). A diversified algal community is perhaps the best guarantee against algal blooming and subsequent crashes. Biological control of phytoplankton by planktivorous fish or by zooplankton has been attempted. Although it has not yet reached a level of excellence, it is progressing. It effectively reduces the overall volume of phytoplankton. The role of fish and zooplankton in the ecosystem are discussed on p. 162 and p. 180. After an algal crash, a supply of green water, enriched with phytoplankton, from an adjacent pond, will help remove the excess nutrients leached from the dead algal biomass (Erez *et al.*, 1990).

Chemical methods for the control of plankton biomass have been discussed earlier (p. 64).

Algal abundance and off-flavours

Another nuisance caused by excessive abundances of algae is off-flavours, which affect fish in some culture conditions and which are often linked to the presence of blue-green algae. It should, however, be pointed out that off-flavours are not only due to blue-greens, as some other taxa also produce them. The actual flavours originate from the decomposition of organic matter by various fungi; the chemical compounds responsible are geosmin and methylisoborneol. These flavours affect fish for several days to several weeks after their release to the water; this renders preventive measures difficult (Smith, 1988).

Phytoplankton life-strategies and behaviour

1. Phytoplankton species have developed life-strategies analogous to r- and K-selection. r-selected species colonize new environments, have high growth and reproductive capacities and are edible, while K-selected species are equilibrium species, with low growth rates, high biomass and lower edibility.

2. The pelagic environment, though seemingly homogeneous, is heterogeneous, made up of an assemblage of microzones or patches, with ephemeral, space-limited conditions. This explains the diversity of phytoplankton species present in water.

3. Algal exudations are releases of organic compounds, which constitute significant supplies of organic carbon to bacteria.

4. An algal bloom is the development of an algal biomass close to the carrying capacity of the environment. The algae prone to these blooms are often inedible.

5. Algal crashes follow the blooms and are a nuisance in fish ponds. The causes for such crashes are as yet undetermined. Possible causes are light or nutrient limitations.

6. Management of the phytoplankton biomass includes proper fertilization practices, and water circulation to avoid the appearance of anaerobic microzones and large-sized inedible phytoplankton species. After an algal crash, addition of green water may help to restore normal conditions and remove excess nutrients.

1.2.5. Influence of nutrient concentration

Primary productivity and nutrient concentration

Limnologists and aquaculturists have searched for relations between primary productivity and nutrient concentration in water (Strabskaba, 1980). An increase of nutrients in water enhances three functions of the aquatic ecosystem. They are: production, decomposition, and recycling of nutrients. However, none of these processes are sensitive indicators of ecosystem responses (Schindler, 1987). Nutrient concentrations are the sole factors on which action is possible, as the light régime cannot be modified and as the volume of autotrophic organisms depends on proliferation of an initial stock, and, thus, on prevailing limnological conditions.

Trophic loading of a water body

The amount of available nutrients determines the trophic load of a water body. According to trophic level, waters are oligotrophic, mesotrophic or eutrophic. In oligotrophic environments, flagellates and Chrysophyceae dominate; in mesotrophic environments, diatoms are predominant; in eutrophic waters, blue-green algae and dinoflagellates are dominant. In lakes of the north-west of U.S.S.R. (9 lakes of the Karelian Isthmus and 8 Latgalian Elevation lakes), algal biomasses less than 0.5 g m^{-3} (< 1.5 mg chl a m^{-3}) characterized oligotrophic lakes. Values of 1-5 g m^{-3} (1.5-10 mg chl a m^{-3}) were considered mesotrophic, 5-10 g m^{-3} (10-25 mg chl a m^{-3}) eutrophic, and > 10 g m^{-3} (> 25 mg chl a m^{-3}) highly eutrophic (Trifonova, 1989). Quite a few water bodies in tropical areas have been classified as oligotrophic on the basis of their nutrient concentrations. Considering the frequencies of algal blooms, net production rates, the extent of macrophyte development and extent of oxygen depletion with depth, it is clear that they are eutrophic, however.

Relations between nutrient concentration and requirements

Nutrient concentrations determine phytoplankton growth. Phytoplankton biomass increases as a function of light, temperature and nutrient supply. Nutrient-limited growth can be modeled using either the Droop model (internal nutrient concentration)

(Senft, 1978; Brown & Harris, 1978) or the Monod model (external nutrient concentrations) (see p. 137).

Sommer (1988) considers that there is common dissatisfaction with the Monod model as a descriptor of the nutritional status of natural phytoplankton. Unsteady state conditions in the field and intracellular storage permit temporal uncoupling between phytoplankton reproductive rates and ambiant nutrient concentrations. The cellular content reflects the concentrations in the water during the recent past and the volume of uptake during the same period (Heyman & Lundgren, 1988). Under such circumstances the intracellular concentration of limiting nutrients (cell quota) is considered to be a better indicator of the nutritional status than ambient concentrations. Ambient nutrient concentrations do not give a true picture of the recent past, and have only a remote relation with plankton biomass at the time of measurement.

In chemostat culture under limiting nutrients, algae belonging to different taxa (chlorophytes, chrysophytes, cyanophytes and dinoflagellates) were shown to have similar internal nutrient concentrations and physiological responses: N- and P-limitation occurred below 70 $\mu g\ mg^{-1}$(DW) N and 5-10 $\mu g\ mg^{-1}$ (DW) P (Healey, 1978; Hecky & Kilham, 1988: 800, Fig. 2).

However, in a natural assemblage *in situ*, internal nutrient concentrations are not a handy deficiency indicator. Also, at low concentrations (especially for phosphorus), dissolved nutrients are not accurately measurable. Yet, numerous plankton species are still capable of absorbing such analytically undetectable quantities.

In climates with little seasonal variation, like the tropics, phytoplankton biomass is high, despite low ambient concentrations. In these conditions, extremely rapid nutrient turnover times occur. A value for phosphate turnover of less than 10 minutes indicates high P deficiency, while values of days would be found when no P deficiency exists (Lean *et al.*, 1988). From studies involving many large African lakes, extending from L. Abhé, Ethiopia to L. Cheski, Zambia, Kilham & Kilham (1990) showed that recycling of nutrients is indeed high and rapid.

Uptake velocity of nutrients was assumed to provide an estimate of the degree of nutrient deficiency. Uptake kinetics, as an index of competitiveness between species, could, however, not be related to growth rates (Grover, 1989a).

In enclosure experiments conducted in Lake Okaro, New Zealand, Lean *et al.* (1989) evaluated nutrient deficiency against phytoplankton growth, and uptake rates in response to nitrate and phosphate additions. The uptake rates for phosphate, ammonium and nitrate normalized per unit particulate P or N were compared to that of C assimilation per unit C. The remarkable ability of phytoplankton to store phosphorus, as compared to nitrogen, was hereby illustrated (see also p. 108). Under extreme N deficiency, the maximum uptake velocity for ammonium and nitrate was only twice as high as when no N deficiency existed, while for phosphate, the velocity increased by about 100 times when the population became P-deficient (Lean *et al.*, 1989).

Algal growth rate is a function of internal concentration of the limiting nutrient, i.e. the one present in the cell in the smallest amount relative to the need of the alga (Droop, 1975). Synchronous multiple nutrient limitation has not been shown for any unialgal culture. This should not be expected, since macronutrients cannot substitute each other in biochemical functions. However, nutrients can compensate, within limits, for physical limitation, light and temperature, or vice-versa, hence colimitation by a physical factor and a nutrient (Hecky & Kilham, 1988) does occur. In multi-species chemostats, multiple nutrient limitation of phytoplankton growth has been demonstrated: species differ in their optimum nutrient ratios for growth. Different species dominate at varying Si:P ratios. At intermediate ratios, the species coexist since each was either limited by Si or P. Nutrient-loading ratios can have a strong selective effect on natural communities of phytoplankton, and can affect the biomass yield for the nutrient limiting most species.

In non-steady state (pulsed nutrients), even more species can be supported, as the kinetics of nutrient uptake is then uncoupled from growth and allows species that are limited by the same nutrient to persist. In a natural phytoplankton assemblage, one or more species are limited by a common nutrient (for which they compete), different nutrients, light, temperature, light or temperature and a nutrient, or any other combination.

In natural phytoplankton, the focus is on total biomass of phytoplankton rather than on one species. The relevant questions therefore are: can the growth of natural phytoplankton be increased significantly by the addition of one or more nutrients? Which nutrient has the greatest effect? These questions have focused interest on complex variables like primary productivity and algal biomass, and their response to nutrient enrichment (Hecky & Kilham, 1988).

Relations between intracellular and extracellular nutrient concentrations and limiting concentrations

Benthic algae in milkfish ponds in Taiwan contain on average 3.8 times more energy in fertilized ponds than in non-fertilized ponds. For blue-green algae, this ratio is 1 to 8 (Fong & Ju, 1987). The effect of nutrients is obvious when dealing with algal communities, which, previously deprived of nutrients, react to nutrient-enriched water. In biomanipulation experiments using plankton from three small lakes in Michigan, U.S.A., where nutrients are limiting, Elser & MacKay (1989) observed significant increases of growth and photosynthetic capacity following nutrient enrichment. Improvement of the physiological status of algae leads to an increase in herbivorous zooplankton biomass. Loss of algal biomass by grazing is thus compensated by an increase in growth rate, due to nutrient regeneration by animals.

Limiting nutrients, and essential differences between nitrogen, phosphorus and silicon cycling

When the cycling of the macronutrients is compared, essential differences appear (Table 21). Nitrogen and carbon cycles are open to the atmosphere (Smith *et al.*, 1989), which is a sink and a source. Phosphorus is immobilized by the sediments, which are a substantial sink, though under some conditions, they also can release phosphorus. The phosphorus cycle is open to the sediment. The main pool of nutrients associated with living organisms is actively exchanged between producers, consumers and decomposers and is maintained in the active portion of the cycle of the living community.

In a study of four small prairie lakes of different trophic levels in south-western Manitoba, central Canada, Barica (1990) found that nitrogen was supplied more easily, being fixed from the air when required; this is why more was wasted, being denitrified when in excess. The appearance of nitrogen-fixing algae was induced by a nitrogen deficit relative to phosphorus. The inputs of nitrogen into the environment by these nitrogen-fixing algae were, however, only temporary, as the accumulated nitrogen could not be stored. It was removed by denitrification, as soon as the algae perished (Barica, 1990). Phosphorus, which is rarer and more difficult of access, is quickly taken up, when present, as shown by a comparison with the uptake kinetics of nitrogen.

Silicon, supplied by run-off water, is a rather inert nutrient, of which a low number of chemical species occur (see p. 122). It is easily lost to sediments within sedimenting diatoms. It is limiting after a period of intense production only. The role of silica is considered, by some authors, as essential in determining the phytoplankton assemblage, mainly through the silicon-phosphorus proportion (Si:P ratio) (Tilman *et al.*, 1986) or by specific requirements of phytoplankton species for amounts available in the water (Hecky & Kilham, 1988).

Relations between phosphorus and nitrogen requirements

Nutrient requirements are usually assessed on the basis of the Redfield ratio. For the phytoplankton, the ratio for C:N:P is 106:16:1. When the ratio of N:P is below 29:1, nitrogen-fixing blue-green algae are favoured, while, above this value, blue-green algae are outcompeted by other algal groups.

In a study analyzing the nutritional status of phytoplankton in a shallow, hypertrophic, brackish, coastal lake in northern Germany (Großer Binnensee) involving stoichiometric analysis of seston and bio-assay enrichment experiments, Sommer (1989) showed that the most limiting nutrient was nitrogen, then phosphorus, and finally silicon.

In Colorado mountain lakes, Morris & Lewis (1988) showed that the dissolved inorganic nitrogen to total phosphorus ratio represented effectively those fractions of nitrogen and phosphorus really available for phytoplankton and, by using this ratio, possible nutrient limitations were detected. Barica (1990) compared different nitrogen to phosphorus ratios (particulate, total, total dissolved and inorganic dissolved N:P), to establish which forms are representative of systems and may be used to predict system evolution. The ratios were based on data collected over a period of several years in four

Table 21. Comparisons between phosphorus, nitrogen and silicon cycling

	Phosphorus	Nitrogen	Silicon
Intracellular stocks	high	no	low
Natural turn-over	high for the biogenic P, sometimes only a few minutes between uptake and excretion, but millions of years for the geochemical cycle	high, varying between tens to fractions of hours	generally low, but shorter when redissolution of the diatom shells occurs in the euphotic zone. The dissolution of silica is accelerated by the presence of salt.
Regeneration	high recycling, up to 70 % of the inputs derive from P mineralization	low recycling, losses around 50 % of the cycling pool	low regeneration rate
Uptake capacities	high	high	limited uptake, not flexible and restricted to cell division and cell wall formation
Supplies	from the soil geochemistry	from air (by N-fixation)	from run-off water (and upwelling in deep lakes)
Losses	to sediment	to the atmosphere	to the sediment

Central Canadian (south western Manitoba) lakes, and were representative of some 100 lakes of that region. The lakes were eutrophic to hypereutrophic. The total and particulate nitrogen to phosphorus ratios were least variable and closest to intracellular values; the total inorganic N:P ratio had the greatest amplitude of variation, and its use to characterize a water body was not convincing. The evolution of the N:P ratio with time, and with season, showed low values, around 5:1 and less; these, and not the average seasonal values, are most likely liable to lead to the appearance and stimulation of growth of nitrogen-fixing blue-green algae.

Chronic or temporary and changing limitations

The problem of nutrient limitation cannot be pinned down to one major nutrient; it is a subtle balance between varying supplies of different essential nutrients. Successive, opportunistic limitations occur, for the relative abundance of one nutrient renders the nutrient which is at the lowest concentration limiting. This is shown by experiments of Yusoff & McNabb (1987) in Malaysian fish ponds. They calculated ratios between inorganic nitrogen and orthophosphate, before and after fertilization (triple superphosphate (TSP) at 1.4 kg P ha^{-1} $week^{-1}$ and urea at 16.6 kg N ha^{-1} $week^{-1}$), and observed that the ratio shifted from conditions of limiting phosphorus (N:P of 36 in control and 44 in ponds fertilized with TSP and urea) to conditions of limiting nitrogen (N:P of 2 in ponds fertilized with urea only). The evolution of this ratio after fertilization demonstrates the frail balance between the correct amounts of nitrogen and phosphorus required to achieve optimal production.

In conditions of low concentrations of phosphorus

and nitrogen, the supplementation of phosphorus to water quickly entails a limitation by nitrogen. The amounts of phosphorus required are, however, more vital than those of nitrogen, and the effect of nitrogen supplementation is less apparent. Chemostat cultures of *Scenedesmus sp.* with nitrogen and phosphorus additions in a ratio of N:P of 5:1 to 80:1 show successive limitation by nitrogen, when the ratio is below 30:1, and by phosphorus when above 30:1. At the same time, cellular nitrogen content remains constant until the nitrogen-phosphorus ratio reaches an optimum value, and increases proportionally (linearly) afterwards. Cellular phosphorus content remains high at low N:P ratios, when nitrogen is limiting, but decreases quickly when the N:P ratio reaches an optimum. At high N:P ratios, intracellular phosphorus remains low (Rhee, 1978).

Ponds react to changes in phosphorus-nitrogen-carbon ratio by a shift in the phytoplankton species present: an increase in phosphorus supply leads to a community of nitrogen-fixing blue-green algae, while a decrease of carbon dioxide favours algae adapted to low concentrations of carbonate (Schindler, 1987). A simultaneous nitrogen and phosphorus limitation rarely occurs, like in Lake Okaro, New Zealand (White *et al.*, 1989). Additions of the most limiting nutrient lead to only slight modifications in algal growth. The prevailing phytoplankton community is the result of inter-specific competition between species for nutrients. It cannot exactely be predicted in the actual state of our knowledge (see p. 157). Nutrient availability depends on supply, which, in turn, depends on recycling processes, specific utilisation of nutrients, and transfer across trophic levels (Elser *et al.*, 1988).

Control model for carbon, nitrogen and phosphorus

In the controversy whether nitrogen or phosphorus are the prime control of organic production, Smith *et al.* (1989) developed a conceptual model, on evidence gathered in Tomales Bay, California, U.S.A., according to which carbon cycling controls nitrogen cycling, but is itself controlled by phosphorus cycling. It is based on the assumption that assimilation and release of nitrogen and phosphorus follow closely carbon cycling, i.e. production (C fixation) and respiration (C oxidation). In most shallow ecosystems, there is a balance between primary production and respiration, and the sum of carbon fluxes of production and respiratory oxidation tends toward zero. The difference between these fluxes is expressed in nitrogen flux units (Fig. 46). Excess respiration or surplus production is compensated by denitrification or nitrogen fixation. This capacity of adjusting the nitrogen balance is potentially present in most systems, though the actual rates of denitrification or N-fixation may be low. The trophic relations between nitrogen-fixing blue-greens, releasing carbon exudates to bacteria, and bacteria providing phosphorus to algae through phosphate exchange can be viewed as a coupling of the cycling of the three nutrients (Bloem & Bär-Gillisen, 1989).

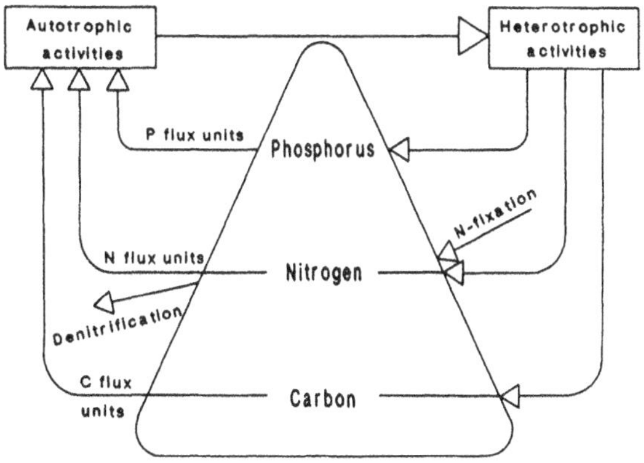

Fig. 46. Schematic representation of the relations between autotrophic and heterotrophic activities and the control of autotrophic activities by the macronutrients through regeneration processes.

In autotrophic systems, net assimilation of dissolved inorganic phosphorus is possible, as long as there is some source of nitrogen. Primary production is limited by biomass (standing crop) and specific growth rate, which in turn is limited by nutrient availability. As long as there is a phosphorus supply, nitrogen-fixing will supply the required amounts of nitrogen for the elaboration of organic matter. Many communities show high primary production, though dissolved inorganic nitrogen concentrations in water may be low. This high primary production results from high rates of nitrogen fixation, which meet the nitrogen needs. Phytoplankton production is significantly correlated to autotroph biomass, but there is only a low correlation with dissolved inorganic nitrogen; this suggests that biomass is ultimately more important than growth rate in the regulation of primary production.

A supply of dissolved inorganic nitrogen will have an immediate stimulatory effect on primary production, provided phosphorus is available. However, along with stimulation of primary production goes an almost proportional increase of respiration; dissolved inorganic nitrogen is released, which stimulates denitrification, bringing the levels of dissolved inorganic nitrogen down to former levels. Hence, inorganic nitrogen loading by a point source enhances production locally and temporarily, but there is no long-term or large-scale effect.

As organic matter is oxidized, nutrients, i.e. phosphorus, nitrogen and carbon, are recycled, thus stimulating primary production. The fate of phosphorus, immobilized in sediment or by primary production, determines that of nitrogen. A nitrogen surplus in relation to phosphorus availability will be denitrified, while a nitrogen deficit relative to available phosphorus is eliminated by nitrogen fixation.

If light supply to a system which relies on autotrophic production increases, autotrophic biomass will increase only as long as phosphorus is available. If nitrogen is required, nitrogen fixation will be stimulated. If available phosphorus is limiting, an increase in organic production can be achieved by a faster recycling and regeneration, but without a change in net production.

Only situations in which phosphorus is supplied externally to the system appear to have long-term impacts on biomass. Systems, with large supplies of organic matter, have an average net heterotrophic metabolism. In these systems, benthic regeneration of organic matter here results in an increase of the relative importance of denitrification, causing a nitrogen-limited growth rate. However, the primary production of such a system remains high, because of the high autotrophic biomass, controlled by phosphorus, regenerated from organic matter.

The control of the production system by nitrogen or phosphorus can be explained by a change in nutrients available or in the N:P ratio. There is a shift from phosphorus to nitrogen limited growth (Smith *et al.*, 1989).

Reaction of the environment to nutrient limitation

Another way to approach the problem of nutrient limitation is to consider how the environment copes with the available amounts of nutrients, or which is the phytoplankton assemblage prevailing under the range of nitrogen and phosphorus fluxes present. A phytoplankton community is composed of a number of species, whose nutrient requirements correspond to the range of nutrients available in the environment. This can be shown by enrichment experiments in limnocorrals with fertilizer applications of various nitrogen-phosphorus ratios. In lake Okaro, New Zealand, the phytoplankton assemblages displayed major changes depending on the nature of fertilizer applications, involving appearance and disappearance of various taxonomic groups (Pick, 1989). Some algae in two lakes (Kennedy and Sproat lakes, Vancouver Island) of British Columbia, Canada, were thriving under limiting nitrogen conditions, while others sustained limiting phosphorus conditions (Suttle & Harrison, 1988). As in Kennedy lake, fertilizer application gave rise to a low N:P ratio, nitrogen limitation promoted the appearance of nitrogen-fixing blue-green algae (Stockner & Shortreed, 1988).

Specific nutrient requirements of algae have been demonstrated by Blomqvist *et al.* (1989), in enclosure experiments in the acidified lake Njuffatet, central Sweden. They experimentally modified the inorganic nitrogen sources and, irrespective of the dominant species at the start of the experiment, the addition of ammonium to the culture medium lead to the dominance of *Merismopedia*, while addition of nitrate lead to dominance of *Peridinium*. Algae will therefore only grow in an environment in which their specific physiological needs are met.

Changes in the supply ratio of silicon and phosphorus also affects the algal community. In semi-continuous cultures of a natural algal community (originated from a Minnesota lake), at Si:P ratios below 20:1, green algae (Chlorophytes) were dominant, while at ratios of around 80:1, diatoms became abundant, and at ratios of around 160:1, siliceous Chrysophytes (silicoflagellates of the genus *Mallomonas*) took over (Grover, 1989b). Diatoms were capable of competitive differentiations, by reducing the size of their cells, when the Si:P ratio decreased. A greater variety of species was associated with variability in phosphorus supply (Grover,

1989b).

Zooplankton affects phosphorus sedimentation; the relation between phosphorus cycling in water and the abundance of zooplankton is discussed on p. 164.

Ways of determining nutrient limitations

Three approaches have been used, with a view to understanding the nutrient requirements of plankton: the first attempts to establish direct correlations between nutrient concentrations and phytoplankton biomasses. Seymour (1980) applied multiple regression (discriminant analysis), to the relation between nutrient concentrations and plankton biomass in a fish pond in the South of France, and obtained significant negative correlations between phytoplankton volume and ammonium and organic nitrogen concentrations in the course of one production season, and between phytoplankton volume and nitrite, nitrate (highly significant) and orthophosphate concentrations, during the following production season. The negative correlation was due to the inverse relation between nutrient abundance and the development of a plankton bloom. At the die-off of the bloom, an important amount of nutrients was released. Jana & De (1983) similarly established a correlation matrix between the physical and chemical variables of a polyculture fish pond in West Bengal, India, and its gross primary productivity. They concluded that productivity was positively correlated with water temperature, and with free carbon dioxide, dissolved oxygen, ammonium, nitrite, nitrate and phosphate concentrations.

A second way to approach the problem is to determine intracellular nutrient concentrations and to compare these with the critical content. These studies are, however, not possible with field populations.

The third type of studies attempts to understand algal nutrient requirements by means of selective enrichment bio-assays, examining phytoplankton responses (especially growth), to additions of one or more nutrients (Sanders *et al.*, 1987). Growth may be an increase in biomass, in the number of cells, in chlorophyll a content, or in particulate nitrogen (White *et al.*, 1989).

Comparative results of these three approaches seem contradictory (Gerhart & Likens, 1975). This is due to differences between field and laboratory studies. Studies of nutrient enrichment have the best potential to identify the limiting nutrient. In the field, there is probably not just one limiting factor, but rather a combination of nutrients, whose concentrations are close to limiting thresholds.

Most mathematical models developed until now consider light, rather than nutrient concentration, as the principal determining factor.

Reliability of predictions on the appearance of algal groups

As indicated previously, predicting the appearance of a given group of phytoplankton, following enrichment experiments with fertilizers, is still unreliable. In simultaneous, comparative trials in adjacent ponds, different phytoplankton associations may appear. For example, in two ponds stocked with the same fish (silver carp, common carp, bighead carp, grass carp, hybrid buffalo) and with similar fertilization (pig dung at 39 and 66 hogs ha^{-1}), different phytoplankton taxa evolved differently. In the first pond (39 hogs ha^{-1}), Euglenophyta were dominant one month after stocking, followed by Chlorophyta the month after, then by Chrysophyta the month after and by blue-green algae a month later. In the other pond (66 hogs ha^{-1}), Euglenophyta were dominant at the same time as in the first pond, but Chlorophyta took over faster, and remained dominant during all subsequent months, while blue-green algae were never present in sizeable amounts. The total amount of phytoplankton in the samples varied over a 10-fold range (Buck *et al.*, 1978). The factors which regulate such succession dynamics are not well understood; the successions observed are probably the result of different nutrient requirements on the part of the various phytoplankton species, and of opportunistic exploitation by the species present. Limnological differences at a micro-environmental scale certainly exists, but are not easily accessible to investigation. These may prove to be major determining factors in the evolution of phytoplankton assemblages.

Seasonal succession is probably less a species replacement than the regular appearance of a sequence of species over the year, related to changes in the environment, which are repetitive (Evans, 1988).

Plankton group-succession or temporal modification of plankton assemblage structure

For a long time, limnologists have been aware of successions in dominance of plankton groups in the course of a year. In temperate waters, this succession has been described by a model (the PEG-model), based on 24 statements about the sequential events which occur in the course of the season (Sommer *et al.*, 1986). Succession is dictated by seasonal imperatives, linked to modifications of the régimes of temperature, nutrient concentration, and of grazing pressure (Vanni & Temte, 1990).

At the end of winter, when temperature becomes favourable, a high availability of nutrients leads to an abundance of small, edible diatoms, with high growth rates. Diatoms require silicon, which is abundant at that timeand of which they exhaust the reserves. In a microcosm experiments on plankton from Schöhsee, northern Germany, nutrient recycling by zooplankton returned proportionally more phosphorus than silicon to the environment and was disadvantageous for diatoms and benefitted green algae (Sommer, 1988). Following this change, herbivores with short generation times appear, for instance rotifers, followed by cladocerans, whose abundance increases quickly and which graze intensely on the algae present. These species are r-selected (Komárková, 1989, observations in the moderately eutrophic Řimov reservoir, Czechoslovakia).

Grazing quickly exceeds algal reproduction, and results in a decrease in the number of algal cells, followed quickly by a decline of the grazers. This period is designated as the clearwater phase. Reduced grazing pressure and increasing recycled nutrients lead to a restoration of phytoplankton biomass, the amount of which is inferior to the initial colonial biomass, but richer in species and functionally more diversified. The species, mostly green algae, are now more resistant to grazing, being larger in size or colonial, or protected by gelatinous sheaths or thickened cellular walls. "Canopy species" appear. These summer species are K-selected (Komárková, 1989).

Differential recycling of phosphorus and nitrogen leads to an imbalance in nutrient supplies. A nitrogen depletion promotes the appearance of nitrogen-fixing blue-green algae, at a period when mixing depth is reduced. Large-sized herbivorous zooplankters are replaced by small-sized species, less vulnerable to predation by fish. Autogenic succession is mainly determined by the mixing depth.

The maintenance of a phytoplankton population is dependent on two prime factors: solar light, as the energy source for photosynthesis, and nutrient concentrations in the water. Nutrient supplies during the year come from regeneration and from microbial activities, which are at maximum at a certain depth below the main phytoplankton biomass.

Algae which sink and sediment are colonized by decomposers and release nutrients there. Some algae migrate vertically to gain access to such nutrients, while others depend on physical mixing to carry nutrients upwards from deeper water (Gawler *et al.*, 1988, on Lake Geneva, Switzerland; Vincent & Dryden, 1989, on lake Okaro, New Zealand). The reduction of the amount of light in autumn restricts photosynthetic activities, which limits algal volume.

The population at any given moment reflects a balance between instantaneous conditions, the recent past, and the phytoplankton species best adapted to these conditions, species being selected from a spectrum of available types (Houk, 1989, on Zelivka reservoir, Czechoslovakia).

In marine environments (Mediterranean and Caribbean Sea), phytoplankton succession has been divided into three main stages in relation to changes in the environment (Smayda, 1980). The first stage is one of nutrient rich water, with algal cells characterized by a high ratio of cell surface area to cell volume, i.e. high nutrient uptake capacities, high rates of potential increase, with 1 to 2 cell divisions per day, and individual population densities reaching 100 to 1000 cells per ml. Such algae belong to the diatoms, various flagellates and Chrysophytes. The second stage is a more mixed community, with larger cell sizes, lower growth rates, cell divisions occurring every second day or even less, and population densities of 20 to 100 cells per ml. These algae are diatoms and dinoflagellates. The third stage appears when nutrient concentrations have reached a minimum, and the water is well stratified. The algae now have lower cell surface area to cell volume ratios, and population densities are less than 10 cells per ml. The most common species are nitrogen-fixing blue-greens.

The evolution of plankton communities in fish ponds grossly follows the "marine" pattern. There is considerable availability of nutrients at the start of production, soon after impoundment (Fig. 47). Nutrients are released from the sediments and by the fertilizers applied at the beginning of production. Ponds are usually drained at harvest time and left dry for some time. During this time, accumulated organic matter decomposes and the resulting nutrients readily dissolve, when the pond is impounded again (see also p. 24).

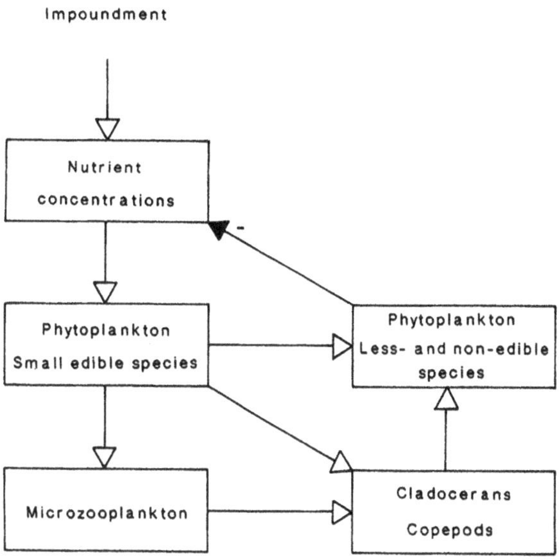

Fig. 47. Schematic representation of the sequence of events occurring within a pond after impoundment. The black arrow indicates a feed-back effect.

Indirect evidence of the relation between nutrient concentrations and primary production was provided by Legendre *et al.* (1987), who examined the evolution of plankton populations and biomass in coastal fish ponds in Côte d'Ivoire. During the first days after impoundment, preceded by a drying period and by liming, they noticed a considerable increase of nutrients in the water. These nutrients stimulated the growth of a large phytoplankton population, leading to the appearance, after 8 to 10 days, of a zooplankton population, mainly composed of rotifers, reaching 34000 ind. l^{-1} or 2.5 g (DW) m^{-3}. The phytoplankton biomass varied widely (up to 15 g C m^{-3}) during the initial phase, but stabilized after a period of 20 to 30 days at a level less than the initial one (around 3 g C m^{-3}). This decrease was due to a drop in average concentration of nutrients, especially ammonium, and also to zooplankton grazing. A second zooplankton population (700 ind. l^{-1} or 1.4 g (DW) m^{-3}) appeared after 13 to 20 days, consisting of crustaceans. The decrease of the phytoplankton was accompanied by an increase in bacterial biomass. The zooplankton biomass stabilized after 25 to 30 days around 0.3 g (DW) m^{-3}.

Spatial heterogeneity of nutrient concentrations

In all observations of physical and chemical variables in water, measurements are made on samples, but results are considered to be representative of the whole lake or pond. However, within a water mass, there is a continuum of conditions, with microzones of varying nutrient concentrations. Samples collected less than 2 m and few seconds apart can show differences in dissolved inorganic nitrogen concentrations of three orders of magnitude. The higher the volume of a water sample, the more such local variations are masked (Smith, 1986).

The development and proliferation of a given phytoplankton species was earlier explained by the presence of such micropatches. This implies that there is no hope to explain the relationship between major nutrient concentrations in water and observed primary production, by conventional sampling procedures.

The presence of a heterogeneous nutrient distribution, resulting from zooplankton excretion and distribution, and, in particular, its effect on phytoplankton growth is another controversial subject (Currie, 1984; Lehman & Scavia, 1984). Still another form of heterogeneity comes from pulsed nutrient supplies, varying in the course of time. Trials of phytoplankton cultured in conditions of pulsed supplies have resulted in the coexistence of various species, competing for the same resource (Sommer, 1984).

In attempts at mapping horizontal, vertical and diel variations of particulate phosphorus, carbon and nitrogen concentrations in ponds in Oahu, Hawaii, Wei & Laws (1989) suggested mechanisms which modify the distributions. About 75 % of the variance of a recorded value of particulate carbon and nitrogen was explained by position within a pond; for dissolved reactive and total dissolved phosphorus, position explained 35 % of the variance, and for particulate N:C ratio 7 %.

Horizontal variability was reduced in conditions of vertical homogeneous mixing (isothermy); determination of temperature difference between surface and bottom of a pond was recommended as an index of mixing conditions. During conditions of homogeneous mixing, depth explained not more than 12 % of the variance among measurements. Time of day was responsible for more than 50 % of total variance; this was ascribed to active migrations and assimilation of plankton, which modified nutrient concentrations in solution. The increase of particulate carbon at the surface between morning and afternoon was the result of photosynthesis. Simultaneously, nutrient concentrations decreased, as a consequence of uptake by phytoplankton.

Total dissolved phosphorus was at a maximum in the morning, and decreased about 7 % in the afternoon, but dissolved reactive phosphorus did not show such a change.

Wind had an appreciable effect on horizontal distribution (see p. 54). Microscale variability of particulate material in water at the bottom of the pond was due to patchy distribution of microzones with intense degradation or to activity of benthic macroinvertebrates. Spatial and temporal variations are not easily distinguished, and the processes have phases of different amplitudes and frequencies. Small scale, high frequency processes sometimes hide those of large scale and low frequency.

In lake Rotongaio, New Zealand, spatial heterogeneity was considered the rule in planktonic life and was the result of the interaction of various factors, including water movement induced by wind action, and zooplankton grazing (Horne & Commins, 1989a). This heterogeneity was temporally unstable and generally lasted from few hours to one day. It was related to the distribution and vertical migration of algae. These migrations were induced by photosynthesis and depended on carbohydrate ballast of the cells. Algae in bad physiological condition did not migrate vertically (Horne & Commins, 1989b).

Temporal heterogeneity of nutrient concentrations

Uptake of nutrients occurs rapidly. This has been shown, for example, by homogeneity tests made near a point source supplying water to a marine fish pond in Israel. At 1 m beyond the water supply, the concentration of nutrients in the water was similar to that of the pond and considerably different of that of the water supply. This homogenization could only result from immediate uptake of nutrients by phytoplankton (Krom *et al.*, 1989a).

Control of excess nutrients

The control of excess nutrients, deleterious to water quality, like in sewage, must also be mentioned. In Israel, Avnimelech & Lacher (1979) studied the recycling of nutrients from surplus feed, which accumulates on pond bottoms. They observed a bottom accumulation of 25 %, 75 % and 80 % of organic carbon, nitrogen and phosphorus respectively, the origin of which was the feed, either directly or indirectly after transformation by algae. When they established the budgets of various nutrients, they found that carbon was lost in the process, through release of carbon dioxide. Small amounts of nitrogen were also lost, but phosphorus was gained, probably through resuspension of particulate material from the sediment. Deterioration of water quality brings about environmental stress. Organisms with high reproductive rates and high dispersion capacities, such as some phytoplankton species, respond quickly to this stress. The most vulnerable species disappear first. Among animals, benthic invertebrates are particularly affected, because accumulation of polluting and harmful substances is most severe in sediments. Anatomical malformations amongst invertebrates are indicators of pollution (Schindler, 1987).

Inorganic nitrogen in excess can be controlled, for instance, by production of unicellular algae (single cell protein), themselves consumed by fish. However, this production needs the presence of other nutrients in adequate amounts, especially inorganic carbon (Avnimelech *et al.*, 1989). In experiments conducted at an aquaculture farm in Singapore, the macrophyte *Elodea densa* removed nitrogen compounds from water and reduced turbidity. A total amount of 25 % of the inorganic nitrogen was found to be removed, 26 % being nitrate, 50 % being nitrite and 17 % ammonium, i.e. 61.36 mg, 2.27 mg and 9.09 mg of N m^{-2} d^{-1} of nitrate, nitrite and ammonium, respectively. At the same time, total alkalinity and total organic carbon increased by 5 %, while dissolved phosphate increased by almost 300 %. The phosphate released originated from the sediment (Ng *et al.*, 1990). The water hyacinth (*Eichornia crassipes*) is a macrophyte which can also be used for the same purpose (Liu & Seki, 1988).

Influence of nutrient concentrations

1. Primary production depends heavily on nutrient availability, but the relation between primary production and nutrient concentrations is complex. Nutrient fluxes and intracellular nutrient contents are more important than the actual concentrations. Algal growth depends on specific uptake capacities of the algal assemblage present.

2. Phosphorus uptake capacities of algae are usually larger than nitrogen uptake capacities and vary to a wider extent between deficiency and normal conditions.

3. Differences in the cycling of nutrients are apparent: phosphorus is readily stored and kept in active cycling, while nitrogen in excess of the immediate needs is wasted by denitrification. Silicon is not recycled and is progressively lost to the sediment.

4. Nutrients control primary production sequentially according to the following model: phosphorus controls carbon, which, in turn, controls nitrogen. Nitrogen concentrations in the environment are adjusted by nitrogen-fixation when in deficit or denitrification when in excess. Phosphorus determines long term modifications of biomass, as it can be stocked; supplies result in a gradual increase in biomass. Nitrogen cannot be stored and any supply immediately increases respiratory activities.

5. Phytoplankton adapts to prevailing N:P ratios in the water through changes in species assemblage, which promote species with specific nutrient requirements. Minimum N:P ratios appearing in the course of time may be more important in governing the evolution of the assemblage than average ratios.

6. Phytoplankton succession is essentially an adaptation of the plankton community to changing water quality. Appearance and disappearance of algae depends on major modifications in nutrient availabilities. The zooplankton community changes accordingly. The phases of the succession which are of most interest to the fish farmer are those which involve a large growth of small edible algae, in a nutrient-rich, changing environment.

7. The major obstacle to an understanding of the dynamics of nutrient cycling in water is the heterogeneity, at a microenvironmental scale, of nutrient distribution in water. Furthermore, this patchiness in distribution is often ephemeral.

1.2.6. Influence of zooplankton

Zooplankton composition

Zooplankton is subdivided in microzooplankton, composed of flagellates, ciliates and rotifers, and macrozooplankton, which consists of planktonic crustaceans, cladocerans, copepods, and less frequent groups, such as typhloplanid flatworms, Acari, and Medusae, and insect larvae (mainly *Chaoborus*).

Microzooplankters are mostly detritivores and bacterivores. They develop abundantly when water is loaded with suspended organic matter, and are consumed by rotifers and small crustaceans (Hul, 1987).

Most but not all cladocerans (also classified as mesozooplankton) are herbivores. In fact they are the most important group of herbivorous crustaceans. Small- and large-bodied species are found, all promptly consumed by planktivorous fish. They are sensitive to predatory pressure by planktivorous fish. As they are numerous, of large size and easily cultured, they have been widely studied and used in biomanipulation experiments. Their relations with phytoplankton are complex and not completely unraveled, which makes predictions about them difficult (McQueen & Post, 1988).

The food of cladocerans is not restricted to planktonic algae. They also consume bacteria, fungi, detritus, organic aggregates, and sometimes even ciliates and rotifers (Burns & Gilbert, 1986).

Cladocerans select their food on the basis of size and by chemoreception. However, great differences in the feeding of even closely related species exist (Matveev & Balseiro, 1990), some being selective, some not (Fulton, 1988).

Copepods (classified as macrozooplankton *sensu stricto)* are subdivided in calanoids, cyclopoids and harpacticoids and comprise species with various feeding regimes, often modified in the course of development. Calanoids are microphages, herbivores or predators. Cyclopoids are omnivorous, herbivorous, or carnivorous. Fish consume cyclopoids (late-stage copepodites and adults), though not so eagerly as other prey. All life stages of harpacticoids are detritivorous or herbivorous (Dussart, 1980).

Copepods in general and calanoids in particular are less preyed upon by fish, as they have better swimming, hence evasion, capabilities than Cladocera (O'Brien, 1987). Cyclopoids have an erratic jumping motion, which makes then more conspicous than the gliding motion of calanoids. The calanoid/cyclopoid ratio is considered to give a reflection of the predation pressure by visual planktivorous fish. The ratio decreases with increasing predation pressure (Hrbáček, 1962; Brooks & Dodson, 1965).

Swimming behaviour of copepods has evolved in close relation to their feeding behaviour. Carnivorous copepods possess mechanoreceptors enabling them to detect vibrations and water displacements created by swimming prey (Strickler, 1970 & 1975; Kerfoot *et al.*, 1980), while herbivorous copepods possess chemoreceptors enabling them to discriminate among different algal prey (Stemberger & Gilbert, 1987). They are considered to be selective feeders (Lampert, 1987).

Zooplankton distribution within ponds

Zooplankton distribution is non-homogeneous. In Dutch fish ponds, some species were mainly found in the littoral, while others selected limnetic waters (Verreth, 1990). Such distributions are probably related to food availability and avoidance of predators.

Distribution varies with time, due, in particular, to vertical migration, which is an almost universal phenomenon. The archetypical form of vertical migration is a swimming movement to deep water during the day, and an active ascent towards the surface during the night. However, there are so many exceptions, including pronounced reverse migration, that generalizations are hazardous. This diel movement from the warm, food-rich surface layers to the cold nutritionally poor deep waters has evolved in response to benefits to the community: (1) vertical migration results in predator avoidance by zooplankton; (2) fluctuations of the grazing pressure owing to vertical migration enables unimpeded growth of the algae during daytime. This results in higher rates of algal production, compared to those obtained in conditions where zooplankton graze continously on the same algal biomass. The disavantages of migration for the zooplankton are: (1) a reduction of energy intake; (2) an increase of energy expenditure required for swimming up and down; (3) a reduction of the rate of development of the eggs carried by migrating animals (Lampert, 1987).

For cladocerans, vertical migration is a compromise between avoiding predatory fish, by staying in deeper water, and the need to feed at the surface. When the amount of available food is insufficient, the degree of migration is reduced. The animals stay longer at the surface and use more time to satisfy their food requirements, defying predation (Gliwicz, 1985; Johnsen & Jakobsen, 1987). The deep waters act as a refuge for zooplankton: low visibility impedes visual prey pursuit and prevailing low oxygen concentrations are avoided by predatory fish. In lake Rotangaio, New Zealand, some zooplankters even survived in deeper water at dissolved oxygen concentrations of less than 0.2 mg l^{-1} (Forsyth *et al.*, 1990).

Influences of zooplankton on the plankton community

Zooplankton is a link between lower trophic levels and fish. Interactions between different plankton groups have been investigated and attempts have been made to establish a general relationship between zooplankton community structure, especially the relative importance of microzooplankton, and the trophic level of water bodies. In a study involving 12 sites in Quebec, Canada, significant relationships between ciliates, rotifers, cladocerans and cyclopoids, but not calanoids, and a range of trophic levels of lakes was shown. Also a clear relationship between microzooplankton and macrozooplankton biomass versus total phosphorus was established. The relative microzooplankton biomass increased with increasing trophy (Pace, 1986).

Though quantification of the relationships between different plankton groups and nutrient availability is the ultimate aim of the ecologist and the fish farmer, qualitative interactions are the first to be investigated. The interactions between species in a community are direct or indirect, involving trophic linkage, behavioural, and chemical responses (Miller & Kerfoot, 1987). Effects of zooplankton on the plankton community are as follows:

a) Grazing

Zooplankton influences phytoplankton community structure, through grazing. Biomanipulation experiments with zooplankton in limnocorrals have clarified the role of zooplankton in ecosystems. In an oligotrophic Swiss lake, the removal of large-bodied zooplankters (above 95 μm in size), consisting mainly of cladocerans *(Daphnia)*, lead to the conclusion that herbivorous zooplankton reduced, by grazing, algal biomass, but did not affect the size structure of the phytoplankton and the relative percentage of size classes contributing to total primary production (Uehlinger & Bloesch, 1987). Similar experiments in meso-oligotrophic Lake Lucerne, Switzerland, where removal of zooplankton (with a plankton net of 95 μm net size) was followed by fertilization with phosphorus, showed an increase in chlorophyll a, organic particulate carbon and particulate phosphorus (in fact, of total algal biomass). The proportion of nanoplankton decreased to the benefit of larger cells and the settling rate increased. This was, however, not attributed only to selective grazing, but also to differences in algal strategies (Bloesch & Bürger, 1989).

In enclosures in a eutrophic farm pond and an oligo-mesotrophic lake in Illinois, the impact of Cladocera on phytoplankton depended on the body size of the grazer. The replacement of small-sized species of cladocerans by large-bodied species was found to reduce the abundance of phytoplankton by one order of magnitude, and also decreased the density of herbivorous cyclopoids and rotifers (Vanni, 1986).

In enclosure experiments in the moderately eutrophic lake Ros (northern Poland), the impact of large-sized zooplankton, such as *Daphnia magna*, and zooplankton of normal size, on phytoplankton were different: the species of large size controlled the density of phytoplankton of all size classes, while the zooplankters of normal size had no effect on net plankton (above 50 μm). Algae of large size increased, because they were subjected to less competition from small sized algae, in conditions of limiting phosphorus concentrations. The efficiency of control of phytoplankton density by large cladocerans decreased with increasing dominance of phytoplankton of large size (Dawidowicz, 1990b).

The impact of large-bodied cladocerans is sometimes so important that it drastically changes the structure of the algal community. For example, in fish ponds in Regensburg, Germany, the introduction of *Daphnia magna* resulted in disappearance of the phytoplankton and appearance of a population of *Elodea canadensis* (Theiss *et al.*, 1990).

In the moderately eutrophic Schöhsee, northern Germany, the clearwater phase, which appeared in the course of phytoplankton succession in temperate waters, and which was characterized by an important increase in Secchi disc transparency, was due, at least partly, to Cladocera and other herbivores grazing on the phytoplankton (Lampert *et al.*, 1986).

b) Phytoplankton defenses against grazing (see also p. 145 & p.150)

In laboratory experiments, large-bodied cladocerans have been found to control the proliferation of filamentous algae. These observations cannot, however, be transposed to nature: here, the stocking of such cladocerans does not prevent the appearance of algal blooms; the animals may even disappear when algal density becomes high, probably because the Cladocera cannot reproduce sufficiently fast (Gliwicz, 1990).

Observations on high and low densities of *Daphnia pulex* and of rotifers, respectively, during a bloom of the filamentous alga, *Anabaena affinis*, were followed by trials on interactions of various species of cladocerans and rotifers with filamentous blue-green algae. The presence of filamentous blue-green algae influenced cladoceran populations directly and this effect was proportional to algal density. Rotifers, however, were not affected. The sensitivity of cladocerans varied with species and with size of the individuals (Gilbert, 1990).

Similarly blooms of *Microcystis aeruginosa*

negatively influence large-bodied cladocerans; feeding is aselective in large cladoceran individuals, so that toxic, non-edible algae or algae without nutritive value are readily ingested (Fulton & Paerl, 1987).

Another reason why cladocerans do not easily coexist with filamentous blue-green algae is due to mechanical interference by the filaments with the feeding appendages. Their thoracic trunk limbs are easily clogged by algal filaments. Blue-green algae also produce endotoxins (Gilbert, 1990). All species do not have the same sensitivity these toxins; *Bosmina* is less. sensitive than *Daphnia* (Burns *et al.*, 1989). By fragmenting non-toxic blue-green algal filaments into pieces, large-sized species of cladoceran, make these edible for smaller cladocerans (Dawidowicz, 1990a).

c) Changes in sedimentation rate caused by zooplankton (see also p. 144)

Zooplankton modifies the amount of phytoplankton lost via sedimentation, by changing the average size of the algal population. Grazing results in the appearance of algae of large size, which are less edible and sink more quickly. Grazing by Cladocera transforms small algae into a slurry of partially digested gut fluids that dissipate rapidly. Particulate matter in cladoceran faeces is likely smaller on average than the food particles, with sinking characteristics comparable to those of small moribund algae (Kitchell & Carpenter, 1987). Observations in limnocorrals showed that small algae (> 12 μm) contributed to 80 % of primary production. Removal of the large-sized herbivores from the limnocorral increased sedimentation rate of particulate phosphorus and particulate organic carbon. From this, Uehlinger & Bloesch (1987) concluded that small-sized algae contribute in a more important proportion to sedimentation when consumed than when unconsumed. The portion of primary production ingested by zooplankton follows two paths: part is incorporated in the food web, being transformed into zooplankton biomass, while the rest becomes excrement, which settles and/or is decomposed. The supply of particulate matter to the seston pool thus generated is important, as comparatively more material is rejected than assimilated (Gulati, 1990).

d) Zooplankton effect on nutrient cycling and supply to phytoplankton

By feeding on algae, zooplankton recycles part of the nutrients (P and C) contained in the algal food, and reduces sedimentation. (1) Herbivores reject in their excrements about half the phosphorus contained in the algae ingested. Dominance of large-bodied herbivores in a Polish lake reduced the recycling of phosphorus contained within phytoplankton and zooplankton biomass and detritus (Esjmont-Karabin *et al.*, 1989). In the absence of zooplankton, there is no orthophosphate excretion, the inorganic phosphorus pool decreases, and the abundance of algae decreases. (2) Detritivorous zooplankton has a consumption rate of detritus similar to the consumption rate of algae by herbivores. Therefore, the cycling of phosphorus within the trophic foodweb of the water column is always maintained (Esjmont-Karabin *et al.*, 1989).

The specific release rate of phosphorus by cladocerans increased as the P:C ratio of the food increased. For a ratio of 6 to 8 μg P mg^{-1} C, the release rate of phosphorus was nil. At that concentration, all ingested phosphorus was used for growth and reproduction. In eutrophic lake Nesjøvatn, central Norway, phosphorus limitation occurred when the P:C ratio of phosphorus-limited algal cells and of detritus was well below this value (Olsen *et al.*, 1986).

Comparisons between 49 lakes in north-east Poland have shown that the phosphorus regeneration rate by zooplankton increases when phosphorus concentration increases, in such a way that the turnover time of phosphorus is maintained fairly constant, even if the total phosphorus pool is variable. The uptake of phosphorus by algae is directly affected by the supply rate of available phosphorus, zooplankton regeneration rate being more important than total phosphorus concentration. The rate of phosphorus regeneration determines the abundance of algae; the specific composition of the assemblage is more related to specific characteristics of the species present. In some cases, this regeneration favours blue-green algae, partly because they are not consumed (Esjmont-Karabin & Spodniewska, 1990).

While zooplankton graze on phytoplankton, exudations of organic carbon are released into the water (sloppy feeding). This algal carbon is put at the disposal of the bacteria and stimulates their growth (Güde, 1988, on Lake Constance, Switzerland).

e) Predation on zooplankton

The zooplankton community of Mondsee, Austria, was affected in its structure by predation of invertebrates and fish (Dokulil *et al.*, 1990). These aspects will be discussed in paragraph 1.2.8., Influence of fish production.

Influence of macrozooplankton on microzooplankton

In a small lake of the Spanish Sierra Nevada, the removal of zooplankton above 45 μm in size from limnocorrals has been found to result in the disappearance of flagellates and in an increase in chlorophyll a concentration to 4 to 6 times that of the control. The number of phytoplankton cells increased to about 3 to 4 times that of the control. This not only indicates the importance of zooplankton grazing pressure on algae, but also on microzooplankton. The net growth of phytoplankton increases when zooplankton increases in abundance, probably due to regeneration of nutrients, which partly make up for the loss by grazing. The phosphate released by grazers is assimilated by algae or recovered by bacteria. The disappearance of flagellates reduces the internal regeneration rate of nutrients (Carrillo *et al.*, 1990).

Cladocerans influence rotifer and ciliate communities. In laboratory experiments, five species of cladocerans isolated from lakes in Vermont, U.S.A., were tested on their capacity to kill, during their normal feeding behaviour, individuals of the rotifer species *Keratella cochlearis*. It appeared that mortalities inflicted were independent of species and positively correlated with size of the cladocerans (Burns & Gilbert, 1986). The suppression of rotifer species by cladocerans is thus often attributed to mechanical interference. Rotifers and ciliates are sucked into the branchial cavity, where they can get fatally injured (Gilbert, 1989).

Similarly, the selective removal of zooplankton from water causes an increase of the relative importance of ciliates, which are apparently more reduced in number by predators than by their resources. Bacterivorous microflagellates, as well as ciliates, are important for the regeneration of nutrients (Hamilton & Taylor, 1987).

Controversy about the diversity of tropical zooplankton compared to that of temperate zooplankton

The species assemblage of tropical zooplankton is different from that of temperate region. Tropical zooplankton is characterized by the loss of several elements. Among the Cladocera, no large-sized herbivorous or raptorial species are found in tropical areas; cyclopoid and calanoids copepods are also less numerous. Only the number of predatory insects is more important than in temperate waters (Fernando *et al.*, 1990). For instance, the larvae of *Chaoborus* (Diptera Nematocera) are important predators of zooplankton in Lake George (Moriarty *et al.*, 1973). From these observations, it has been concluded that tropical zooplankton is less diverse than temperate one, though a range of pelagic invertebrates uncommon in temperate water are present: medusae, typhloplanid flatworms (Mesostominae), water mites or Acari. Their presence often remains undetected, but they may be a typical feature of tropical zooplankton (Dumont, 1992).

One of the reasons invoked for this allegedly lower diversity is that the efficiency of food capture does not increase in proportion to the increase of metabolic needs with rising temperature (Fernando, 1980). Crustaceans of smaller size are promoted because of their lower volume-surface ratios. In the case of rotifers, parthogenetic reproduction and resting stages are not suitable to environments with low seasonal differences (Lehman, 1988).

Cladocerans which do well in the tropics are small and bilaterally compressed and show morphological or chemical defense features, such as spines, spinules, predator repellents. Pelagic invertebrate are not visible, equipped with nettle cells, distastefull, produce toxic substances or are so jelly-like that they are not worth being eaten by fish (Dumont, 1991).

Considering the differences between tropical and temperate zooplankton, Dumont (1992) advocated that predation is probably the most important driving force shaping zooplankton communities in the tropics. Warmwater fish are important predators of zooplankton (Crisman & Beaver, 1990), as a large fraction of them remains planktivorous throughout life. The role of herbivorous macrozooplankton is partly taken over by phytoplanktivorous fish. The number of invertebrate predators is also increased by novel elements such as medusae, typhloplanid flatworms and water mites. Top-down control overrides the effects of resource competition or bottom-up control, as shown by numerous defense adaptations (Dumont, 1992).

The phytoplankton-zooplankton interface

The phytoplankton-zooplankton interface is the crucial point where changes in the predators at the top of the food web are translated into changes in the properties of the ecosystem, such as primary producer biomass, productivity and nutrient recycling (Elser *et al.*, 1990).

In ecosystems of high or low productivity, the coupling between phytoplankton and zooplankton is weak. The quantitative response of algae to grazing pressure is often not a direct function of grazing, but involves secondary compensatory effects, such as nutrient recycling. The qualitative responses of algae are specific and are, for instance, the appearance of inedible algal species, large-sized or colonial.

A study of three Californian lakes, varying in trophic level, showed that in ultra-oligotrophic environments, nutrient concentrations limit primary production and herbivore populations, while in hyper-eutrophic waters, the proliferation of inedible algae hampers the development of herbivores. In strongly eutrophic lakes, the corresponding zooplankton was unable to produce a detectable grazing effect, even when increased to eight times its natural density. Thus, the strength of coupling between phytoplankton and zooplankton is greatest in waters of intermediate trophic status (Elser *et al.*, 1990).

Types of control in plankton populations

The concept of interface implies an interaction of plankton with its environment. In experiments of selective removal of large-sized zooplankton, followed by fertilization of the water, increases in total algal biomass occurred. This indicates control of the phytoplankton not only by the grazers (top-down), but also by the resources (bottom-up), as the increase in resource availability stimulated phytoplankton production (Fig. 48). The structure of the zooplankton community depends on the specific characteristics of the phytoplankton assemblage with regard to edibility, and of the morphological capabilities of grazers to make use of this food source.

Analysis of data from three north-temperate lakes, bioassays, nutrient enrichment experiment and mesocosm experiments lead Elser *et al.* (1988) to conclude that the sequence of complex interactions

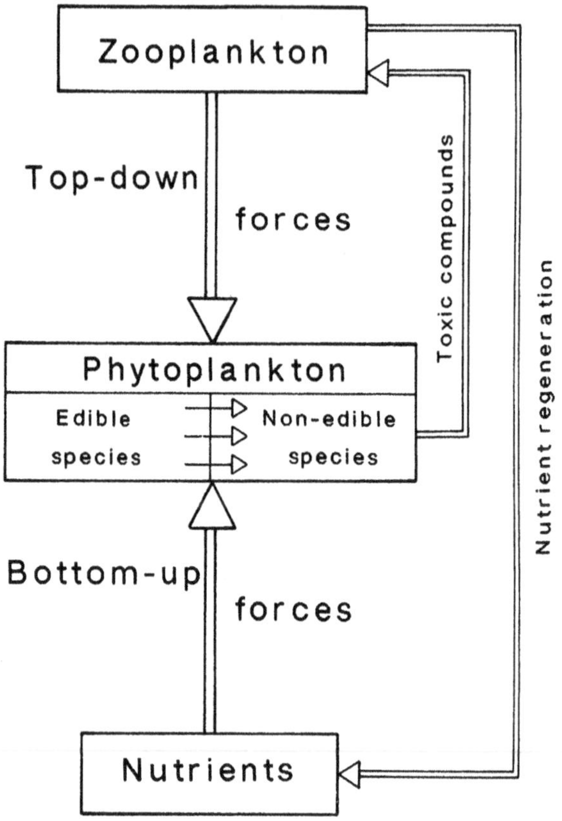

Fig. 48. Schematic representation of the types of controls exercised on the phytoplankton. Right-hand arrows indicate feed-back controls of zooplankton on phytoplankton.

between nutrient concentrations and the phytoplankton assemblage, through interspecific competition for nutrients, is another type of bottom-up force (producer controlled). The edibility of the phytoplankton assemblage affects the zooplankton community, especially the selective zooplankton grazers. This creates a feed-back whereby selective grazing promotes less edible and better protected algal species, and finally a change in nutrient recycling, mediated by zooplankton. The various direct and feed-back effects are not easily separated, which makes it difficult to establish a hierarchy of causes.

One of the specific mechanisms considered in connection with the modifications driven by the zooplankton is that the P:C ratios of nutrient requirements by zooplankton are lower than those of algae; zooplankton recycles greater amounts of nitrogen than phosphorus. This results in changes in relative availability of nitrogen versus phosphorus, and modifications in zooplankton biomass and/or size structure occur. Also, differences exist between the P:C ratios of substances which are ingested and those which are excreted or metabolized by zooplankton; nutrients present in excrements are more directly available for uptake than those present in algae or in zooplankton. The path followed for the reincorporation of nutrients is differential, being a function of the transformation and uptake they have been subjected to (Elser *et al.*, 1988).

In microcosm experiments on a natural plankton suspension enriched with netphytoplankton from the epilimnion of the moderately eutrophic Schöhsee, northern Germany, the Si:P ratio was modified by grazing and subsequent recycling of nutrients. As proportionally more phosphorus than silicon was returned to the water, species requiring more silicon, such as diatoms were outcompeted by green algae (Sommer, 1988).

Population controls do not only temporarily shape community structure, but also trigger micro-evolutionary changes. An important feature of the bottom-up and top-down forces is that they induce genetic changes within a community. The species evolve in response to availability of resources, but also with respect to predator-triggered defenses (environmental cues originating from predators rather than from prey (Havel, 1987)) by developing antipredator traits (Sih, 1987). The selection of traits within a population acts upon genotypes, with identical morphological phenotypes, as in the case of vertical migration (Dumont & De Meester, 1990). This ensures genetic heterogeneity within populations. Conservation of genetic heterogeneity is needed, if a population is to adapt to future changes in the environment.

Controls and regulations on plankton populations

It is important to understand the regulatory mechanisms, including feed-back actions, which promote or inhibit the development of various communities.

Do buffer mechanisms exist, which enable the maintenance of an assemblage of plankton communities (Moss, 1990)? Or are some equilibrium or successive equilibrium stages reached, which are maintained within some limits of changes in external factors, after which there is a sudden shift towards another equilibrium (Levitan, 1987; Scheffer, 1990)? Are trophic interdependence and stability of ecosystems imperatively linked (Paine, 1988)? A change in light regime in a shallow eutrophic lake, which resulted in a radical transformation of the structure of the ecosystem (Lammens *et al.*, 1990), would tend to confirm this contention.

Some uncoupling of the links between zooplankton and phytoplankton has also been observed (McQueen & Post, 1988), indicating that the controls of these trophic levels are numerous and their respective effects sometimes balanced and compensating each other. The effects observed in experiments with single species are seldom additive when species are in polyculture under natural conditions.

Interaction and exploitative competition

Interactions between organisms of similar status in the food web are manifold, but can be summarized as competition for the exploitation of common resources. A resource is here understood as all that is needed by an organism to satisfy its requirements of space, food, oxygen, etc. The tactics of organisms for obtaining such resources can be reduced to r- and K-strategies, i.e. species which have high adaptation potentialities, and high growth and reproduction rates, occurring in temporary habitats, and specialized species, with low growth rates, efficiently using all resources, and living in stable environments.

At the physiological level, this results in different adaptations for the uptake of nutrients or food, based either on velocity of uptake (fast versus slow) or on the affinity of uptake (high versus low concentration thresholds). Species can be distinguished on the basis of these properties. Calanoid copepods exploit aquatic environments with low levels of food, while some rotifers with high food requirements, along with high growth rates, have an advantage over other organisms in conditions of high food abundance. The concepts of transient dominance and mutual competitive exclusion apply to the case. They partly explain seasonal succession between species with fast growth rates, dominant when resources are plentiful, and species with lower growth rates, successful when resources are scarce (Rothhaupt, 1990).

Other hypotheses which explain competition between species have been discussed by Rothhaupt (1990) and are related to such characteristics as metabolism, feeding behaviour and predator escape responses. It is clear that a better understanding of the ecophysiology of the various plankton species will reveal a range of capabilities which allow species to respond to the various constraints and opportunities of the environment.

Interspecific competition acts at different levels: on minimum threshold concentrations required for a given resource, on partitioning of the resource within the environment, on abilities of organisms to substitute or take up alternative food, and on their responses to fluctuations in resource availability. The responses of species to changes in the food availability are variable and even include temporary halts of growth (resting stages) during life history (Rothhaupt, 1990).

Control by limitation of resources

From 25 whole-lake experiments in different countries of Europe, Benndorf (1990) concluded that top-down control on lower trophic levels decreased with increasing distance from the level on which it directly acted. The maximum potential biomass which could be reached by each trophic level was determined by the availability of resources, but the actual biomass reached was the combined effect of control mechanisms which were not only a function of resource availability but also of predation. The potential productivity of a trophic level is determined by nutrient supply, the action of predators, or a combination of these two factors. Actual productivity depends on nutrient recycling and nutrient partitioning between various populations with different growth rates (Gophen, 1990).

An effect on a given trophic level or on a group of organisms is quickly dissipated within the food web. Feed-back actions, which are not always directly apparent, must also be taken into consideration. For example, a reduction of predatory pressure of fish on zooplankton leads to an increase in zooplankton biomass. Beyond a certain zooplankton biomass, their living conditions deteriorate, however. The increase of zooplankton biomass amplifies the availability of nutrients to the surviving phytoplankton, not only by a reduction of phytoplankton biomass, but also by increased mineralization of nutrients by the zooplankton. This nutrient availability benefits phytoplankton species with high nutrient requirements associated with high growth rates, or species which can escape zooplankton grazing. However, increased grazing on the phytoplankton usually leads to higher water transparency, which stimulates photosynthetic activity. This results in some cases in colonization of the environment by macrophytes (Benndorf, 1990).

Some predatory effects are independent of or poorly dependent upon mechanisms of resource control. Reduction of top-down control rebalances the phytoplankton-zooplankton relationship. Its leads to the dominance of large-bodied herbivorous zooplankton and induces changes in the structure of the phytoplankton community, with selection of algae protected against grazing or which have characteristic high growth rates. Other effects are largely dependent on mechanisms acting on the resources. Average long-term biomasses are reduced by predation only if a certain level of availability of the resources is not exceeded. The probability of a mass development of inedible algae increases when resource (for instance P) limitation decreases. The mechanisms of control of the resources rely for their action on the availability of phosphorus. A phytoplankton standing crop can only be reduced, if the availability of phosphorus is decreased.

Herbivore tactics to escape predation

Zooplankton adaptations to escape predation are manifold, ranging over morphological, physiological, chemical and behavioural features (Sih, 1987). In some cases, defenses are predator-induced (Havel, 1987). Amongst crustacean prey, defenses include hardened integuments, spines on different parts of the body, for instance around the thoracic and cephalic zone (Williamson, 1987), size reduction, as in some cladocerans (Stenson, 1987), or unwieldy sizes or shapes rendering the handling process of the prey difficult for the predator (Sih, 1987). Among rotifers, the presence of spines, of a rigid lorica, of mucus sheats, protect them from predation (Stemberger & Gilbert, 1987). A small size entails a reduction of visibility and retention by gill rakers of fish, which is even more reduced by a lateral compression of the body as in cladocerans. Absence of pigmentation or hemoglobin (transparency), and reduction of the size of the eye spot lessen the visibility and enhance survival, though several disavantages go along with these features. Camouflage, mimicry, warning coloration have also evolved as a means of defense. Some prey, such as mites, are distasteful or toxic to predators. Behavioural modifications include reduced mobility, capacity to detect water currents (rheotaxis of rotifers) and escape them, vertical migration (O'Brien, 1987), and aggregation (Folt, 1987) or coloniality (Stemberger & Gilbert, 1987). The evolution of antipredator traits were divided by Sih (1987) into two main categories based on whether they act before or after the encounter with the predator. This choice affects the life-style of the species.

Avoidance of predators results either from features enabling the non-detection by the predator (avoidance of entering the sensory field of the predator) or from a hiding behaviour. Cladocerans benefit from the presence of refuges and provision of artificial refuges was found to increase their abundance (Irvine *et al.*, 1990). Species living in temporary or stressful habitats are considered to have evolved traits to decrease encounter rates with predators and competitors. The difference between species based on resource utilization, which has lead to the distinction between species with r and K strategies, can be extended to antipredators traits (Sih, 1987).

Evolving defenses results in costs for the prey, such as lower feeding, growing or reproductive rates than prey without these defenses. Sometimes they are constraining the prey's overall life-style. On the whole, behavioural traits appear to less constraintful than morphological ones, as they inflict lesser costs on feeding and mating (Sih, 1987). Riessen & Sprules (1990) have quantified the costs of these defenses for *Daphnia pulex*: individuals with spines in the neck region reached sexual maturity 8.4 to 14.6 % later than individuals not developing neck spines.

Zooplankton predator-to-prey proportions

Based on data from a range of freshwater habitats in Britain and North America, Jeffries & Lawton (1985) considered that an invertebrate predator to prey species ratio of 1:2 to 1:3.5 seems to be optimum. In a small enclosed ecosystem in Villefranche, France, the predation of copepods on ciliates and of ciliates on flagellates reached an equilibrium stage, whereby biomass increase compensated predation (Sheldon *et al.*, 1986).

In the Řimov reservoir, Czechoslovakia, the size structure of zooplankton was related to fish biomass and its predation, indicating the importance of their planktivory (Seďa *et al.*, 1989). The predatory pressure of fish is far more intense than that of all other groups.

Influence of zooplankton on the plankton community

1. Zooplankton is subdivided, according to size, into microzooplankton, mesozooplankton and macrozooplankton. Microzooplankton are the smallest organisms, consisting of flagellates, ciliates and rotifers, mostly detritivorous and bacterivorous, though rotifers also consume small algae. Mesozooplankton is comprised of cladocerans, and macrozooplankton of copepods (or cladocerans and copepods). The dietary range of zooplankton organisms is often wide.

2. The influence of zooplankton on phytoplankton is manifold: (1) zooplankton herbivores graze the algae and reduce the overall volume of the algal biomass. The algal community reacts to grazing pressure by an increase in proportion of large-sized and less edible algae. This, in turn, reduces the stock of grazers; (2) the zooplankton modifies the sedimentation rate of algae, by its grazing pressure. It releases faecal pellets of the algae consumed, which would otherwise have been kept in suspension. They also promote algae of greater size, which sediment faster; (3) they promote the recycling of nutrients in the water column. Nutrients are regenerated, recycled, and kept in constant turn-over, owing to the presence of zooplankton. This results from grazing of the zooplankton, whereby organic compounds (via sloppy feeding) and faecal pellets are released into the water, and subsequently decomposed, these nutrients thereby again becoming available for phytoplankton.

3. The diversity of tropical zooplankton compared to that of temperate regions is a controversial subject. The role which zooplankton plays in temperate regions may be taken over in the tropics by phytoplanktivorous fish, but predation pressure by zooplanktivores is another factor, and tropical zooplankton is enriched in unconventional elements with defenses against vertebrate and invertebrate predators.

4. The control of phytoplankton by zooplankton is quite loose and there is often no strong coupling between phytoplankton and zooplankton. The forces controlling such populations dynamics are a combination of bottom-up and top-down controls.

5. The relations between communities of organisms depend on trophic links, which command the establishment of equilibria between groups of organisms. But probably other types of relations besides trophic levels must be taken into consideration, such as competition for space, partitioning of resources within the environment (spatial heterogeneity in the distribution of resources), and resource limitation. Where grazing is not a limiting factor to the development of phytoplankton, limitation of available resources (especially phosphorus) must be taken into account.

1.2.7. Contributions to secondary production

Limitations of the concept of primary production

Primary production is conceptually confined to the autotrophic production of microphytes and macrophytes and designates the elaboration of particulate organic material at the lowest level of the trophic pyramid. However, in fact, this definition also includes heterotrophic bacterial production, unless the definition of the elaboration of organic material is limited to carbon fixation. The extension of primary production to microbial activities allows the inclusion of algal 'losses', such as organic carbon exudations and secretions, and a better assessement of primary production. However, the differences in significance of the processes of autotrophic and heterotrophic production are too wide to allow us to collect them under the same label (Flynn, 1988).

The microbial food loop

Many aquatic environments are characterized by a trophic dependency of much of the animal biomass on dead plant material. Sometimes, consumers directly ingest large amounts of plant detritus, but often there is no such direct link. Heterotrophic bacteria and fungi are then the intermediates between detritus and primary consumers (Miller, 1987).

Transfer of primary production to secondary production is also possible, and involves the microbial loop (Fig. 49). This designates the energy and nutrient flow from algae through bacteria and flagellates to zooplankton (Heyman & Lundgren, 1988); it constitutes a rechannelling of carbon and energy through the microbes into the planktonic food web (Bloem & Bär-Gillissen, 1989, in lake Vechten, Netherlands). The term bacterial food loop was coined by Azam *et al.* (1983) to designate trophic links between bacteria, heterotrophic flagellates and the microzooplankton in marine water. This loop should, however, be considered as an integral part of the food web of any ecosystem (Sherr & Sherr, 1988), and not as a source or sink of carbon. Bacteria and protozoa channel dissolved organic matter back into the food web by repackaging parcels of fixed carbon or recovering fixed carbon. The repackaging designates the uptake of bacteria by mixotrophic algae, while recovering is done by heterotrophic flagellates ingesting and assimilating polysaccharides of high molecular weight at relatively low concentrations (Sherr & Sherr, 1988).

Relations between bacterial populations, living plankton and dead cells

Bacteria recover organic matter with energetic value and release nutrients. Bacteria are present wherever there is accumulation of planktonic cells, which sink and decompose, and where algae exudations occur. In stratified lake Vechten, Netherlands, Bloem & Bär-Gilissen (1989) observed that important amounts of carbon and energy released as organic material byphytoplankton was processed by bacteria and returned to the 'main' food web through the microbial

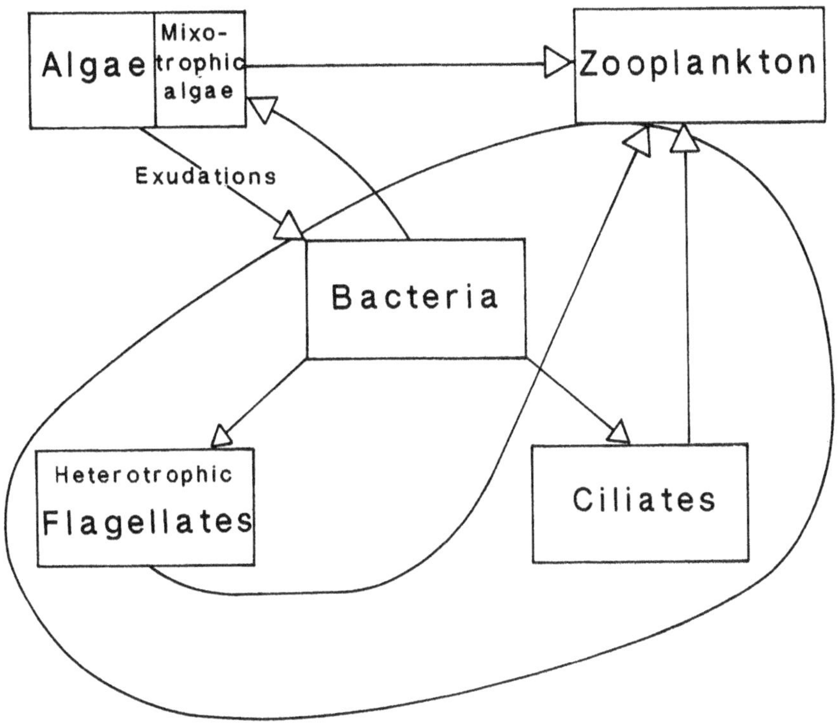

Fig. 49. Schematic representation of the bacterial food loop (encircled portion).

loop. Coupling between bacteria and heterotrophic nanoflagellates was such that about 10 to 70 % of the water column was daily cleared of bacteria. Efficiency of the process depended on the yield, i.e. the gross growth efficiency, of the grazers.

In the metalimnon of the lake, chlorophyll concentration and primary production rate were high, because of deep penetration of light and an upward flux of nutrients. Primary production along with detritus accumulation also promoted bacterial production. Bacteria were consumed by heterotrophic flagellates and ciliates, which themselves were consumed by microzooplankton. Ciliates (two *Coleps* species and one unidentified spherical ciliate) were found in great numbers at the oxic-anoxic boundary, where they were feeding on bacteria. There, ciliate predators were avoiding these oxygen-poor zones, while ciliates were either shifting to nitrate respiration, as an alternative energy source, or having a microaerophilic metabolism. High bacterial activity coinciding with high chlorophyll concentration indicated a link between phytoplankton and bacteria, phytoplankton exudations and organic material from decomposing algae cells, providing carbon and energy to the bacteria.

In eutrophic lake Nesjøvatn, central Norway, an algal die-off caused an increase in bacterial abundance, which constituted up to 30 % of primary production (Vadstein *et al.*, 1989). Methods of measuring microbial production are still controversial. Two main methods are used, namely thymidine and Karl's adenine, which give results varying by orders of magnitude (see, for example, the discussion between Moriarty and Costa-Pierce, 1990). The biomass and production of bacterial populations increase with nutrient concentrations. This makes the relation between primary production and nutrient concentration difficult to grasp, as nutrient supply occurs partly through the decomposition cycle.

Differences in organic material

Organic material in the water column is either dissolved (DOM) or particulate (POM). DOM is usually material freshly released by organisms, of greater nutritional value, consisting, for example, of amino acids which can be directly taken up (Mann, 1988). The organic substances released by algal cells during their active growth are photosynthetic products, made of small organic molecules, easily assimilated by bacteria, such as hexoses. Algal decomposition products are macromolecules, probably polysaccharides. The products exuded during active algae growth inhibit and repress the synthesis of bacterial decomposition enzymes, while the decomposition products induce their synthesis (Chrost, 1989).

Organic particles or aggregates (POM) are also released by organisms, as part of dead organisms, and become substrata for decomposer microorganisms. In the process, nutrients get regenerated and transferred between living communities on and around particles in an almost closed circuit. This is what Mann (1988) called the aggregate spinning wheel concept. Bacteria also aggregate around phytoplankton cells, using their DOM excretions and regenerating nutrients.

The organic material in suspension or solution is regenerated as far as it releases nutrients or energy for the organisms which use it as food, as a source of nutrients, or as substrate for extracellular enzymatic attack. The importance of regeneration is therefore function of the nutritional value of the organic material. The complexity of the organic compounds present in water is great and has been discussed in the section on organic phosphorus compounds (p. 97). Most particulate organic carbon in the detritus of a humic lake, nearby Oslo, Norway, was of low nutritional value for bacteria (Hessen *et al.*, 1990).

In laboratory experiments with humic water, and in a study involving eleven oligotrophic Swedish lakes with differing humic contents, Transvik & Sieburth (1988) showed that bacteria use humic material, when it is partially flocculated or dissolved. As such, humic material could contribute positively to secondary production as the flocs of organic aggregates were colonized by attached or free-swimming bacteria, on which free-swimming or attached flagellates fed.

Bacterial types and decomposition in water

Dissolved organic matter is used as a carbon and energy source by heterotrophic bacteria. The amounts of organic compounds transferred from algae to the bacterial loop are largely undetermined; it has been estimated that 10 to 50 % of the carbon fixed by photosynthesis might pass into the bacterial loop (Berman, 1988; Pomeroy & Wiebe, 1988). Largely undetermined are also the types and populations of bacteria occurring in the water column. In lake Constance, Switzerland, bacteria were separated into small (0.2 to 1 μm) and large sizes (1.0 to 3.0 μm), and into free swimming and attached types. Small free-swimming bacteria were estimated to represent two-thirds of secondary bacterial production. Attached bacteria are usually of less importance. The bacterial biomass and production were predominantly controlled by supplies of organic matter for decomposition (Simon, 1987).

The attraction of bacteria to material to be decomposed is chemosensitive and allows pelagic bacteria to remain in contact with particles (Jackson, 1989). Bacterial activity results in the presence of enzymes in water. The activity of glucosidase is linked to the presence of heterotrophic bacteria, forming part of a coupled system of hydrolysis and assimilation. This enzymatic activity is more important in the 0.2 to 3.0 μm size fraction. Synthesis and exocellular excretion of ß-glucosidase is induced by the presence of macromolecules stemming from the decomposition and autolysis of algal cells. Other exudates, such as hexoses, are directly assimilated by algae (Chrost, 1989).

The microbial food loop : source or sink of energy?

Pomeroy & Wiebe (1988) consider the bacterial food web as an energy sink, because detritivorous metazoans and bacteria compete for the same resource. However, through this loop, dissolved organic carbon, mainly polysaccharides of high molecular weight at low concentrations, oozing out of algae, is recovered by bacteria and is returned to the main food web. These organic compounds are repackaged by bacteria which are themselves consumed by mixotrophic algae and heterotrophic flagellates (Sherr & Sherr, 1988).

Nutrient regeneration by bacteria is in essence a repackaging of nutrients released by organic material into new bacterial tissues. But bacteria do not use carbon, nitrogen and phosphorus in the same ratio as phytoplankton, i.e. the Redfield ratio. The value of organic material, assessed on the C:N:P ratio, therefore depends on the nutritional requirements of the organisms making use of it. Different organisms have different nutrient requirement ratios. Nutrients taken up are used for cell respiration or growth and each process releases or keeps nutrients in specific ratios. Therefore, nutrient composition of organisms does not reflect nutrient intake. From this, it can be concluded that at each trophic step, some nutrients are wasted, being in surplus of the ratio required. For instance, Goldman *et al.* (1987) showed in a batch culture that marine bacteria do not use nitrogen efficiently.

If the efficiency of energy transformation across different trophic levels is considered, losses appear important. In Israeli fish ponds, Schroeder (1987) assessed the loss factor of carbon by microbial transformation to about 3; this, nevertheless, is much less than if the organic material was not recovered.

Trophic links between organisms

Between different groups of organisms, links are established. Bacterial populations appear there where organic detritus and dead cell accumulate. Bacterial production is mainly grazed upon by heterotrophic nanoflagellates (2 to 20 μm in size), which are themselves consumed by microzooplankton, especially ciliates (20 to 200 μm in size). Populations of ciliates, flagellates and bacteria therefore develop accordingly to the trophic links binding them.

In lake Vechten, Netherlands, a high density of blue-green algae induced high densities of bacteria and of heterotrophic nanoflagellates, the main consumers of these bacteria (Bloem & Bär-Gillisen, 1989).

The predators of bacteria are flagellates (almost exclusively at some well-defined depth), rotifers, and ciliates. Heterotrophic flagellates in fresh water are able to feed solely on bacteria. Ciliates are important predators of flagellates, consuming bacteria less readily. In Lake Constance, flagellates and ciliates feed on picoplankton, nanoplankton and detritus (Weisse, 1990).

The impact of feeding of some mixotrophic chrysophytes, such as *Dinobryon*, on bacteria is important and at some depths, even exceeds that of zooplankton. In fact, these algae depend more on bacteria for their metabolism than on photosynthesis (Bird & Kalff, 1987). Phagotrophic protozoans form an interface in the regulation of carbon and energy fluxes of aquatic systems. Their rapid growth, the grazing they are subjected to and which they exert on bacteria, and their nutrient regeneration capabilities are responsible for this. Fast feed-back mechanisms regulate these transfers (Sherr *et al.*, 1988). The mixotrophy of flagellates is an opportunistic response, controlled by environmental conditions. They may use either particulate and dissolved organic material, or the photosynthetic process (Porter, 1988).

In a study involving zooplankton of an eutrophic lake of Georgia, U.S.A., Sanders *et al.* (1989) showed that cladocerans inflicted only little mortality to bacterial populations, but actively consumed heterotrophic flagellates. They ingested particles of 2 to 8 μm of size, but also consumed ciliates of 100 to 200 μm in size.

The trophic pyramid or trophic food web concept

The trophic interrelations between the species in the pelagic and benthic assemblages show that the ideal image of a trophic pyramid is unrealistic. The diversity and flexibility of feeding regimes of the various species making up such communities do not warrant simple pigeonholing of species or groups of species into producers, consumers, and decomposers. In fact, the situation ressembles a web, where the entanglement of production and consumption does not allow a distinction of separate trophic levels. Organisms, such as cladocerans, considered as strict herbivores, appear, at close scrutiny, unclassifiable as to their feeding regime. In laboratory experiments, they feed on planktonic algae, detritus, bacteria, heterotrophic flagellates (Sanders & Porter, 1990), ciliates, and rotifers, thus crisscrossing the traditional trophic levels and short-circuiting them.

In the same context, the transfer of material and energy through a food web should not be viewed too rigidly or too schematically. A detailed knowledge of the feeding regimes of all organisms present in an ecosystem is required to understand the functioning of the food web (Paine, 1988).

The complexity and diversity of the trophic relations in this part of microbial production is so complex that some do not hesitate to call it a bacterial maze rather than a bacterial loop (Berman, 1988). The energy and nutrient fluxes in the microbial loop are still largely unexplored; in the case of intensive culture systems, they are quite important: they constitute a nutrient regeneration system benefitting production.

The bacterial food loop: contribution to production

1. The microbial food loop is the feeding-back of nutrients into the trophic food web via the bacteria-flagellates-ciliates pathway.

2. The organic material in solution and suspension in water, produced by phytoplankton (algal exudations and lysis) is decomposed by bacteria, which support a population of flagellates and ciliates. Even some mixotrophic algae feed directly on bacteria. The energy and nutrient content of these organic compounds are thus recovered for natural productivity.

1.2.8. Influence of fish production

Effects of fish production on primary production and on water quality

Lastly, the role of fish in the process of primary production must be examined. Fish release various waste products, such as carbon dioxide, ammonium, and organic material containing nutrients, while they remove oxygen and particulate organic material (plankton and bacterioplankton) from the system.

According to the effect of fish on the plankton community and on water quality, their overall influence will be positive, i.e. stabilize the food web and maintain a steady production, or negative, leading with increasing fish biomass, to a deterioration of the production factors. Examples are polyculture manure-treated ponds, like the traditional ponds in China, as opposed to high-density artificially fed monoculture crops, which show deteriorated water quality, at the end of the culture period.

Effects of fish on artificial environments

In Israel, Van Rijn *et al.* (1986) compared physical, chemical and biological conditions in cement tanks used for intensive culture of tilapia and eel (Fig. 50). Daily dissolved oxygen was low in tanks with tilapia (Fig. 50a), a consequence of the consumption of phytoplankton by the fish (Fig. 50e). As a result, the water became enriched with ammonia (Fig. 50g). The relative increase in carbon dioxide enhanced the buffer capacity of the water and stabilized pH (Fig. 50c). Bacterial activities became more important (Fig. 50b, d & f) owing to the presence of more detritus and the relative decrease of photosynthesis. The influence of feeding also clearly appears (Fig. 51). The effect of a temporary cessation of feeding was reflected in the physical and chemical condition in the tanks. Feed polluted the water directly and indirectly: left-overs of feed, as well as excrements produced by fish, supplied organic matter to the water, which, on decomposition, consumed oxygen (Fig. 51a) and released ammonia (Fig. 51e). Bacterial activities paralleled the amount of organic material present. Over a period of six weeks, a progressive deterioration of the culture conditions was clear (Fig. 52), with an increase in ammonia content (Fig. 52e) and a progressive alkalinization of the pH, which rendered the proportion of toxic free ammonia threatening. Bacterial activities increased in a similar way.

The effects of two planktivorous fish, the filter-feeding gizzard shad (*Dorosoma cepedianum*) and the visual-feeding blue-gill (*Lepomis macrochira*) were investigated in tanks for one month by Lazzaro *et al.* (1991a & b). Gizzard shad is an omnivorous planktivore, whereas blue-gill is carnivorous planktivorous and benthivorous. An increase of fish biomass lead to increasing primary production: higher chlorophyll a, turbidity, more unicellular green algae and pennate diatoms. The two planktivores were affecting the water independently. Fish biomass reduced particulate phosphorus in the size fractions > 200 μm (macroplankton), but increased it in the size fraction of 20-200 μm (microplankton)

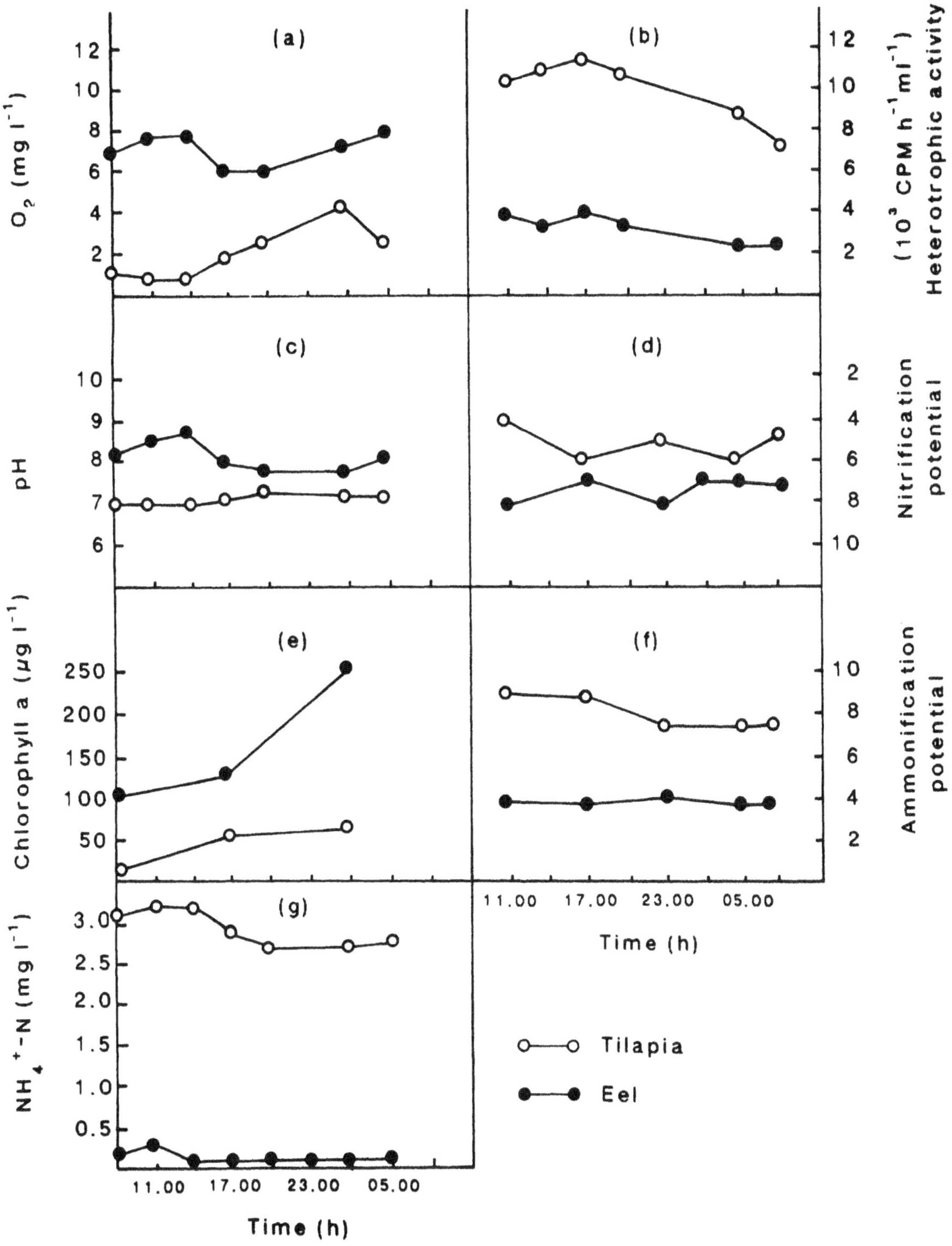

Fig. 50. Comparison of diel changes in physical, chemical and biological production parameters in two concrete ponds (50 m^2 x 0.8 m), stocked with tilapia (625 kg and 1250 kg of total biomass) and eel (less than 200 kg) respectively. Water was supplied at a continuous rate of 1 l s^{-1}. Ammonification potential is expressed as daily ammonium production (mg NH_4^+ d^{-1}). Nitrification potential is expressed as the number of days needed to convert half of the added ammonia to nitrite (d^{-1}). Heterotrophic activity is expressed in counts of radioactivity by a scintillation counter (from Van Rijn *et al.*, 1984) (with kind permission of Prof. Van Rijn on behalf of authors).

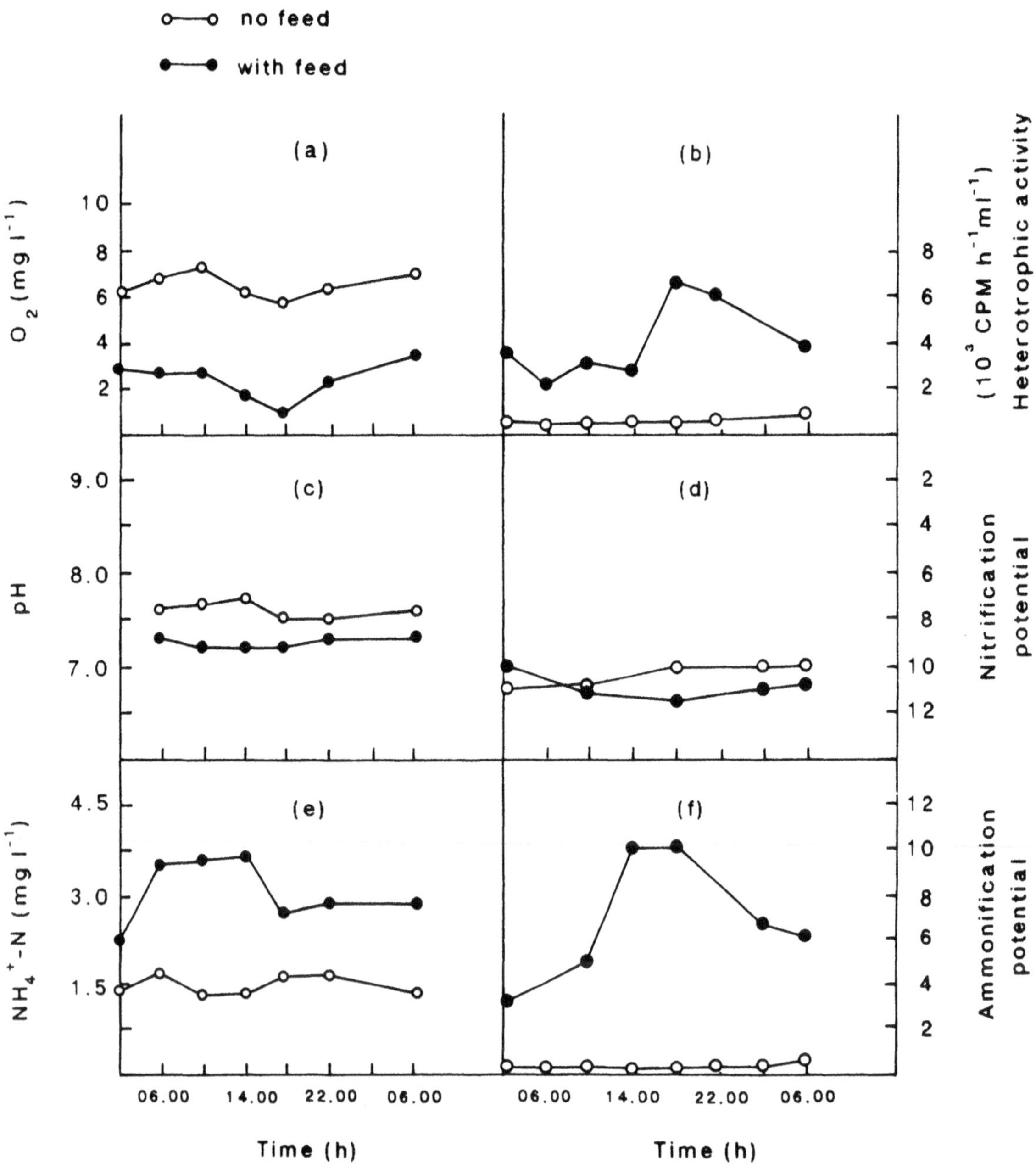

Fig. 51. Comparison of diel changes in physical, chemical and biological production parameters in two concrete ponds, stocked with tilapia, fed (3 % of the biomass, daily) and non-fed (ca. 1200 kg of total fish biomass). The water supply is the same as in Fig. 50, so are as well the units of ammonification, nitrification potentials and heterotrophic activity (from Van Rijn *et al.*, 1984) (with kind permission of Prof. Van Rijn on behalf of authors).

and 12-20 μm (large nanoplankton), which is consistent with the consumption of zooplankton and the resulting increase in phytoplankton. Total phosphorus increased with increasing fish load, while total nitrogen remained unchanged.

Effects of fish on ponds

Tucker *et al.* (1979) monitored changes in physical, chemical and biological factors in ponds used for culture of American catfish. The culture period extended over 10 months and three different stocking densities were used. The ichthyomass was important

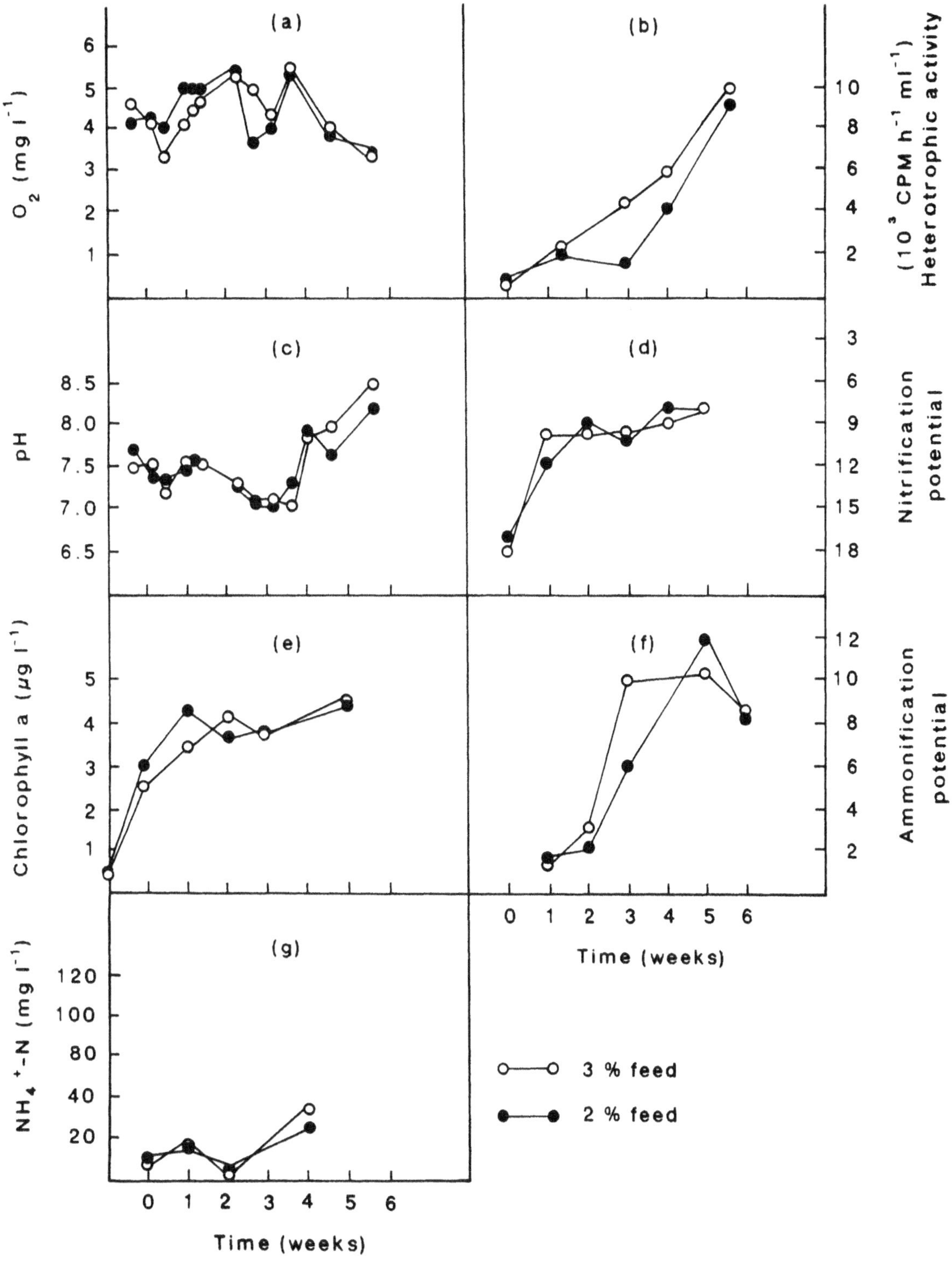

Fig. 52. Effects of two different feeding rates, 2 and 3 %, on the modification of physical, chemical and biological production parameters in two concrete ponds, stocked with tilapia (ca. 1000 to 1400 kg of total fish biomass) during a period of 6 weeks. The water supply is the same as in Fig. 50, and so are as well the units of ammonification, nitrification potentials and heterotrophic activity (from Van Rijn *et al.*, 1984) (with kind permission of Prof. Van Rijn on behalf of authors).

and increased in the course of time, stimulating decomposition of organic material, as well as photosynthesis. This is illustrated by increases in ammonium concentration (Fig. 53e), in COD (Fig. 53a), in chlorophyll a (Fig. 53c), and in turbidity (Fig. 53a). The increase of photosynthesis was a consequence of the decomposition and recycling of nutrients, boosted by fish excrements and feed leftovers.

Cole & Boyd (1986) confirmed these results, by establishing correlations between physical and chemical features of water and amounts of feed added to ponds, which varied between 0 to 200 kg ha^{-1} d^{-1}. The ponds were used for the culture of American catfish. Dissolved oxygen was negatively correlated, while chlorophyll a, chemical oxygen demand, carbon dioxide and ammonium concentrations were all positively correlated with the amount of food added. The use of feed at a rate above 112 kg ha^{-1} d^{-1} resulted in a decline in fish

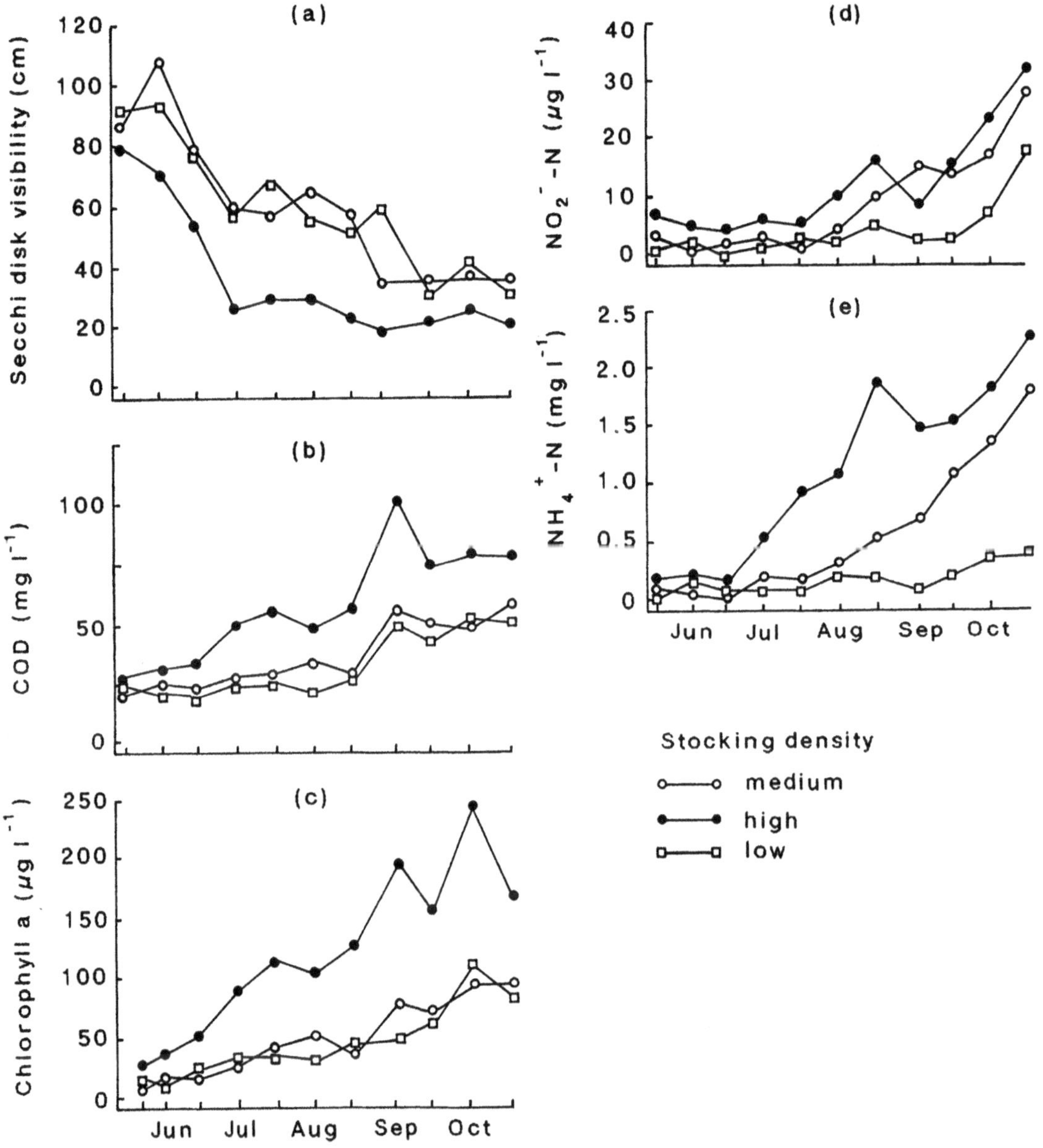

Fig. 53. Effects of three different stocking rates on the physical, chemical and biological production parameters of channel catfish ponds. The three different stocking rates are: 4942 fish ha^{-1} (low), 10007 fish ha^{-1} (medium), and 20385 fish ha^{-1} (high). (COD = chemical oxygen demand) (from Tucker *et al.*, 1979) (with kind permission of Prof. Boyd on behalf of the authors, & Am. Fish. Soc.).

production, which could not be alleviated by the use of an aerator. The authors did not find an explanation for this decline, but it was probably associated with accumulation of metabolites.

Specific effects on ponds of some groups of fish

Planktivorous fish have a major influence on the structure of the whole plankton: they modify the density and size structure of communities (Hrbáček, 1962; Brooks & Dodson, 1965), and the nature of detrital seston, as the sizes of settling particles are determined by the zooplankton (Søndergaard *et al.*, 1990).

The most obvious effects of fish on primary production are direct consequences of their feeding, i.e. the removal of organisms from the pond. Besides these, other effects appear in cascade. They diffuse through the production system and reveal the intricacy of the interrelations. But do links between organisms exist other than trophic ones?

Main effects of fish on plankton communities result from direct and indirect impacts of predation, depending on feeding mode. Fish show three different modes of feeding: (1) particulate or visual feeders visually select their prey; (2) pump filter feeders use rhythmic suction to engulf water straining off the plankton, while swimming slowly or remaining stationary; (3) tow-net feeders swimm rapidly with their mouth fully agape, engulfing water and capturing prey. In each case, capture depends on retention ability of prey by the predator. In the case of filter-feeding fish, the sieving structures are the gill rakers, the microbranchiospines, the pharyngeal teeth, assisted by the secretion of mucus. The mode of feeding of a species can change within life-history or be seasonal, due to changes in prey abundance. With increasing abundance, fish shift from visual feeding to filter feeding. Visual feeding is also modified by learning processes (Lazzaro, 1987).

Main direct and indirect effects on the environment are listed hereunder.

Effects of benthophagous fish, especially common carp

The impact of benthophagous fish on the trophic structure of temperate ecosystems is quite important (Lammens *et al.*, 1990) (Fig. 54). Their effects on the sediments are direct and indirect. They establish a contact between the benthic and pelagic compartments, which are otherwise fully separated (Richardson *et al.*, 1990):

a) predation modifies the structure of the benthic community:

The common carp feeds on benthic macroinvertebrates, such as Chironomidae, Oligochaeta, and Mollusca. Its mode of feeding is mainly tactile and olfactory; it ploughs through sediments in search of food, ingesting sizeable amounts of detritus. Its degree of selection is low and practically all organisms are suppressed. Other benthophagous species, which feed relying on visual stimuli, select prey based either on size or on apparent visibility. They modify the structure of the benthic community towards organisms of smaller size and cryptic behaviour.

b) predation on benthic organisms reduces bioturbation:

By reducing the number of benthic organisms, fish cause a reduction in bioturbation, which is of particular importance for aeration of the sediment.

c) transfer of benthic resources to water is promoted:

Fish transfer particles of sediment to the water column. Some become incorporated to fish biomass or are released as excrement and made available to pelagic bacteria.

d) by resuspension of sediments, fish modify the cycling of nutrients, especially of phosphorus:

The resuspension of sediments makes it possible for nutrients, accumulated in pond mud, to become dissolved in the water (see p. 50). This mainly concerns soluble substances from interstitial water. Sedimentary phosphorus, soluble or labile, is thus transferred to water. The mechanical mixing of sediments with water does not dissolve phosphorus (Qin & Threlkeld, 1990). The importance of phosphorus transfer during sediment disturbance is dependent on sorption properties of soil, amounts of labile phosphorus, and concentrations of interstitial dissolved phosphorus. The influence of benthophagous fish, especially common carp, on planktonic production has been shown by Lamarra (1975).

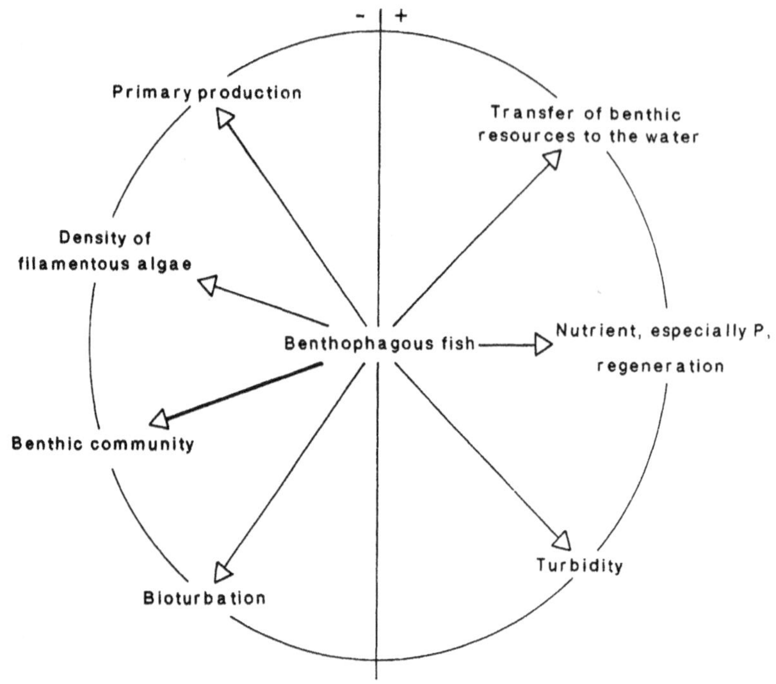

Fig. 54. Schematic diagram of the effects of benthophagous fish on a pond. The positive or enhancing effects are on the left-hand side, while the negative or decreasing effects are on the right-hand side. The bold line indicates the main effect.

Similarly, herbivorous fish (*Scardinius erythrophthalmus*), feeding on aquatic plants removed and transferred phosphorus from sediments to the pelagic compartment, as shown in aquaria (Hansson *et al.*, 1987).

e) the resuspension of sediments affects turbidity:
The common carp, feeding on the bottom of a pond, brings mud into suspension and makes the water turbid. Carp ponds are notorious for their turbidity, especially if pond soil contains clay. A reduction in density of benthophagous fish has resulted in an increase in water transparency in Dutch fish ponds (Meijer *et al.*, 1990b).

f) benthophagous fish cause a decrease in primary production by modifying the light regime in the water:
The increase in mineral turbidity in the water results in a decrease of primary production, entailing a decrease in dissolved oxygen, sometimes leading to oxygen depletion (Szumiec, 1989).

g) the density of filamentous algae is reduced:
Benthophagous fish reduce the density of filamentous algae, directly by feeding on them, and indirectly by changing turbidity and light penetration.

Effects of zooplanktivorous fish on ponds

There is an abundant literature on effects of planktivorous fish on plankton communities, initiated in the thirties by research in fish ponds (Lazzaro, 1987). The main effects are here summarized using a selection of recent papers. Richardson *et al.* (1990) conducted experiments in 18 large outdoor tanks at Kingston, Oklahoma, U.S.A.; Mazumder *et al.* (1990) performed *in situ* experiments, in enclosures (8 m in diameter and 15 m deep), in Lake St. George, Toronto, Canada. The presence of zooplanktivorous fish changed the plankton community as follows (Fig. 55):

a) a decrease in density and biomass of large filtering herbivores (Brooks & Dodson, 1965):
In experiments using a combination of planktivorous and benthophagous species (Richardson *et al.*, 1990), or only planktivores (Mazumder *et al.*, 1990), a decrease in density of large and medium-

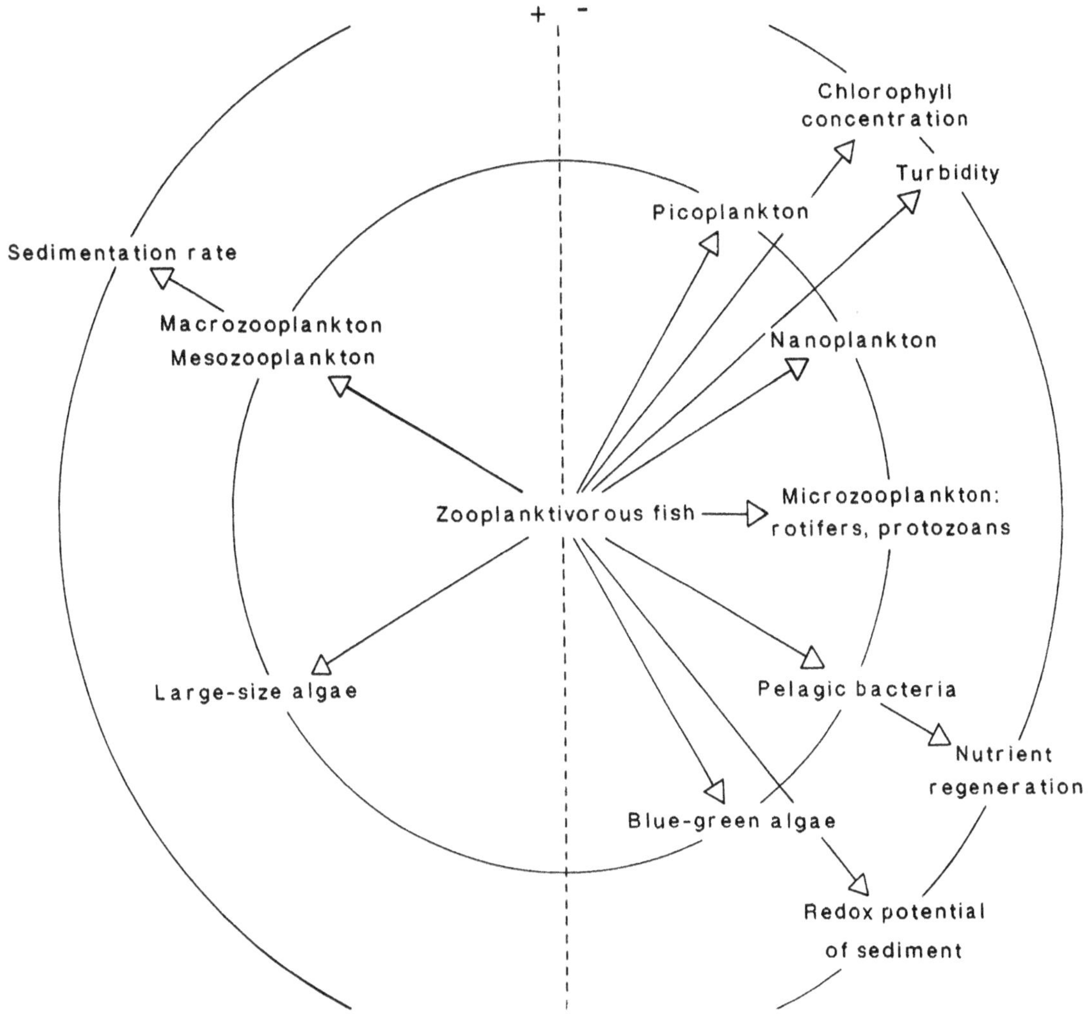

Fig. 55. Schematic diagram of the effects of zooplanktivorous fish on a pond. The positive or enhancing effects are on the left-hand side, while the negative or decreasing effects are on the right-hand side. The bold line indicates the main effect. The inner circle groups the direct effects on the living organisms, the outer circle the indirect effects.

sized zooplankton (copepods and Cladocera) was observed, which themselves prey on microzooplankton and on nanoplankton. In lakes in southern Norway, removal of larger zooplankton resulted in an increase of microzooplankton, in particular of rotifers and protozoans. An increase in density of zooplankton, especially herbivores, and of grazing pressure on algae, following the removal of planktivorous fish, was found by Lyche *et al.* (1990). A decrease in average size and size at maturity of Cladocera has also been observed (Gliwicz *et al.*, 1981). In some cases, predatory cyclopoids increase in density, fecundity, biomass and production as a result of the increase of microzooplankton (Hillbricht-Ilkowska & Weglenska, 1973; Kerfoot, 1987).

b) increase of the picoplankton and nanoplankton: The presence of planktivorous fish alters the structure of the planktonic assemblage, increasing the relative importance of picoplankton and nanoplankton. Their losses, by grazing of zooplankton, are reduced and their biomass increases (Richardson *et al.*, 1990; Mazumder *et al.*, 1990). This results in an increase of chlorophyll and of turbidity (biotic turbidity) (Threlkeld, 1988; experiment in 5 outdoor tanks in Kingston, Oklahoma). This increase of nanoplankton, composed of green algae of less than 10 μm of size, is also observed in the presence of silver carp (phytoplanktivorous) and bighead carp (omnivorous).

c) this increase in algal biomass is observed in the presence or absence of sediment:

The increased algal biomass has been considered to result from nutrient enrichment, following resuspension of sediment by benthophagous fish. The impact of a possible transfer of nutrients from sediment to water has been tested by comparing the increase of algal biomass in the presence and absence of sediment. In a study involving indoor and outdoor tanks of different capacities (900 l and 7000 l) in Kingston, Oklahoma, Qin & Threlkeld (1990) observed that, in mixed cultures of common carp and planktivorous fish, an increase of phytoplankton occurred in the presence and in the absence of sediment, regardless of whether fish had access to the sediment or not. Their interpretation is that the increase of phytoplankton resulted from the presence of the planktivorous fish, which fed on large-sized herbivorous zooplankton, thus causing a reduction in the grazing pressure of the zooplankton on the phytoplankton. In eight enclosures in Lake St. George, Toronto, Canada, the grazing of zooplankton and predation of fish on zooplankton promoted growth of small algae, which would otherwise be subjected to competition with large-sized algae and to nutrient limitation (Mazumder *et al.*, 1986). Increased nutrient regeneration may also be caused by smaller zooplankton, which excrete nutrients at a greater rate (per unit biomass) than large zooplankton (Bartell, 1981; Vanni, 1987b).

d) planktivorous fish reduce the density of large-sized algae:

The removal of about 50 % of planktivorous fish was found to alter the planktonic community towards an increase in large-sized diatoms, green algae and cryptophytes in Lake Søbygård, Denmark (Søndergaard *et al.*, 1990).

e) increase in microzooplankton:

Microzooplankton (rotifers, nauplius larvae, ciliates, and heterotrophic flagellates) develops better when predatory pressure by zooplankton crustaceans is reduced (Richardson *et al.*, 1990; Mazumder *et al.*, 1990). Such an increase in zooplankton has been observed in a eutrophic lake near Wuhan city, China, in the presence of silver carp and bighead carp (Miura, 1990). In enclosures in a small humic lake, nearby Oslo, south-eastern Norway, Hessen *et al.* (1989) noticed an inverse relation between heterotrophic ciliates and crustacean zooplankton, the abundance of the ciliates increasing as that of crustaceans declined. Zooplankton filter-feeding on small suspended particles and bacteria also benefits from the increasing density of detritus resulting from fish feeding. In such conditions, fish may thus actually cause an increased zooplankton production (Hillbricht-Ilkowska & Weglenska, 1973).

f) specific increase in rotifers:

Rotifers increase in the presence of fish, if fish increase rotifer resources; the particles filtered by herbivorous rotifers are partly similar to those consumed by cladocerans. This leads to exploitative competition between both groups, in which cladocerans have the advantage. Therefore, suppression of cladocerans and an increase of algae results in an increase of rotifers (Richardson *et al.*, 1990; Mazumder *et al.*, 1990). The reverse is also true. In Lake Søbygård, Denmark, the removal of about 50 % of the planktivorous fish shifted the plankton community from a dominance of rotifers to large-sized cladocerans (Søndergaard *et al.*, 1990).

g) increased release of organic waste products in the presence of fish:

Organic matter released to the water from excrements, residual feed and sloppy feeding of fish, act as substrata for pelagic bacteria (Richardson *et al.*, 1990; Mazumder *et al.*, 1990). This enrichment of the environment by fish has been called ichthyeutrophication.

h) increase in pelagic bacteria:

As a consequence of increased dissolved organic load, pelagic bacteria increase in biomass; this boosts the microzooplankton, which is mainly bacterivorous. Bacteria are important in the bacterial food loop owing to their role in organic matter decomposition and regeneration of nutrients. Their activity controls the quality of water (Zweig, 1989). The predation pressure on bacteria is not reflected in their population size, because of their rapid growth, especially if availability of substrata is non-limiting (Hessen *et al.*, 1989).

i) increased nutrient regeneration:

The increase of suspended organic material entails a proliferation of decomposer bacteria, hence an increase of nutrient recycling rates and nutrient availability. This increase of nutrient availability favours efficient nutrient absorbers such as the picoplankton and nanoplankton. At the same time, bacterial regeneration of nutrients supplied by the feed and excretion leads to an increase in productivity (Richardson *et al.*, 1990; Mazumder *et al.*, 1990). Qualitative modifications of nutrient concentrations in the water, resulting from changes in grazing pressure in Norwegian lakes, were as follows: the selective retention of nutrients by grazing organisms changed the nutrient proportions in the water; the supply of nutrients was modified from a pulsed mode, when fish were dominant, to a steady mode, where zooplankton was dominant. The phytoplankton assemblage was therewith modified (Lyche *et al.*, 1990). In conditions of phosphorus limitation, increased grazing pressure results in a decreased algal biomass, which is accompanied either by a reduction of the phosphorus pool within the phytoplankton, or by a reduction of the intracellular C:P ratio of the phytoplankton. The removal of fish from the small eutrophic lake Mosvatn, Norway, reduced its internal phosphorus loading (Sanni & Waervagen, 1990), while the presence, for example, of yellow perch reduced phosphorus turn-over time by a factor 3 and increased phytoplankton biomass by a factor 1.5.

j) decreased sedimentation:
Small-sized algal cells settle less quickly than large-sized algae. They decompose in the water column and the nutrients released remain in the pelagic compartment (Richardson *et al.*, 1990; Mazumder *et al.*, 1990).

k) effect on the seston:
The size structure of the seston depends on its mode of formation. Part of the particles are zooplankton excrements. Modifications in species assemblage lead to changes in the types of excrements released and entail modifications in the cycling of nutrients within the water column. The dynamics of particulate and dissolved carbon and phosphorus in the water column and the dynamics of nutrients from the sediments are thereby also affected. In Lake Søbygård, Denmark, the removal of about 50 % of the planktivorous fish was found to reduce the internal loading of phosphorus, enhance the redox conditions of the sediment, and decrease sedimentation rate (Søndergaard *et al.*, 1990).

l) modifications at the sediment level
The decrease of the sedimentation rate enhances redox conditions and decreases the oxygen demand of the sediment, with a favourable effect on the benthic fauna. In Lake Søbygård, Denmark, the removal of 50 % of the planktivorous fish was found to stimulate the production of microbenthos and macrophytes (Søndergaard *et al.*, 1990).

m) long-term dominance of blue-green algae
Uncontrolled development of nanoplankton is considered to be responsible for long-term dominance of blue-green algae in lake Ringsjön, Sweden (Bergstrand, 1990). However, it seems that generalized conclusions cannot yet be drawn. Experiments involving a reduction of planktivorous fish populations in eutrophic lakes have resulted in effects differing between lakes: blue-green algae developed at different densities and showed various tendencies to form blooms. This disparity stems from initial differences in nutrient concentrations, mixing depths and water renewal rates (residence time) between lakes. Such parameters induce differences in the plankton community prevailing after fish reduction, e.g. through discrepancies in specific loss rates of algae.

The impact of planktivorous fish also decreases with increasing mean depth. Lastly, lakes show differences in their ability to create a reduction in dissolved phosphorus, relative to their phosphorus loading (Jeppesen *et al.*, 1990b), e.g. by the establishment of a macrophyte belt (Jeppesen *et al.*, 1990a).

Impact of phytoplanktivorous fish

The effect of phytoplanktivores on plankton communities is less well documented than that of zooplanktivores. Phytoplanktivorous fish are filter feeding, and tilapia are essentially pump filter feeders. Their food selection is passive, size-selective, retaining all material, larger than the interraker spacing, contained in the engulfed water. (A discussion on retention capacities of the filtering apparatus is found in Lazarro, 1987:116). Their effect on plankton is therefore different of that of visual feeders. Filter-feeding fish consume net phytoplankton, enhance nanoplankton and thus induce better water quality (Lazarro, 1987). Their impact on phytoplankton depends on fish feeding rate and algal growth rate. By removing large-sized algae, with slower growth rates, they favour small algae. However, environmental conditions affecting decomposition rates and nutrient regeneration, such as water temperature, mixing pattern, residence time, surface level fluctuation, and initial trophic state, tend to obscure the impact of filter-feeding planktivores.

As mentioned earlier (p. 165), the role of herbivorous invertebrates is less in tropical waters, and the importance of phytoplanktivorous fish on plankton communities (and on trophic state of the water body) is greater here.

a) on zooplankton

In pond experiments conducted in U.S.A., *Oreochromis aureus* reduced rotifers and increased copepodids and adult stages of a large calanoid copepod. The effect on plankton communities was to enhance large evasive zooplankters (Drenner *et al.*, 1984b). Similar effects were obtained with gizzard shad, which is a pump filter-feeding omnivore (Drenner *et al.*,1982 & 1984a)

b) on phytoplankton

The omnivorous silver carp was used in Poland to improve water quality and overcome algal blooms in Lake Warniak (Mazurian lakeland, Northern Poland) (Kajak *et al.*, 1975). Silver carp kept in pens within the lake reduced by 4.5 times the average biomass of phytoplankton regardless of density (450 or 1350 kg ha^{-1}). At the highest densities (1350 kg ha^{-1}), it reduced blue-green algae, but enhanced dinoflagellates and nanophytoplankton. The grazing of plankton cleared the water and the resulting fish faeces enriched the sediment benefitting the benthos. This effect was stronger in stratified lakes than in well-mixed or shallow waters. Silver carp resorts to feeding on bottom sediment during periods of low plankton or at high abundance of blue-greens. Filamentous blue-green algae, of the species *Oscillatoria agardhii*, or *Aphanizomenon flos-aquae* are not digested and if these species dominate, silver carp suffer high mortalities, if access to the lake bottom is denied. The increase in nanoplankton results from the silver carp's consumption of herbivorous zooplankton; the fish do not feed on nanoplankton, having access to particles larger than 20 μm only (Lazarro, 1987).

O. aureus (in the size range of 12.5 to 19.0 cm SL) reduced standing crops of the largest algae ($> 150\ \mu$m), but did not affect smaller algae ($> 15\ \mu$m), probably because of their higher growth rate which masked the impact of grazing. Smallest algae ($< 8\ \mu$m) were enhanced, owing to inefficient filtration, nutrient regeneration by fish, higher growth rates, algal enhancement during gut passage, and modification of the herbivorous zooplankton. Tilapia in the size range of the experiment were feeding on particles larger than 25 μm (Drenner *et al.*, 1984b).

Plankton consumption by fish

Our knowledge of the consumption of plankton by fish is still fragmentary.

Trophic links between fish and plankton are often loose, (1) because plankton is a changing assemblage of pelagic organisms of variable nutritional value. Some of the changes occur as sequels to the grazing or the predation, while other are linked to environmental abiotic changes (seasonality and nutrient availability). (2) The predatory pressure of fish on plankton changes (for a species) with age, with density, and with the presence of other species with similar feeding regimes.

As mentioned earlier (p. 145 & 168), some plankton species are inedible or of poor nutritional value, while some other species are readily consumed.

Blue-green algae are often regarded as a nuisance in ponds, because indigestible to fish and reference was made to observations of Fish (1955) that *Microcystis* is not digested. It has been shown that indigestibility of the alga by fish resulted from the fact that diel rhythms of digestion were not appreciated (Trewawas, 1983). The non-consumption of blue-green algae has also been contradicted with regard to *O. niloticus* by observations of Moriarty (1973), Moriarty & Moriarty (1973), Moriarty *et al.* (1973), Bowen (1982). Tilapia and silver carp are indeed capable of digesting and assimilating all but some toxic blue-green algae (Gophen, 1990, in Lake Kinneret, Israel; Leventer & Teltsch, 1990; Miura, 1990).

Moreover, how selective is fish feeding on plankton? The impact of fish on plankton depends on the selectivity of its feeding, which in turn depends on its mode of feeding. Visual feeders select larger preys, while filter-feeders capture non-evasive preys, within a certain size range.

When discussing selectivity of feeding, feeding efficiency of fish must be considered under the angle of maximization of net energy intake per unit time and type of prey eaten, and the cost of prey capture must be assessed in terms of energy and time spent (Lazarro, 1987).

As discussed earlier (p. 158), changes in the plankton assemblage induce concurrent modifications in the volume of edible food for fish.

Further, many fish are facultative planktivores, feeding also on other material, such as suspended organic material (detritus), periphyton, macrophytes, and benthos. Many are moreover opportunistic feeders, switching feeding mode, according to prey densities and environmental conditions (light or competitors). For example, Eggers (1982) found that sockeye salmon in Lake Washington preferred large non-evasive preys as long as they were available, but resorted to evasive preys in their absence.

Predation on zooplankton or phytoplankton induces reaction within the community. Appearance of predator-induced defenses among zooplankton, including inedible or distasteful species of animals or algae.

For example, silver carp, an obligate planktivore, feeds on particles above 20 μm (i.e. larger than nanoplankton), reducing the abundance of blue-green algae and total phytoplankton biomass. Introduced into some reservoirs of Israel, it was able to prevent the development of many blue-green algae. It acted as a biological filter for the ecosystem, removing important quantities of suspended particulate material, feeding on algae, bacteria, zooplankton, and detritus, even ingesting inert clay particles. However, a side-effect was that dinoflagellates and nanoplankton increased in abundance. and silver carp became unable to control these algae. In lake Kinneret, at the end of summer, silver carp even become zooplanktivorous (Gophen, 1990). In experiments conducted in small carp ponds (320-2500 m^2) in Königswartha, Germany, Barthelmes (1989) found a positive relationship between silver carp density and phytoplankton and a negative one with zooplankton density. These relationships are, however, not extendable to any kind of situation and might be only valid during a restricted period of time in that given situation.

Synergism between aquatic organisms

Changes in plankton assemblage resulting from the presence of planktivores are illustrated by the following example. In a eutrophic lake near Wuhan city, in China, the abundance of blue-green algae of the genus *Microcystis* decreased in the presence of silver carp (Miura, 1990). The reduction of blue-green algae as a result of increased grazing pressure enhanced the green algae biomass, resulting in an increase in food supply through a path of the food chain connecting the green algae to the bighead carp via the zooplankton. The increase of small green algae was favourable to herbivorous zooplankton and eventually to the bighead carp. The bighead carp benefitted of this reduction of blue-green algae biomass and its growth was fastened. This improvement of water quality beneficial to the bighead carp is an example of positive synergism (Leventer & Teltsch, 1990).

Interactions between fish feeding regimes

In comparison to facultative planktivores, *Tilapia rendalli*, *Lepomis macrochirus* (the blue-gill), and *Colossoma macroponum*, the same silver carp had a different impact on plankton in limnocorral experiments in a reservoir in Brazil. The silver carp now considerably reduced the abundance of copepod nauplii, cladocerans and rotifers; if tilapia and blue-gills were also present, the density of rotifers increased. Blue-green algae increased in the presence of all fish, except silver carp (Starling & Rocha, 1990).

Another interaction between species was that tilapia spent more time collecting surface-floating prey than filter-feeding in midwater in the presence of other fish (Starling & Rocha, 1990).

Influence of fish on the decomposition cycle

The presence of a large biomass of fish must be compensated by a biomass of decomposers sufficiently large to recycle the amounts of organic matter released. Without these decomposers, water quality will inevitably deteriorate and production will be affected. Moreover to sustain algal growth and compensate for grazing a steady supply of nutrients is required. Decomposer activities is therefore essential for the maintainance of production. The autotrophic and heterotrophic cycles are henceforth complementary (Gophen, 1990). Fortunately, bacterial decomposer populations respond quickly to internal organic matter loadings, provided their oxygen requirements are met. Their nutrient recycling starts up plankton production, the level of which will correspond to nutrient availability.

Biomanipulation and relations within the aquatic environment

The network of interrelations between the organisms present highlights the fundamental role of nutrients in pond production system. Changes in the assemblages of aquatic organisms eventually backfire on nutrient fluxes and their relative proportions. The complexity of such couplings is best illustrated by biomanipulation experiments.

Biomanipulation, as termed by Shapiro *et al.* (1975), is a tool which has only recently been used by aquatic biologists. Biomanipulation is the management of aquatic communities with a view to improving water quality (Gophen, 1990), or manipulations of aquatic organisms (and their habitats), in order to elicit responses beneficial to the environment (Shapiro, 1990). The effects of mixing fish with different feeding regimes, such as benthophagous and planktivorous fish, on the structure of plankton communities has been the subject of biomanipulation and mesocosm experimentation (Qin & Threlkeld, 1990; Richardson *et al.*, 1990; Post & McQueen, 1987; Mazumder *et al.*, 1990; Benndorf *et al.*, 1989; Vostradovský *et al.*, 1989; Barthelmes, 1989; Brandl *et al.*, 1989). A variety of interactions between fish with different feeding regimes and the plankton were revealed. The ultimate objective is to predict the evolution of a plankton population under natural conditions. The cascade of effects is sometimes surprising. For instance, the impact of fish on water quality had never been seriously considered (Gophen, 1990).

Responses to biomanipulation are determined by the trophic state of the water body; this explains some differences in the results of biomanipulation on food webs (Gophen, 1990).

According to Post & McQueen (1987), biomanipulation of planktivorous fish as a technique for the control of the abundance of phytoplankton is largely ineffective. Meijer *et al.* (1990a), in experiments in ponds in the Netherlands, support this statement, estimating that zooplanktivorous fish have no significant effects on algal biomass, nor on total nutrient concentrations. However, Lazzaro *et al.* (1991) consider that control of phytoplankton biomass by fish depends on the fish - phytoplankton biomass relationship. If the relationship is linear, a reduction in zooplanktivores at high biomasses produces improvements in water quality. If the relationship is sigmoid, a reduction in zooplanktivores will not result in a commensurate reduction in phytoplankton biomass.

Control of planktivorous fish by piscivorous fish

The use of piscivorous fish at a reasonable density is beneficial to production. It prevents planktivorous fish from preying too heavily on zooplankton and it rebalances the whole food web (Shapiro *et al.*, 1975; Bennsdorf *et al.*, 1989). In pond culture in Central African Republic, the African catfish (*Clarias gariepinus*), an omnivore, has been used to control tilapia population. A good proportion of catfish to tilapia has been found to be 14 to 20 % (in biomass); tilapia (*Oreochromis niloticus*) initial stocking weight were 10 to 25 g and that of catfish was 6-10 g. In these conditions total fish yield reached 8000 to 12000 kg ha^{-1} yr^{-1} and no (or only few) tilapia fingerlings were found at the harvest (Delincé, unpublished data).

Piscivorous fish are grouped into a feeding mode which is gape-limited and one which is not. This aspect must be considered in the choice of predatory fish for the control of forage fish (Zaret, 1980; Gophen, 1990).

Importance of fish on plankton production

1. The effect of fish on artificial environments is deleterious for water quality. The capacity of water for neutralizing metabolites, carbon dioxide and ammonia, released by fish, as well as

the left-overs of feed, is limited. This process of nutrient supply to ponds and natural environments, with a certain autotrophic capacity, leads to ichthyoeutrophication.

2. The impact of fish on the environment depends on its mode of feeding: benthophagous, phytoplanktivorous or zooplanktivorous. The effect on a group of prey organisms diffuses and becomes dissipated in a cascade of effects through the network of interrelations within the foodweb.

3. The effects of fish on the environment are, in first instance, the result of feeding activities. These effects are difficult to investigate because the diets of fish change over time, are often plastic, a great variety of food being consumed, and because predation or grazing induces changes in the dynamics of the prey populations.

4. The interactions of communities of organisms within the environment reveal the existence of positive synergisms between groups of organisms, such as the grazing on blue-green algae by fish, benefitting overall water quality. Fish species with similar feeding regimes have to share the available food resources and their feeding regimes will adapt to reduce direct competition.

2. Fish production in a pond

2.1. Stocking, management practices and production

Types of management operations in a fish pond

To conduct fish production in a pond, several operations must be performed: stocking with fingerlings, feeding the fish, and fertilizing the pond, monitoring fish growth, and harvesting the fish. The farmer must decide on the choice of the species of fish to culture, the stocking density, the type of feed and feeding rates, the type of fertilizers and fertilization rates, and the length of the culture period.

Fertilization has already been discussed; the choice of artificial feed and the feeding rates will not be discussed here. Selection of fish for fish farming includes scrutinizing available species for an array of desirable characteristics, including fast growth rates at high stocking densities leading to high fish yields, resistance to manipulation and to diseases, ease of spawning in captivity, high fecundity, wide feeding regime and ready acceptance of artificial feed, good food conversion rates, good market value, etc. The species currently most cultured in Africa are tilapia and catfish (Table 22).

The best tilapia species in terms of growth rate is *Oreochromis niloticus*. It is also the most commonly cultured species, though in some countries it is not available or some other species are cultured, because they are better adapted to local farming conditions, such as brackish water (*Tilapia guineensis*, *Oreochromis mossambicus*). Information on distribution of tilapia species in Africa is found in Philipppart & Ruwet (1982) and in Trewawas (1983).

Tilapia hybrids are used to avoid some inherent problems of tilapia culture. Unselected strains of tilapia in high density culture conditions show precocious maturation and prolific breeding. This results in reduced individual growth rate and a production

Table 22. Species names of most common farmed fish in Africa.

Tilapia	Catfish	Other potential species	Introduced species
Oreochromis niloticus *O. aureus* *O. mossambicus* *O. hornorum* *O. macrochir* *Sarotherodon galilaeus* *Tilapia guineensis*	*Clarias gariepinus* *Heterobranchus longifilis* *Chrysichthys nigrodigitatus*	*Heterotis niloticus* some species of *Labeo*, *Citharinus*, and others.	*Cyprinus carpio* (common carp) *Ctenopharyngodon idella* (grass carp) *Micropterus salmoides* (black bass) *Salmo trutta* (trout)

of fish of low average weight. Food ingested by female fish is diverted to production of gametes and this energy is lost to weight increase (resource allocation principle). To curb the breeding activity of the fish, different techniques exist: (1) fish are are selected on the basis of fast growth rates and late maturity. It has been shown that all the traits of gonadal development and body development of *O. niloticus* are genetically controlled and mutually independent. Selection for late maturity can therefore be efficiently combined with selection for fast growth (Kronert *et al.*, 1989; Oldorf *et al.*, 1989); (2) All-male tilapia are produced essentially by three techniques: hand-sexing of adult fish (> 30 g of body weight), hormonal treatment, and hybridisation. (Castration or neutering of fry by chemicals or radiation has also been attempted). The hormonal treatment is a sex-inversion, successfully achieved with various androgens (methyltestosterone) and estrogens and to which fry are submitted after yolk-sack resorption. Hybridisation is achieved by crossing two pure species which produce 90-100 % male hybrids. *O. niloticus* x *O. aureus* combines fast growth and hardiness, but only 80 % are males; *O. niloticus* x *O. hornorum* and *O.mossambicus* x *O. hornorum* gives 100 % males, of which the *hornorum-niloticus* cross gives better growth performances. The number of hybrid crosses available for all-male hybrid production is limited. It depends on obtaining pure species and on an ability to maintain this purity. The sophistication of this technique is adding costs to the culture as well (McAndrew, 1988).

Tilapia hybrids are currently produced in some private farms in Africa, e.g. in Ibadan, Nigeria, in Mombassa, Kenya, but to a much larger extent in farms in Israel, in the Philippines, in Taiwan, in Hong Kong, and in U.S.A. To start with this technique, a batch of pure strain tilapia of the species mentioned as well as holding facilities which will ensure isolation of the fish are required. A live collection of pure strains of tilapia are held at the Aquaculture Institute of the University of Sterling, in Scotland. Farming hybrid tilapia in Africa considering local facilities, will need the production, within the farm itself, of the hybrid fish, as purchase of the fish from the outside will imply high costs.

Other species with high production potential are the catfish species, *Clarias gariepinus* (formerly known as *Clarias lazera*) *Heterobranchus longifilis* and *Chrysichthys nigridigitatus*, more adapted to brackish water farming. All these species were difficult to breed in captivity, but breakthroughs for mass breeding these fish have been achieved in Bangui, C.A.R, for *C. gariepinus*, and in Layo, near Abidjan, Côte d'Ivoire for the two other species.

Species with high potential include some *Labeo*, *Citharinus*, *Bagrus*, and others and more research on their real growth rates is needed. Among the introduced species, the common carp is the most commonly cultured. In some countries, Chinese carp have been also introduced.

The monitoring operations of the farmer must be directed at maintaining a good water quality, while maximizing fish biomass in the pond. Control of water quality aims at keeping the fish in optimum ecological conditions, i.e. meeting its oxygen requirements, avoiding exposure to harmful ammonia, and ensuring optimum limnological conditions for maximum primary productivity. Control of fish growth involves the adjustment of feeding rates and the length of the culture period. To be monitored are the increase of individual weight of fish (growth rate), its increase in biomass, and the utilization of feed by the fish.

The role of the fish farmer is primarily to adjust the conditions to his aim of obtaining a high production in the most profitable way. Each intervention of the fish farmer must carefully weigh the supplementary cost involved against the extra yield or benefit gained.

Importance of growth rate and production rate

The absorption of food leads to an increase in weight of individual fish, measured by the growth rate, the increase in weight per unit time (usually per day). The specific growth rate is the daily increase divided by the weight of the fish. Growth is a specific characteristic of fish and varies within a given range. Growth rate is usually greater in young animals and decreases with age. Fish of low initial average weight grow proportionally faster than fish of larger size, but, in absolute terms, their increase in weight is lower. Differences in growth also exist between strains of the same species. Growth can thus be improved by genetic selection. Graphically, growth is represented by an S-shaped curve: at the beginning, the rate is low, then steeply increases until it reaches an inflexion point, and levels off afterwards.

At the population level, the biomass stocked represents a given weight per unit surface area, and the increase of cumulative individual weight represents fish production. It appears that, at low loading, the increase is slow, increases rapidly for a given biomass/volume ratio and levels off when fish biomass gets too large for the available space. The maximum fish biomass for a given space is called the carrying capacity (Fig. 56). The graphical representation of the increase of biomass is also an S-shaped curve. The carrying capacity concept was brought forward by Chen & Prowse (1964) and appears to be the limiting factor at high fish density (Edwards *et al.*, 1981). This phenomenon is controlled by a so-called crowding factor, which combines different physiological and behavioural factors affecting growth, defying strict physiological interpretation (Brett, 1979). Growth rate is affected by the rate of exchange of oxygen and carbon dioxide and by feeding rates, but also by behavioural factors. In territorial species, they include social interaction involving dominance, restricting free movement and access to food. In schooling species, fish feeding is promoted by social stimulation, which is lacking at low densities, resulting in poor growth.

2.1.1. Feeding operations

Importance of feeding

Fish production at or below carrying capacity of the pond can be considered a function of the food available to the fish. This food can be subdivided into three categories: the natural food, produced by the natural productivity of the pond, the extra food produced by fertilization of the pond, and the artificial feed cast into the pond.

The fish farmer has a variety of artificial feeds at his disposal. The efficiency of its transformation into fish depends on the qualitative value of the food, i.e. the quantity of food consumed per unit of increase expressed by a feed conversion ratio (FCR), i.e. the

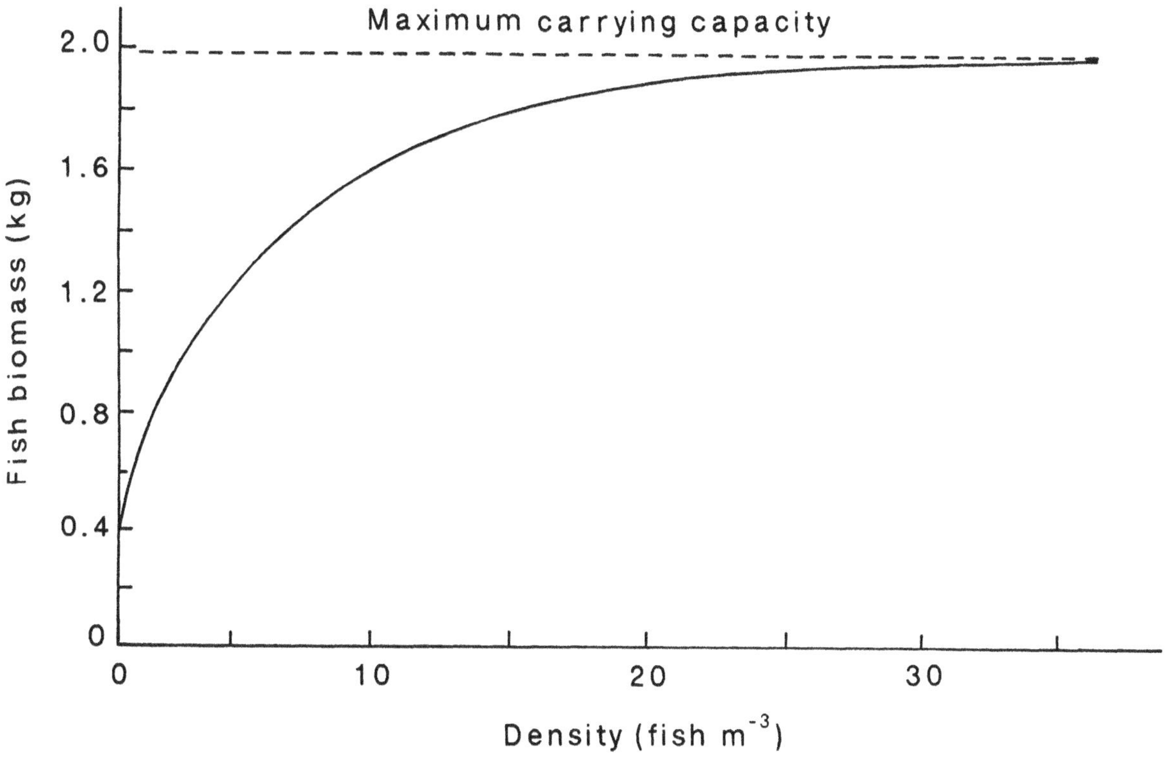

Fig. 56. Tilapia biomass at harvest as a function of fish density in concrete ponds. The fish biomass reached at the highest densities is close to the carrying capacity of the culture system (modified from Edwards *et al.*, 1981).

ratio of the amount of feed offered over the increase of biomass. The food should not only be available in sufficient amounts, but should also suit the nutritional requirements of the species cultured. Usually some amount of natural food is available and artificial feeding is thus supplementary. The artificial feed is never completely consumed and the left-overs fertilize or even pollute the pond. Abundant consumption of feed by fish produces an important volume of excreta, which also fertilize the pond.

Influence of feed on growth rate

The influence of feed on the growth of different fish species is illustrated by experiments of Moav *et al.* (1977) in Israel. In polyculture experiments, they fed silver carp, tilapia (*O. aureus* or hybrids) and common carp with high- and low-level protein feed, at varying stocking densities. The results (Fig. 57) are measures of the growth potential of the three species in the different culture conditions. The growth of common carp was significantly increased by high-protein feed, while tilapia and silver carp met their protein requirements using the natural food available in the pond. A negative effect of density (crowding) was more evident in common carp than in silver carp or tilapia.

The fish communities increased in weight in the course of the culture period and the daily weight increases were regular (Fig. 58). The end of the growing period was reached when a decrease in the average daily increase occurred. The study of Moav *et al.* (1977) showed a sharp increase in FCR at the end of the culture period (Fig. 59). This was explained by a deterioration of culture conditions and an arrestation of growth. In the context of intensive feeding, other aspects, such as frequency, nature and form of feed, i.e. pellets, dough, meal, must also be taken into account.

In experiments conducted in small containers, Opuszyński *et al.* (1989) noted that fingerlings of various species reacted differently to food composed mainly of zooplankton: the common carp and the bighead carp grew quickly when fed zooplankton *ad libitum*, while in the same conditions silver carp and grass carp showed low growth rates.

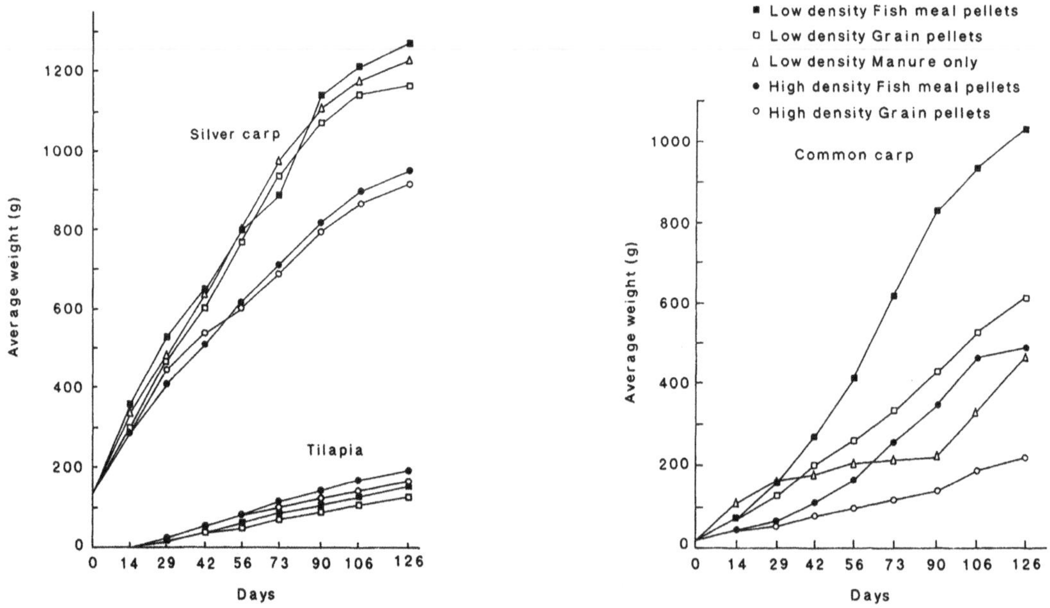

Fig. 57. Comparative growth curves of common carp, silver carp and tilapia submitted to five different treatments. The average size of the ponds was 400 m^2, total stocking rates were 19700 fish ha^{-1} (high) and 6380 or 6130 fish ha^{-1} (low), i.e. 11450 fish ha^{-1} (high) or 3300 fish ha^{-1} (low) for common carp, 2500 fish ha^{-1} (high) or 1250 fish ha^{-1} (low) for silver carp, 5000 fish ha^{-1} (high) or 1500 fish ha^{-1} (low) for *Oreochromis aureus*. The remaining fish were grass carp (from Moav *et al.*, 1977) (with kind permission of Dr. Wohlfarth, on behalf of the authors, & Aquaculture).

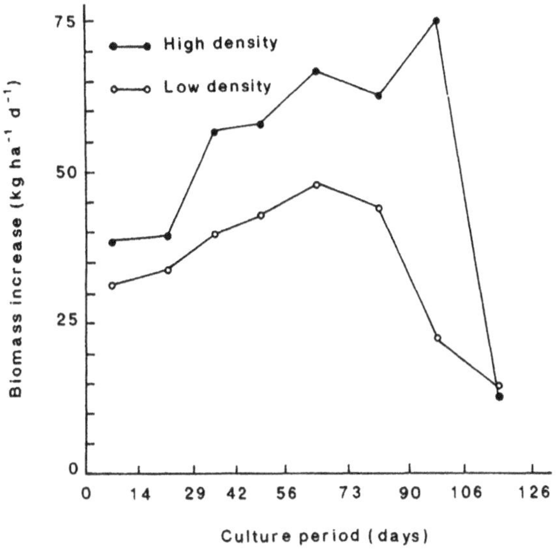

Fig. 58. Calculated average daily yield increments at high and low fish densities in the course of a production period. The stocking rates and other production parameters were the same as in the experiment of Fig. 57. The fish were fed predetermined feed ratios with high-protein fish meal pellets (from Moav *et al.*, 1977) (with kind permission of Dr. Wohlfarth, on behalf of the authors, & Aquaculture).

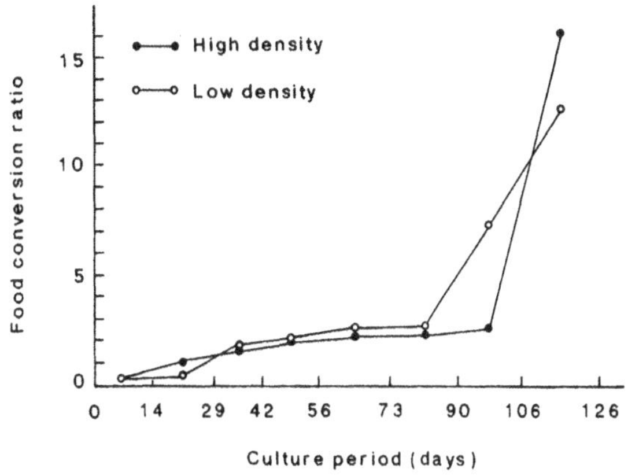

Fig. 59. Evolution of feed conversion ratios at high and low fish densities in the course of a production period. The pond conditions and feeding rates are the same as in Fig. 58 (from Moav *et al.*, 1977) (with kind permission of Dr. Wohlfarth, on behalf of the authors, & Aquaculture).

Relations between feeding rate and production rate

Production depends directly on the amount of available feed, as shown by the experiments of Edwards *et al.* (1981). In trials conducted in cement tanks using *O. niloticus* at high densities (25 fish m^{-2}), the weight of the fish harvested after 12 weeks of culture was directly dependent on the amount of phytoplankton available (Fig. 60). The fish were fed this phytoplankton at concentrations ranging between 20 to 70 mg l^{-1}. At low fish densities, when average plankton concentration increased from 50 mg l^{-1} to 80 mg l^{-1}, the amount of fish harvested did not change much, indicating an excess of available food. At higher densities, the relationship was more obvious. When feed was abundant, fish production became more or less constant; when the fish density was low, individual growth was high, and, with increasing fish density, the decrease of individual growth was compensated by an increase of the biomass at harvest. This range of phytoplankton, 20 to 70 mg l^{-1}, corresponded to a range in increase in fish biomass of 1 to 3 (Fig. 60). Continuous changes in phytoplankton composition, but without appearance of blue-greens (*Microcystis*) euglenoids (*Euglena*) or

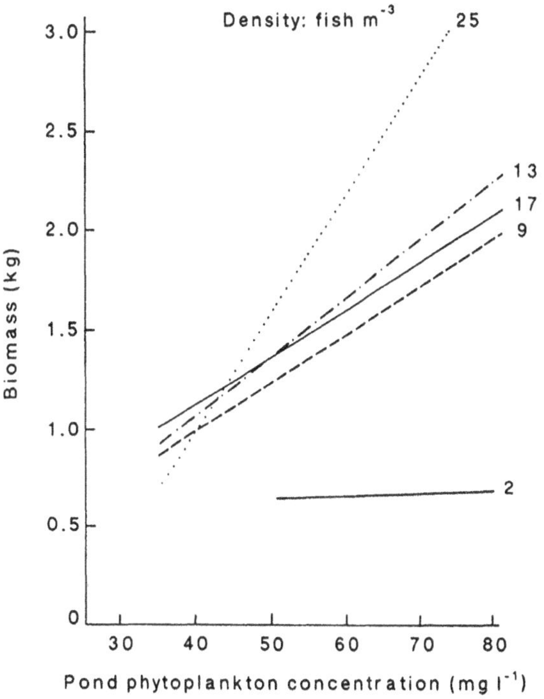

Fig. 60. Tilapia (*Oreochromis niloticus*) biomass at harvest as a function of different pond phytoplankton concentrations and different fish densities. The tilapia were held in concrete ponds (4 m^3), fed with algal-laden high-rate stabilization pond effluents, for a period of 12 weeks (from Edwards *et al.*, 1981) (with kind permission of Prof. Edwards, on behalf of authors, & Aquaculture).

cryptophyte species (*Cryptomonas*), did not affect the feeding of tilapia. Phytoplankton composition comprised 32 genera, 17 green algae, 6 blue-green algae, three euglenoid algae, three diatoms, one dinoflagellate and one cryptophyte.

The use of feed was more efficient at high densities. The estimated feed conversion ratio (FCR) varied between 1.3 and 2.5 (Edwards *et al.*, 1981).

2.1.2. Stocking density and growth

Definition of stocking density

The optimum stocking of a fish pond is that amount of fish released into the pond at the start of the production period, which guarantees the highest possible yield of fish on harvesting. This depends on pond productivity and on the specific potential of the fish species used.

Density is expressed as the number of fish per unit surface area or volume in a pond (m^{-2}, ha^{-1} or m^{-3}). Some authors distinguish between stocking density and population density. By definition, the first designates the initial density, while the second considers the actual density at any time of the culture period. Differences between these two densities result from mortalities, which occur during the course of the culture or are a consequence of stocking manipulations, and reproduction occurring during the culture period. The farmer is concerned with the stocking density, which he decides on, and the density at the harvest.

Relations between density, growth rate and final mean weight

Edwards *et al.* (1981) quantified the effect of fish density on mean average weight at harvest of *O. niloticus* cultured in cement tanks of 4 m^3 and fed phytoplankton originating from sewage stabilization ponds. Average individual weight of the fish after 12 weeks of culture at 1, 10, 20, and 30 m^{-3}, decreased with density (Fig. 61). At a density of 1 fish m^{-3}, the average weight at harvest (after 12 weeks) was little above 120 g, while at 10 fish m^{-3} the average weight remained around 20 g and less for higher densities. Grouping the data according to increasing density illustrates the exponential decrease of mean weight at harvest (Fig. 62) which results.

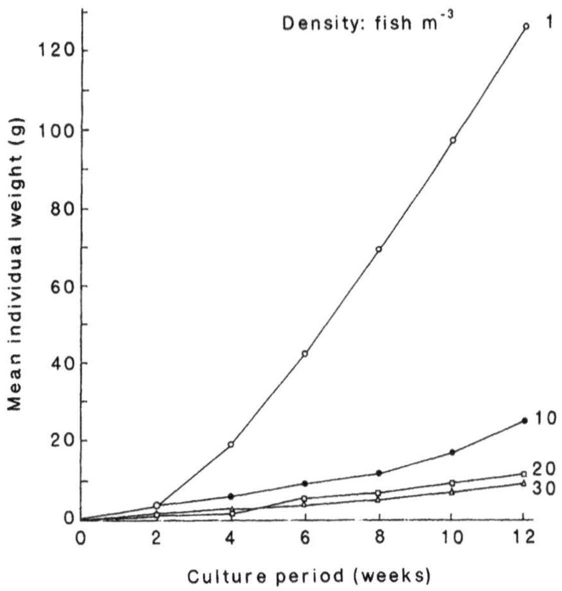

Fig. 61. Evolution of average weight of tilapia (*Oreochromis niloticus*) at two weeks intervals in the course of a culture period as a function of different fish densities and under pond phytoplankton concentrations corresponding to ca. 40 mg l^{-1}. The tilapia were held in concrete ponds (4 m^3) (modified from Edwards *et al.*, 1981).

Individual growth and fish production are in direct relation with population density (Fig. 61, 56, 62 and 63). At low density, the individual growth rate is high and production low, while at high density, the individual growth is low, but production is high. The fish farmer must compromise between a good production and an acceptable individual growth rate, resulting in larger amounts of marketable table-size fish.

Table 23 lists various data on growth of tilapia obtained at different densities in production ponds. It aims at showing the influence of density on growth in real production conditions. The influence of an increase in density can be counterbalanced by artificial aeration, as shown by experiments of Sin & Chiu (1983) obtained in Hong Konk, but growth rates varied widely between experiments indicating a strong influence of food. In indoor tanks experiments, in Belgium, Mélard (1986) explained 80 % of the variance in growth of inbred *O. niloticus* by the following intrinsic and extrinsic variables: initial average weight, gender of the fish, food, feeding frequency, dissolved oxygen concentration, water temperature, and illumination.

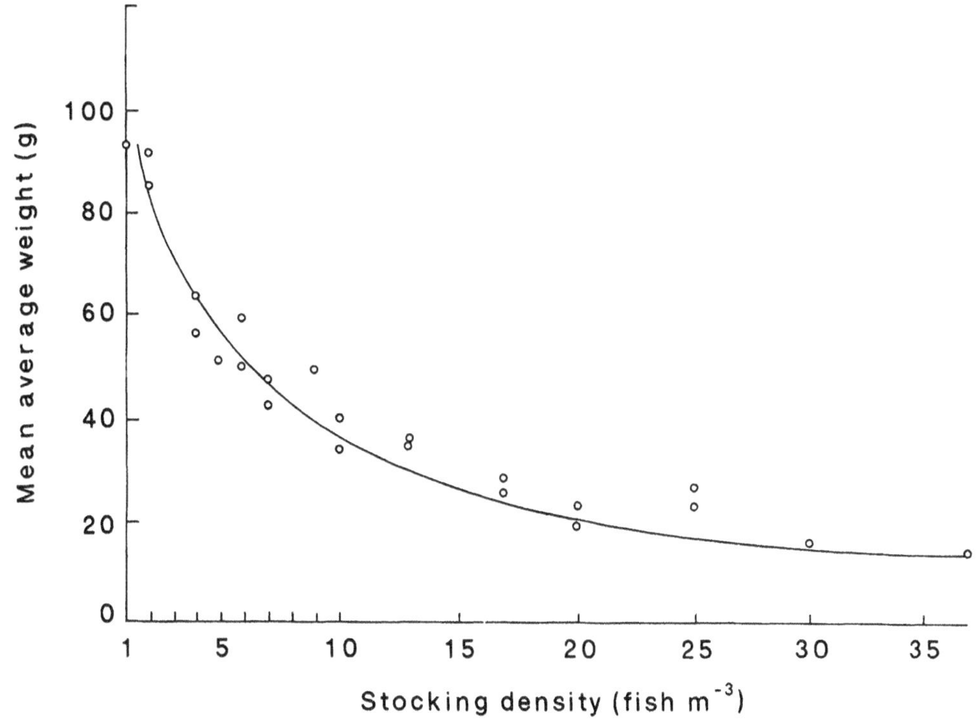

Fig. 62. Mean average weight of tilapia (*Oreochromis niloticus*) at harvest as a function of stocking density. The fish are harvested after a culture period of 12 weeks in concrete tanks (4 m^3) and held under pond phytoplankton concentrations varying in a range of 55 to 60 mg l^{-1} (modified from Edwards *et al.*, 1981).

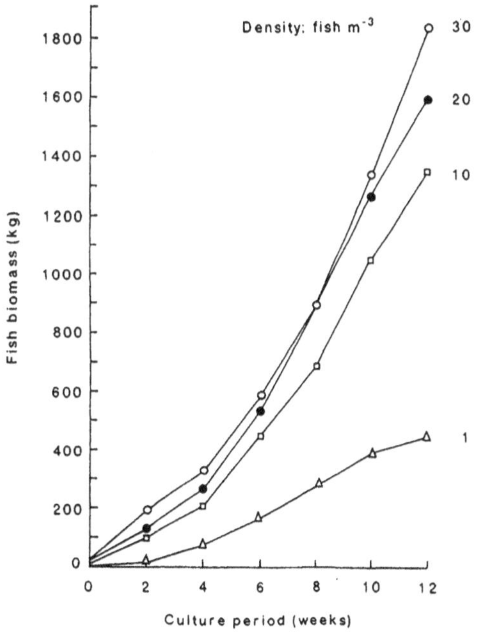

Fig. 63. Evolution of total weight of tilapia (*Oreochromis niloticus*) stocked in concrete ponds (4 m^3) and sampled at two weeks intervals during a culture period as a function of different fish densities and under pond phytoplankton concentrations corresponding to ca. 55 mg l^{-1} (modified from Edwards *et al.*, 1981).

Density and reproduction of tilapia

In tilapia culture, farmers hope that an increase of stocking density will lead to a reduction in reproductive rate of the fish, and thus solve the problem of uncontrolled reproduction (see p. 188). Yet, even at very high densities, this fish continues to reproduce. Edwards *et al.* (1981) observed spawning in tanks of 4 m^3 at densities of 2 to 25 *O. niloticus* m^{-3}, for culture periods of 12 to 18 weeks. Mélard (1986) noticed spawning in growth trials of *O. niloticus* in indoor tanks, outdoor cages and small ponds, in Tihange, Belgium at every density below 500 fish m^{-3}. Densities above that value are not used in farming.

2.1.3. Stocking density and production

Influence of density on production rate

An increase of total biomass with increasing density

Table 23. Daily average growth rate of tilapia, *O. aureus* and hybrids, at different initial densities and stocking weights.

Initial density (ha^{-1})	Average initial weight	Daily weight increase ($g\ d^{-1}$)	Species *Oreochromis*	Remarks and references
59000	6.6	1.01	hybrids:	Sin & Chiu (1983)
122000	5.0	0.69	*mossambicus* x *niloticus*	
30000	111.7	1.40	hybrids:	no aeration
111000	97.7	1.30	-idem-	aeration
				same authors
20000	16.7	1.12	hybrids:	Verani *et al.*
30000	14.6	0.85	*niloticus* x	(1983)
40000	15.0	1.00	*hornorum*	
50000	14.0	0.80		
20000	fingerlings	2.25	hybrids:	Sarig & Arieli
100000	fingerlings	1.19	*niloticus* x *aureus*	(1980)
		At harvest		
460000	0.8	0.32	hybrids:	Sarig & Arieli
700000	0.8	0.15	-idem-	(1980)
20000	279	2.40	males of *O.*	Sarig & Arieli
30000	265	1.96	*aureus*	(1980)
60000	240	3.39		
100000	243	1.48		
1800	100	3.8	*O. aureus* in	Sarig & Arieli
3000	190	2.8	polyculture with	(1980)
4000	175	3.3	carps	
4900	150	2.4		

has been shown in experiments by Edwards *et al.* (1981). *O. niloticus* were cultured in cement tanks of 4 m^3 and fed phytoplankton. The densities tested were 1, 10, 20, and 30 m^{-3} (Fig. 63). A correlation between total biomass at harvest and density was established, shown in Fig. 56. Total biomass increased exponentially with increasing density until the carrying capacity of the cement tanks (2.0 kg) was reached. In the experiment, under apparently non-limiting food conditions, a density of 10 fish m^{-3} permitted an extrapolated production of 11043 kg ha^{-1} yr^{-1} of fish to be obtained. Higher densities (20 and 30 fish m^{-3}) increased the fish yield to 12240 kg ha^{-1} yr^{-1} and 13679 kg ha^{-1} yr^{-1}, respectively.

Details of pond production trials conducted in Israel and Taiwan at various densities are listed in Table 24.

Fish used were hybrid tilapias or *O. aureus*, in monoculture or polyculture, at densities varying between 0.3 and 10 fish m^{-2}. Sin & Chiu (1980) also tested the influence of artificial aeration, which was beneficial and allowed, in fact, a doubling of production. Information with regard to artificial aeration of ponds in Israel is not clear.

Table 24. Daily average weight increase, per fish and per surface area, of tilapia, *O. aureus* and hybrids, at different densities.

Initial population density or at harvest (ha^{-1})	Daily weight increase per fish (g d-1)	per ha (kg d^{-1})	Species cultured *Oreochromis*	References and remarks
59000	1.01	52	hybrid tilapia:	without aeration
122000	0.69	67	*mossambicus* x	with aeration
			niloticus	Sin & Chiu (1983)
30000	1.40	41.7	-idem-	without aeration
110000	1.30	133.4		with aeration
				same authors
20000	2.25	45	hybrid tilapia:	Sarig & Arieli (1980)
100000	1.29	129	*niloticus* x	
			aureus	
18000	2.75	50.3	*O. aureus*	male monoculture
20000	2.40	46.0		Sarig & Arieli (1980)
30000	2.50	75.6		
30000	1.96	59.0		
59000	1.75	103.0		
60000	2.39	143.4		
80000	2.00	163.0		
100000	1.48	146.0		
3000	2.80	8.5	*O. aureus*	polyculture with
4000	3.30	13.2		common carp
4900	2.40	11.1		Sarig & Arieli (1980)
5740	2.53	14.3		
7000	3.10	22.0		
8720	2.87	24.6		
9000	3.30	29.0		
10000	5.50	55.0		
20000	3.30	66.0		

For Israel, yield versus stocking density is analyzed graphically on data from Table 24. The relationship is an exponential increase (Fig. 64).

Appropriate densities

In the past, a density of 2 m^{-2} or 20,000 ha^{-1} was recommended for *O. niloticus*. The results obtained in Israel, Hong Kong, and other places, indicate that this density can be notably increased, with intensification of production conditions. Under a regime of abundant feeding, with high-protein pellets, regular fertilization, and intensive organic fertilization, the stocking densities to be recommended are 3 to 6 fish m^{-2}. With supplementary aeration of the ponds, at least at night, 6 to 10 fish m^{-2} can be stocked. The intensification of production conditions should be accompanied by a selection of either monosex male fish, hybrids or, at least, superior strains.

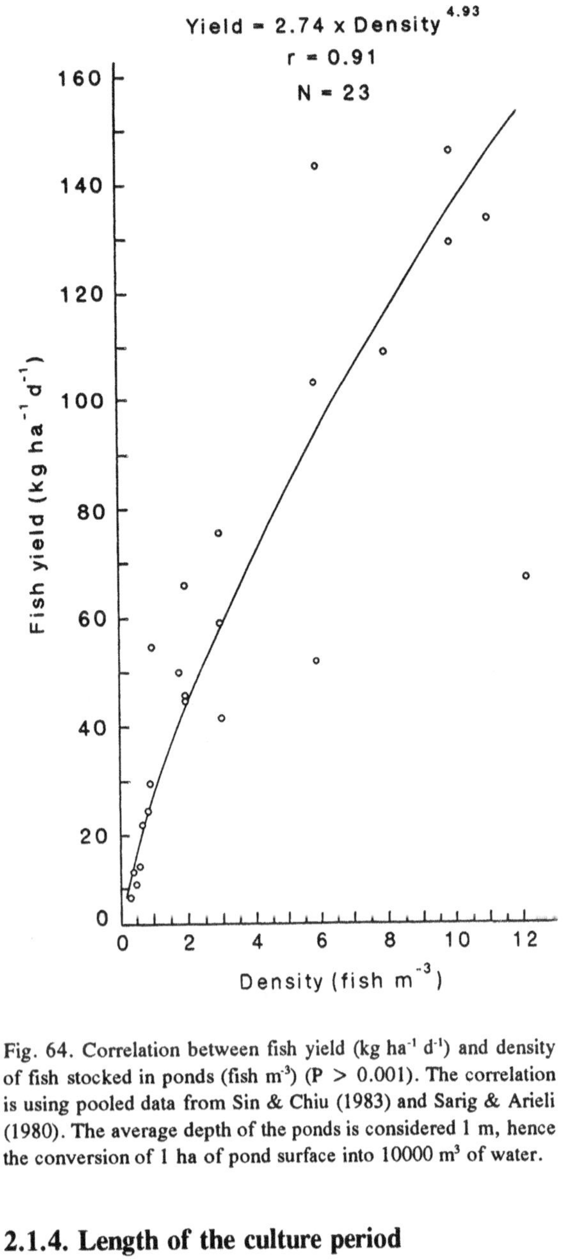

Fig. 64. Correlation between fish yield (kg ha^{-1} d^{-1}) and density of fish stocked in ponds (fish m^{-3}) ($P > 0.001$). The correlation is using pooled data from Sin & Chiu (1983) and Sarig & Arieli (1980). The average depth of the ponds is considered 1 m, hence the conversion of 1 ha of pond surface into 10000 m^3 of water.

2.1.4. Length of the culture period

The length of the culture period in a pond is determined by initial stocking density and the growth rate of the fish. The stocking density should be situated within the range of fast growth rates. Harvest should occur before growth levels off, close to the maximum carrying capacity. The culture period should only be as long as significant increases in biomass occur. If the average size (or weight) of the fish should not have reached a marketable size at maximum carrying capacity, the farmer will have to choose between two possibilities: reduce the fish biomass by selective shinning-out, which will remove the largest fish (intermediate cropping), or transfer the fish to a larger pond. The first solution has the disadvantage of removing the fish with the highest growth rates, leaving only the slow-growers in the pond. Also, if culture conditions have started to deteriorate, the fish remaining in the pond will not fully benefit from this increase in 'breathing space'. The second solution is adopted in the Philippines in the culture of shrimp and milkfish. Here, growout ponds are constructed in modules of three ponds of increasing size, the areas of which vary in a ratio of 1:2:4 or 1:3:9. The fish are transferred from one pond to the other in the course of the culture period by opening sluice-gates and allowing smaller ponds to drain into larger ones (dela Cruz, 1983).

2.2. Fish production and production factors

2.2.1. Production and production factors

Definition of fish production

The fish production of a pond is the amount of fish gained at harvest. Gross yield is the total weight of fish harvested from the pond, and net yield the difference between total weight of fish harvested minus stocking weight. It is conventionally expressed as weight unit per hectare and per day or per year, though it is understood that seldom a production lasts for a whole year. The calculation of yield per ha and per year is extrapolated linearly from harvest size, culture period, and culture surface area.

Benefits to the producer will be maximized when the production of table-size fish is greatest. Therefore, total production should be distinguished from market-size production. Total production includes the production of fingerlings, as most fish, especially tilapia, breed within the pond. Total production and table-size production coincide when the fish do not breed.

Factors determining production

Fish yields are a function of intrinsic factors, related to the genetics of the cultured fish population, and extrinsic factors, peculiar to the culture regime and the nature of the pond. The dependence of individual growth on population density, already mentioned on the preceding pages, operates through competition for space and food. To determine what part of the daily feed is available to each fish of the population, the rules which govern the social behaviour of food procurement should be known. A high level of competition causes an important decrease in individual growth rate, which, in turn, affects final fish biomass and thus yield.

The fish production in a pond depends on farmer's management, including fish care measures. This covers a wide array of factors. Some attempts have been made to quantify these (Piedhrahita, 1988 & 1990), with a view to determining their respective impacts and making recommendations to the farmer as to how he should intervene most efficiently to increase production, but too few publications with a comprehensive description of the culture factors are available. This is one of the greatest handicaps to modelling the production of tilapia. In Africa, for example, the sad reality is that stocking densities, feeding frequencies, and amount of feed cast, for example, often depend on what is locally and momentarily available, and even so, the exact procedures which were followed are not reported. In situations of high yield, it is therefore impossible to reconstruct the culture conditions of the production. One of the few exceptions to this rule is described hereafter.

Production factors in fish farming in CAR

Thirty-five pond production cycles of *O. niloticus*, in monoculture, at the governmental farms in Central African Republic (CAR) (Delincé, original data). were performed between 1974 and 1977, mainly at the Landjia fish farm, Bangui, and yields ranged between 400 to 9300 kg ha^{-1} yr^{-1}.

The feeds were classified on an arbitrary scale, ranging from 1 to 5, according to their nutritional value, 1 standing for no feed, 2 for the least nutritive feed and 5 for the fish-cum-pig culture, which is widely considered the most productive treatment (Fig. 65).

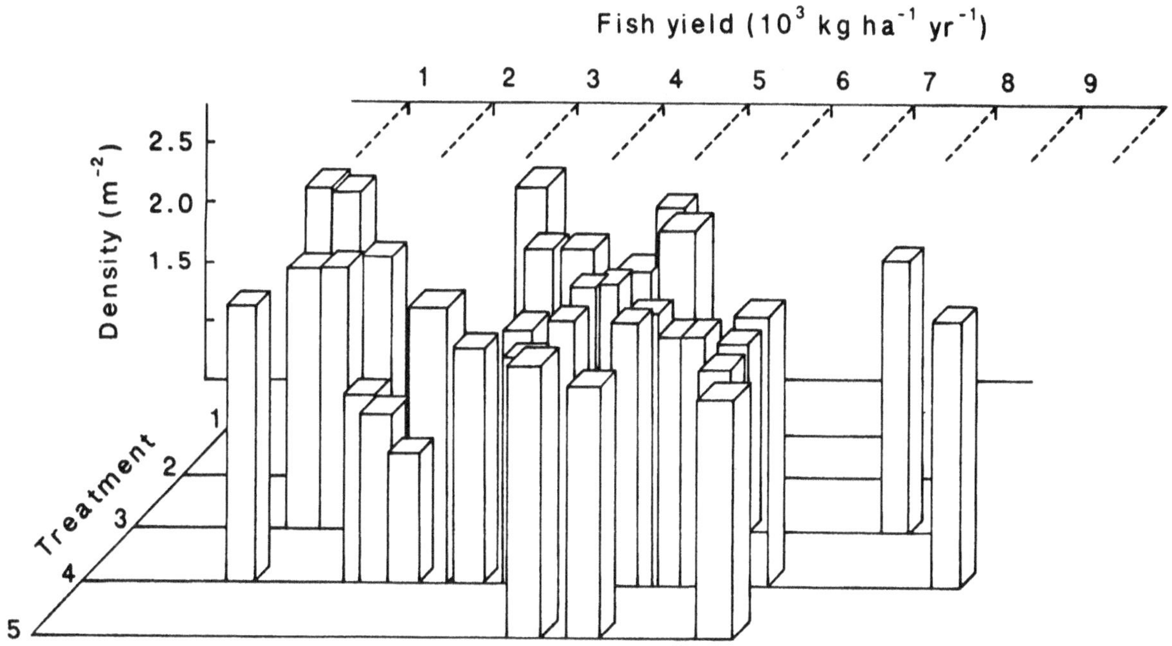

Fig. 65. Fish yield in a monoculture of tilapia (*Oreochromis niloticus*) in earthen ponds in Central African Republic as a function of density and pond treatment. The Y-axis is graded in a relative scale. Detailed explanation in the text.

The various feeds (and/or fertilizers) used were the following: brewery waste, brewery waste and cow dung, brewery waste and stale flours, crushed cotton seeds, crushed cotton seeds and rumen content, pellets, mixed feed unpelleted (same composition as the pellets), duck manure (duck-cum-fish farming), pig dung (pig-cum-fish farming). Production under control conditions, i.e. without feeding or fertilization, was also carried out.

Stocking densities ranged from 1 to 2.6 fish m^{-2}, and mean stocking weight between 8 and 60 g. The culture period extended from 53 to 188 days. Two sizes of ponds were used, the first ca. 350 m^2 in area (16 trials), and the second ca. 3200 m^2 in area (19 trials).

A multiple regression analysis was carried out between production (dependent variable) and the following factors (independent variables): initially stocked biomass, initial average individual weight and density, the feed used, and pond surface area. The multiple correlation coefficient, 0.86, was significant. Individual regressor reports for the initially stocked biomass, the treatment intensity, the initial density, the average individual weight stocked, the surface area of the pond, and the culture period were 0.87, 0.92, 0.90, 0.79, 0.73, and 0.69, respectively. The dependency of fish yield on intrinsic and extrinsic factors of production was therewith confirmed. The relation between production and the explicative variables is expressed by the following equation:

Total production = - 4.00 x stocked biomass (in kg ha^{-1})
+ 862.59 x treatment intensity (scale ranging between 1 to 5)
- 13.44 x culture period (in days)
- 1554.67 x stocking density (m^{-2})
+ 67.53 x average initial stocking weight (in g)
+ 0.14 x pond area (in m^2)

2.2.2. Critical factors limiting production

Deterioration of production conditions

In production trials of increasing intensity, the fish biomass ceases growing at the end of production as a consequence of a deterioration of production conditions. This deterioration has been observed by several authors (Moav *et al.*, 1977; Edwards *et al.*, 1981; Lazard, 1980, in Côte d'Ivoire), who tried to determine its causes. The following factors, all of which were discussed in detail earlier were advocated: a decrease in dissolved oxygen, combined with accumulation of waste products (for example, association of ammonia with suboptimal oxygen concentration in early morning hours); accumulation of inhibitory substances; a decrease in number of organisms consumable by tilapia, and a paralled increase in inedible or toxic organisms, such as euglenoids and blue-green algae (Moav *et al.*, 1977).

Problems with oxygen consumption in ponds

When the biomass of the fish increases, dissolved oxygen concentrations may become limiting. Fish respiration increases, and the environment becomes depleted of oxygen. The fish are not the only oxygen consumers in the pond, however. If the fish load is significant, the amount of excrements released by them is important and the mineralization of this organic matter alone consumes much oxygen.

Jana & De (1983) studied the oxygen consumption of the plankton community in a pond used for polyculture in West Bengal, India. The oxygen consumption of bacteria reached up to 26 g O_2 m^{-2} d^{-1}, with an average of 7-8 g O_2 m^{-2} d^{-1}. This represented about 65 to 100 % of total planktonic consumption of oxygen. In comparison, tilapia (*O. niloticus*) consume between 3.96-4.40 g O_2 kg^{-1} d^{-1} when feeding (fish between 200-400 g, at 25-30° C) (Mélard, 1986). African catfish (*Clarias gariepinus*) consumption was between 3.3 and 7.5 g O_2 kg^{-1} d^{-1} according to feeding level, at 25° C (Hogendoorn, 1983). The oxygen demand of fish ponds depending on organic matter supply amounts to a maximum of 3.5 g m^{-2} d^{-1} (Oláh *et al.*, 1987). Bacterial population sizes of ponds vary little, being about 10^6 cells per ml. In the case of large phytoplankton populations, as found in ponds with high organic fertilization, the autotrophic dark respiration must also be added. The result is that dissolved oxygen levels vary cyclically, and that the concentration often drops to critical values at the end of the night.

Measures to avoid anoxic conditions

Aeration of ponds involves the use of such devices as aerators, diffusers, paddle wheels and air blowers. Extensive discussion on their use can be found in Boyd (1979). The paddle wheel is considered by Boyd & Tucker (1979) as the most efficient device available. In the absence of the above-mentioned devices, aeration by an out-board engine can be a solution. Aeration induces an increase in net photosynthetic activity of the phytoplankton, and alleviates the respiratory demand for oxygen of the pond (Segupta & Jana, 1987).

The use of aerators or other machines to supply air or oxygen to a pond is particularly useful to combat oxygen depletion in the early morning hours. In culture experiments using tanks, aerators are frequently used too. However, their cost is high and their use thereby limited. According to some authors, the solution is palliative, because only a temporary oxygen-enriched refuge for fish is created, and the fundamental problem of oxygen supply to the pond is not solved.

The lack of oxygen also affects the plankton community and the decomposers of the pond; this leads to a die-off of plankton, an additional supply of organic matter, and the establishment of even more acute anoxic conditions. The production conditions deteriorate further.

The replacement of oxygen-depleted stagnant water by fresh oxygen-enriched water is another solution. Unfortunately, such an exchange of water leads to a loss of water fertility and of the natural food present, which is adverse to production.

An uncommon practice in Africa, but which has proved useful in China, is the use of agitators. These devices mix the pond water, like by the movement of a paddle-wheel. This agitation is done during the hot hours of the day, when photosynthesis is active, and helps to oppose thermal stratification, and the establishment of dissolved oxygen gradients. Surface oxygen supersaturation is reduced, as the entire water mass comes into contact with the oxygen-rich upper layers, resulting in an homogenization of dissolved oxygen. This is important, because it builds up a storage of oxygen in deeper strata, to be used during nocturnal oxygen consumption.

In China, in the provinces of Kiangsu and Shanghai, aerators are commonly used. Types found are either blowers powered by a turbine engine (Tapiador *et al.*, 1978) or electrically powered paddle wheels (FAO, 1981). These devices are floating in the center of the pond and are operated 2 to 3 hours a day, around noon, or in the morning by overcast weather. Fish harvest is increased by 14 % by use of these aerators (Tapiador *et al.*, 1978).

Biological control of oxygen production and consumption

Oxygen supply to a pond is dependent on a phytoplankton population of adequate size. One of the most important managemental problems therefore is the control of phytoplankton. Often, an excessive increase of algae results in wide diel fluctuations of dissolved oxygen, which often lead to a sudden die-off of the plankton bloom responsible for these fluctuations (see p. 148). The fluctuations are such that supersaturation of oxygen in the upper layers builds up during the day, followed by a decrease to full depletion during nighttime, when all living organisms, autotrophs and heterotrophs, consume oxygen.

In view of controlling phytoplankton populations by biological methods, Costa-Pierce *et al.* (1985) studied the effect of a polyspecific fish population on primary production in a *Macrobrachium rosenbergii* pond in Hawaii. Six different treatments were analyzed. The ponds were stocked with shrimps at the usual density (11.5 postlarvae m^{-2}), and fed an artificial feed, or fertilized with organic manure, at a rate of 700 kg per hectare and per week. Water hyacinths were added to two ponds, in view of determining their role in removing excess nutrients and controlling phytoplankton. The ponds containing the hyacinths (covering about 25 % of the water surface) were normally fed and fertilized. In two other ponds, grass carp (500 fish ha^{-1}) and silver carp (2000 and 25000 fish ha^{-1}) were stocked, fed, and the ponds fertilized. The water quality of these six types of ponds was compared. Chlorophyll a concentration was more important in the fertilized polyculture ponds, with primary productivity being heavily stimulated and diel oxygen fluctuations reduced. In the other ponds, however, the oxygen reached critical values, potentially dangerous to production. Water hyacinth were of little use in controlling water quality in ponds.

The salient point was that fish polyculture

markedly improved water quality. Turnover time of nitrogen and phosphorus in water was reduced and primary productivity increased. Fish grazing stabilized phytoplankton biomass (lowest coefficient of variation for chlorophyll a).

Means of eliminating deleterious decomposition products

A novel way of removing part of the sediments stemming from the accumulation of left-over feed and settled algae, and thus improve water quality, has been introduced by Israeli farmers (Zohar *et al.*, 1984). They observed that detritus accumulates on the anaerobic sediment and takes much time to decompose and be recycled. Evidently, such material contains metabolites harmful to fish growth. Removal of this detritus is done by a quick flushing of the water of the sediment-water interface, using a pipe laid at the sediment level, and draining the accumulated detritus (Fig. 66). This flush of water from the bottom of the pond is done three times daily for 20 seconds, which is sufficient to remove all the sediment of dark colour.

Analysis of the sludge drained showed that it was rich in organic matter (80 mg l^{-1}) and thus a significant source of oxygen consumption and of metabolites had been removed from the pond.

In flushed fish ponds, with stockings of 15 tilapia per m^2 and with aeration, production amounted to some 50 tons per ha. The FCR of the feed was even decreased, which attests to good production conditions (Zohar *et al.*, 1984). Aeration alone would be detrimental to production, since it would mix sedimented organic material to the water mass, thus even increasing oxygen demand and metabolite concentrations.

In a similar way, bivalves may be used to improve water quality near pond bottoms. In fish ponds in Illinois, Buttner (1986) observed that populations of *Corbicula* filter and remove a great part of the material in suspension. He found significant differences in dissolved oxygen concentrations, in the number of days when dissolved oxygen dropped below 3 mg l^{-1} at dawn, in primary production, turbidity, and ammonium concentration in the water, between ponds stocked with bivalves and those without. The stocking rates ranged around 1000 kg ha^{-1}. However, in some cases, nitrogen detritus was more important in the ponds with bivalves.

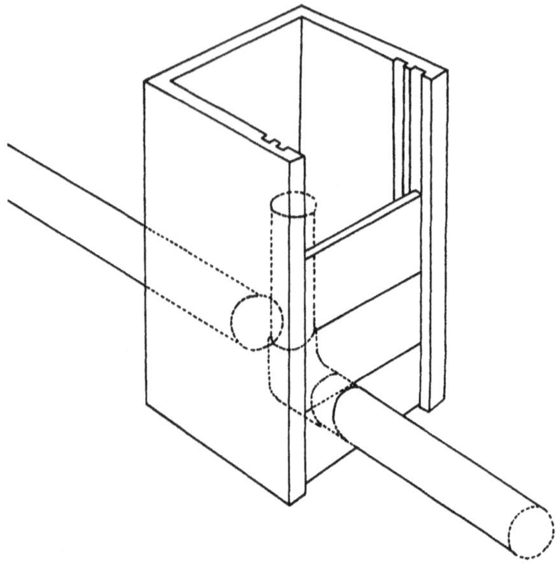

Fig.66. Modified pond monk (drainage device) enabling the draining of the water at the water-sediment interface level. A 15-cm diameter pipe is piercing through the lowest shutter and connected behind the shutter to a removable stand-up pipe, the height of which regulates the water level in the pond. The removal of the stand-up pipe allows the draining of the water at the level of the pipe in the course of the culture period. The extremity of the pipe is screened off to avoid the escape of fish (modified from Zohar *et al.*, 1984).

Successive limiting factors in fish farming

The different components of a pond ecosystem were seen to be complementary and to contribute synergistically to production. Primary productivity acts on fish production, through the quality of the water, primarily via the diel cycling of dissolved oxygen concentration, and through direct use of the phytoplankton as food.

Amongst all environmental variates, water temperature is, of course, a critical factor. Temperature directly affects fish metabolism, but since it influences all other biological and chemical activities within the pond, it is implicitly present in the effects of all other variables.

The factors which influence fish production really act within a network of interrelations. There is not just one limiting factor but a suite of them, each of which is linked to a major increase in production.

This implies that, at each production increment, the bottleneck of that particular limiting factor must be overcome. The three main limiting factors are (1) availability of feed (*ad libitum*), (2) dissolved oxygen concentration, and (3) ammonia concentration of the water. In cultures in artificial environments, these variables can be constantly monitored.

Mélard (1986), in a study of the biological basis of intensive tilapia production in indoor facilities in Belgium, found that the main effect of fish density on the culture environment was to reduce the dissolved oxygen level, so that the effect of fish density on production could be completely masked. Increasing density reduces either the amount of feed available to each fish or reduces dissolved oxygen, which in turn reduces feed intake and consequently growth. Jana & De (1983), in their study of production factors, also found a correlation between gross daily photosynthetic rate and fish production.

Cuenco *et al.* (1985) in a study of environmental factors affecting the population growth of American catfish, showed that, besides available food and oxygen, non-ionized ammonia in the water is an important factor (see Ammonia toxicity, on p. 89). In a culture system with high stocking and feeding rates, daily N-NH_3 inputs are high due to fish excretion and decomposition of excess feed. At alkaline pH, the ammonium/ammonia equilibrium shifts towards unionized ammonia, which is toxic for fish even at low concentrations. Increased fish production was achieved by successive optimalization of growth factors. Stocking densities of catfish were 7000 fish ha^{-1} in all cases. In natural conditions, the yield was 394 kg ha^{-1} (1.1 kg ha^{-1} d^{-1}); with *ad libitum* feeding, the yield reached 3897 kg ha^{-1} or 15.2 kg ha^{-1} d^{-1} (daily feed ratio was computed considering size, temperature, DO and unionized ammonia rather than only body weight); by aeration of the pond, the yield could be further increased to 6141 kg ha^{-1} (25.0 kg ha^{-1} d^{-1}). By keeping non-ionized ammonia concentration low, finally the yield reached 27659 kg ha^{-1} (118.5 kg ha^{-1} d^{-1}).

In ponds, the farmer's art consists in making best possible use of this ecological wisdom, so that each available resource is utilized and does not accumulate beyond a point where it becomes a nuisance. The autotrophic and heterotrophic production cycles must be complementary to ensure a continuous availability of nutrients for photosynthesis and a supply of organic matter to the decomposers. The natural food available to fish then remains continuously high. The diversity of natural production suggests the use of different species of fish with non-overlapping feeding niches in polyculture, adapted to the array of food resources available. Polyculture of tilapia with carp, or with catfish is commendable. Other polyculture associations, with intensive organic fertilization can, however, be established, using other species to widen the resource utilization. In Asian countries, polyculture often involves shrimp and fish, but except for some experimental ponds at the Malindi fish farm, near Mombassa, Kenya, this type of culture is not yet practised in Africa.

2.2.3. Ranges of fish yields

Yield levels according to production intensity

According to yield obtained, four groups of production intensity have been distinguished (Fig. 67) (Balarin & Haller, 1983; Balarin, 1988; Vincke, 1988). Extensive culture produces, on harvest, between 400 to 2500 kg ha^{-1} yr^{-1}, or 1 to 7 kg ha^{-1} d^{-1}. The degree of fertilization is low, as is stocking density, being about 0.3 to 2 fish m^{-2}. Subsistence ponds in rural areas of Africa (most countries in West, Central, East and Southern Africa, in all about 40 countries) indulge in this type of culture. The ponds are usually of small size (less than 0.5 ha), rain-fed or fed by the water table, lacking a water control structure. They depend on fingerlings provided from public or support services and the fish are fed domestic wastes. The ponds are under water for 6 to 12 months and angling occurs regularly. The fish harvested at the end of the culture period consist often of small-sized tilapia, due to the removal of all large-sized fish in the course of the production and due to overcrowding, resulting of poor feeding, prolific breeding and space competition. Fishfarming is practised as a part-time occupation by communities, families or individuals. In that category fall also most of the brackish water farms, such as the many farms in the mangrove areas of Nigeria. Their production is low due to the acidity of the local soil.

The second group is termed semi-intensive (or semi-extensive) and reaches yields ranging between

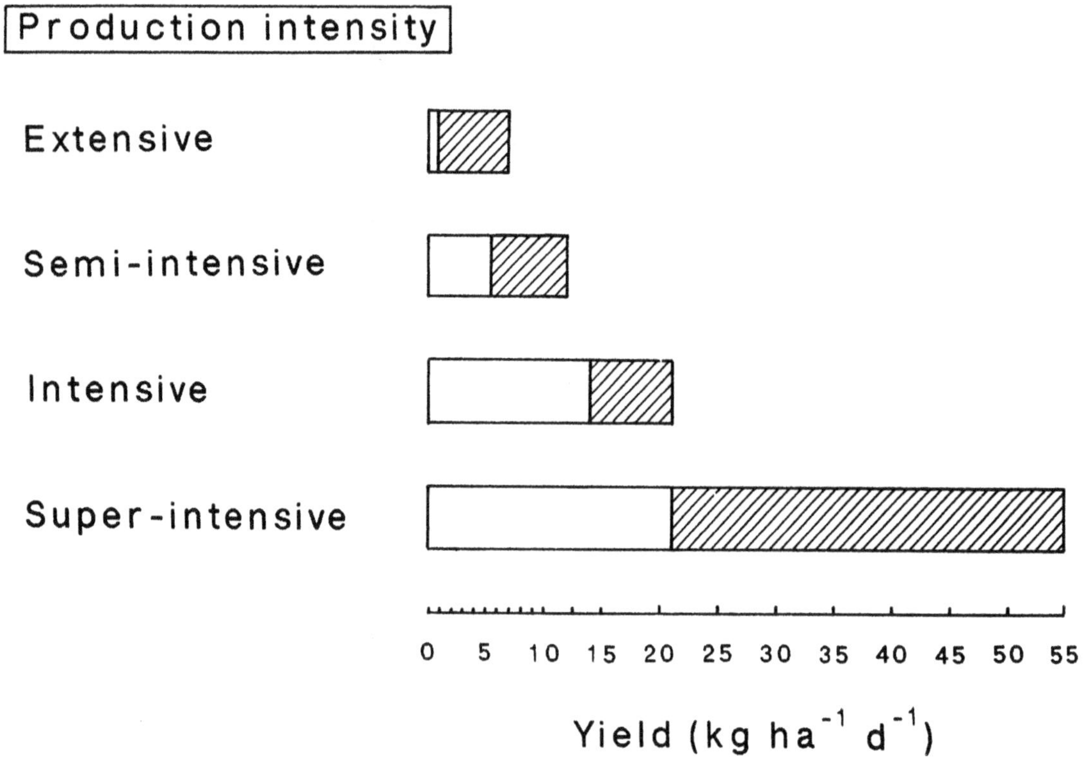

Fig. 67. Ranges of the fish yields obtained at different production intensities.

2000 to 4500 kg ha^{-1} yr^{-1} (or 5.5 to 12 kg ha^{-1} d^{-1}). The layout and the construction of the ponds is properly planned; stocking of the pond is made with tilapia, at 1-3 fish m^{-2}, and the fish are fed and the pond fertilized regularly. The feeds used are single agricultural byproducts or mixed ingredients, with low nutritional value, such as rice or wheat bran, brewery wastes, rumen content, whole or crushed cotton seeds, and molasses. Fertilizers are sometimes used alone. This type of fish farming relies often, but not always on an organised extension service, and is currently practised by individuals or families owning a few ponds from which they draw their living. This semi-intensive culture is found in many African countries, such as in Zimbabwe, Zambia, Malawi, Central African Republic, Kenya, Nigeria, Ivory Coast, Benin, Cameroon, Congo, Madagascar. In some countries, these farms are managed by small entreprises.

Intensive culture is practised in commercial farms. Farm infrastructure comprises sizeable ponds, fingerling production facilities, feed and fertilizer stores, water supply structures. Adequate management of the farm and entrepreneurial skills are required. The ponds are stocked with tilapia in association with catfish, carp or other species, such as *Heterotis*, *Heterobranchus*, *Hemichromis*, Chinese carp. The feed used are composite, nutritionally balanced, in pellet form or not, accompanied with a light organic or inorganic fertilization. The yields of this type of culture range between 5000 to 7000 kg ha^{-1} yr^{-1} or 14 to 21 kg ha^{-1} d^{-1}. This type of culture is practised in the same countries as the semi-intensive type, though not to the same extent. The major limitation to larger extension of this type of culture is the initial investment, and regular supply of feeds. Competition for agricultural byproducts with live-stock farms is impeding the development of these intensive fish farms.

Super-intensive culture is also practised in commercial farms and makes use of intensive organic manuring, resulting from a livestock crop of pigs or ducks, for example, or uses liquified dung, with or without supplementary feeding, and polyculture. Cultured fish are tilapia, the African catfish, *Clarias gariepinus*, common carp. Temporary aeration is carried out. The harvests range from 7500 to 20000 kg ha^{-1} yr^{-1} or 21 to 55 kg ha^{-1} d^{-1}. This type of culture is often part of a commercial farm and is integrated in a larger farm complex, where different agricultural activities are practised. Examples of this type of farms are found, for instance, in Central African Republic, Nigeria, Ivory Coast.

Pond size and length of production period are especially important for the last two production types. If rich artificial feed is used, a homogenous distribution of the feed across the whole surface of the pond is required, to ensure that all fish have access to it. The feed should not be wasted, as its value as fertilizer is much less than as feed. The unconsumed feed transforms into fertilizer. Ponds of large size, i.e. larger than 5000 m^2, are more difficult to feed this way. In ponds where fertilization comes directly from livestock stables, several pens should be constructed at different locations around the pond to avoid accumulation of manure at just one site, resulting in a polluted area. The location of pens at different sites can, however, increase such hazards as theft, a non-negligeable problem in poor third-word countries.

The harvest should take place when fish biomass has come close to the maximum carrying capacity. At the beginning of the production period, an approximate time for harvest is set on the basis of expected daily fish growth. As the production period is reaching its end, regular sampling of the fish should be done, to monitor biomass and FCR. These two items of information will signal when biomass reaches its carrying capacity.

The two intensive types of culture should use selected fish, such as tilapia of a known strain, in male monoculture, or hybrid tilapia (see p. 187).

Maximum yields obtained in Africa

Examples of yields obtained in intensive and super-intensive productions are summarized in Tables 25 and 26. These data come from the governmental farm Landjia, at Bangui, Central African Republic, and are in the same range as the figures of Moav *et al.* (1977), obtained in Israel, and which are considered very high.

The top productions were obtained in polyculture with intensive organic (pig dung) fertilization. The pigsties were located close to or over the ponds. The two fish species used, the tilapia, *O. niloticus*, and the African catfish, *Clarias gariepinus*, have excellent potential, which, in the case of tilapia, is counteracted by its prolific breeding. The advantage of the intense metabolic rate of tilapia is lost when it is diverted from growth to the elaboration of breeding products. In polyculture, catfish recovers part of this energy nutritionally for its own growth,

Table 25. Data on yields obtained in monoculture ponds of *O. niloticus* in the Central African Republic.

Size of the pond (m^2)	3325	2865	3325	3325	2200	484	450	2800
Period (day)	172	53	102	119	146	63	62	142
Initial density (ind m^{-2})	2.32	1.45	2.28	2.20	1.90	2.00	2.00	2.00
Stocking weight (g)	8.00	50.	9.0	15.2	23.0	24.4	23.4	15.3
Total production (kg ha^{-1} yr^{-1})	8168	6395	7167	5644	6818	9348	6163	7261
Table size production (kg ha^{-1} yr^{-1})	3093	3615	4315	3040	2930	4249	3401	1028
Treatment	Artificial feeding							Pig dung

Table 26. Data on intensive production (polyculture) of tilapia, *O. niloticus*, and African catfish, *Clarias gariepinus*, in ponds in Central African Republic.

Size of the pond (m^2)		3400	3400	250	3400	3400	3400	3400
Tilapia	Initial density (ind m^{-2})	2.6	2.0	2.0	3.88	3.27	2.10	1.98
	Initial weight (g)	24	7.1	8.0	10	10	26	14.8
	Weight at harvest (g)	124	110	93.2	120	-	90.3	71
	Production (kg ha^{-1} yr^{-1})	6465	6003	5856	8609	7152	10317	10461
Clarias	Initial density (ind m^{-2})	0.32	0.25	1.0	1.18	0.2	0.15	0.17
	Initial weight (d)	32.9	8.2	6.4	8.2	6.0	1.17	2.3
	Weight at harvest (g)	471	389	128.4	834	112	1170	1320
	Production (kg ha^{-1} yr^{-1})	667	1595	3824	2126	4035	1552	695
Tilapia (%)		81.4	79.0	60.5	80.2	63.9	86.9	93.8
Clarias (%)		18.6	21.0	39.5	19.8	36.1	13.1	6.2
Period (days)		168	153	90	173	180	148	132
Total production (kg ha^{-1} yr^{-1})		7942	7598	9680	10735	11187	11869	11156
Treatment		Organic manuring (pig excreta)						

by feeding on the tilapia fry and fingerlings. As such, it also alleviates the problem of food and space competition brought about by uncontrolled reproduction of tilapia. With improved strains, selected for late maturity, in all-male culture, or in sterile hybrids culture, even this problem could have been eliminated (see p. 187).

Considering that the production conditions were intensive only from the point of view of organic fertilizer supply, and not from the point of view of stocking densities, which were invariably around 2 fish m^{-2}, or fish used in production, which were unselected tilapia strains, it can be expected that production can be increased even more. Using higher stocking densities, hybrid strains and aeration, it seems reasonable to reach 15 to 20 tons per hectare and per year.

Fish production in ponds

1. Fish production depends on food availability, fish growth rate and density, and available space, which determines carrying capacity.

2. Food, when available at satiation levels, is used more efficiently at higher fish densities. Food requirements are specific, and depending on fish species used. The food conversion ratio indicates whether growing conditions are optimum.

3. Fish production increases, but mean individual weight at harvest decreases, with increasing stocking densities.

4. Production is restricted by dissolved oxygen availability. Measures to increase dissolved oxygen concentrations include the use of aerators, blowers and agitators. The removal of decomposition products which accumulate on the pond bottom is beneficial to production.

5. Successive limiting factors impede the increase of fish production; they are: food, dissolved oxygen and ammonia.

6. The production figures in African culture ponds compare favourably with the maximum production figures obtained in intensive farming outside the region.

References

Adams, D. D., G. Matisoff & W. J.Snodgrass, 1982. Flux of reduced chemical constituents (Fe^{2+}, Mn^{2+}, NH_4^+ and CH_4) and sediment oxygen demand in Lake Erie. Hydrobiologia 91: 405-41.

Agustí, S., C. M. Duarte & D. E. Canfield, Jr., 1990. Phytoplankton abundance in Florida lakes: evidence for the frequent lack of nutrient limitation. Limnol. Oceanogr. 35: 181-188.

Agustí, S., C. M. Duarte & J. Kalff, 1987. Algal cell size and the maximum density and biomass of phytoplankton. Limnol. Oceanogr. 32: 983-986.

Agustí, S. & J. Kalff, 1989. The influence of growth conditions on the size dependence of maximal algal density and biomass. Limnol. Oceanogr. 34: 1104-1108.

Ahlgren, G., 1988. Phosphorus as growth-regulating factor relative to other environmental factors in cultured algae. Hydrobiologia 170: 191-210.

Almazan, G. & C. E. Boyd, 1978. Effects of nitrogen levels on rates of oxygen consumption during decay of aquatic plants. Aquatic Botany 5: 119-126.

Almer, B., W. Dickson, C. Ekström, E. Hornström & U. Miller, 1974. Effects of acidification of Swedish lakes. Ambio 3: 30-36. Quoted by Mulholland *et al.*, 1986.

American Public Health Association, 1976. Standard methods for the examination of water and wastewater, 14th Ed. APHA, Washington, D.C., 874 pp.

Anderson, R. F. & S. L. Schiff, 1987. Alkalinity generation and the fate of sulfur in lake sediments. Can. J. Fish. aquat. Sci 44 (Suppl. 1): 188-193.

Andersson, G., W. Granéli & J. Stenson, 1988. The influence of animals on phosphorus cycling in lake ecosystems. Hydrobiologia 170: 245-266.

Auer, M. T., M. S. Kieser & R. P. Canale, 1986. Identification of critical nutrient levels through field verification of models for phosphorus and phytoplancton growth. Can. J. Fish. aquat. Sci 43: 379-388.

Avnimelech, Y. & M. Lacher, 1979. A tentative nutrient balance for intensive fish pond. Bamidgeh 31: 3-8.

Avnimelech, Y., S. Mokady & G. L. Schroeder, 1989. Circulated ponds as efficient bioreactors for single cell protein production. Israel. J. Aquacult. 41: 58-66.

Azam, F., T. Fenchel, J. G. Field, J. S. Gray, L. A. Meyer-Reil & F. Thingstad, 1983. The ecological role of water-column microbes in the sea. Ecol. Prog. Ser. 10: 257-263.

Balarin, J. D., 1988. Systems in use and species cultivated. In H. R. King & K. H. Ibrahim (eds), Village level aquaculture development in Africa. Proc. Comm. Cons. Workshop on Village Level Aquaculture Development in Africa. Freetown, Sierra Leone, 14-20 February 1985. Commonwealth Secretariat, London: -23.

Balarin, J. D. & R. D. Haller, 1983. Commercial tank culture of Tilapia. In L. Fishelson & Z. Yaron (eds), Tel Aviv University, Tel Aviv, Israel. Proc. Int. Symp. on Tilapia in Aquaculture: 473-483.

Barat, S. & B. B. Jana, 1987. Effects of farming management on the distribution pattern of ammonification rate, protein mineralizing and ammonifying bacterial population in experimental culture tanks. Bamidgeh 39: 120-132.

Barbosa, F. A. R. & J. G. Tundisi, 1989. Diel variations in a shallow tropical Brazilian lake. I. The influence of temperature variations on the distribution of dissolved oxygen and nutrients. Arch. Hydrobiol. 116: 333-349.

Barica, J., 1990. Seasonal variability of N:P ratios in eutrophic lakes. Hydrobiologia 191: 97-103.

Bartell, S. M., 1981. Potential impact of size-selective planktivory on P release by zooplankton. Hydrobiologia 80: 139-146.

Bartell, S. M. & J. F. Kitchell, J. F., 1978. Seasonal impact of planktivory on phosphorus release by Lake Wingra zooplankton. Verh. int. Ver. Limnol. 19: 2478-2484. Quoted by Sternik, 1983.

Barthelmes, D., 1989. Evidence for system dependent interactions of zooplancton-eating fish with phytoplankton. Arch. Hydrobiol. Beih. Ergebn. Limnol.33: 579-586.

Benndorf, J., 1990. Conditions for effective biomanipulation; conclusions derived from whole-lake experiments in Europe. Hydrobiologia 200/201: 187-203.

Benndorf, J., H. Schultz, A. Benndorf, R. Unger, E. Penz, H. Kneschke, K. Kossatz, R. Dumke, U. Hornig, R. Kruspe, S. Reichel & A. Köhler, 1989. Food web manipulation by enhancement of piscivorous stocks: long-term effects in the hypereutrophic Bautzen reservoir. Arch. Hydrobiol. Beih. Ergebn. Limnol. 33: 567-569.

Ben-Tuvia, A. & K. Reich, 1983. Stocking and management of cichlids fishery resources in Lake Kinneret (Lake Tiberias). In L. Fishelson & Z. Yaron (eds), Tel Aviv University, Tel Aviv, Israel. Proc. Int. Symp. on Tilapia in Aquaculture: 18-27.

Bergstrand, E., 1990. Changes in the fish and zooplankton communities of Ringsjön, a Swedish lake undergoing man-made eutrophication. Hydrobiologia 191: 57-66.

Berman, T., 1988. Microbes in a watery world. Hydrobiologia 159: 5-6.

Beveridge, M. C. M., M. Begum, G. N. Frerichs & S. Miller, 1989. The ingestion of bacteria in suspension by tilapia *Oreochromis niloticus*. Aquaculture 81: 373-378.

Bird, D. F. & J. Kalff, 1987. Algal phagotrophy: regulating factors and importance relative to photosynthesis in *Dinobryon* (Chrysophyceae). Limnol. Oceanogr. 33: 277-284.

Björk-Ramberg, S., 1984. Changes in sediment nutrient content in a subartic lake subjected to lake fertilisation. Freshwat. Biol. 14: 157-163.

Björk-Ramberg, S., 1985. Uptake of phosphate and inorganic nitrogen by a sediment- algal system in a subartic lake. Freshwat. Biol. 15: 175-183.

Bjørnsen, P. K., 1988. Phytoplankton exudation of organic matter: why do healthy cell do it? Limnol. Oceanogr. 3: 151-154.

Blauw, T. S., 1988. Nitrogen fixation (acetylene reduction) in the sediments on the Pluß-See. With special attention to the rôle of sedimentation. Arch. Hydrobiol. (Monogr. Beitr.), Suppl. 81 (4): 359-489.

Blažka, P., T. Backiel & F. W. Taub, 1980. Trophic relationships and efficiencies. In E. D. Le Cren & R. H. Lowe-McConnell (eds), The functioning of freshwater ecosystems. Cambridge University Press, Cambridge. International Biological Programme 22 : 393-410.

Bleiwas, A. S. H., P. M. Strokes & M. M. Olaveson, 1984. Six years of plankton studies in the LaCloche region of Ontario, with special reference to acidification. Verh. int. Ver. theor. angew. Limnol. 22: 332-337. Quoted by Mulholland *et al.*, 1986.

Bloem, J. & M.-J. B. Bär-Gilissen, 1989. Bacterial activity and protozoan grazing potential in a stratified lake. Limnol. Oceanogr. 34: 297-309.

Bloesch, J. & H. R. Bürger, 1989. Changes in phytoplankton and zooplankton biomass and composition reflected by sedimentation. Limnol. Oceanogr. 34: 1048-1061.

Blomqvist, B., H. Olsson, H. Olofsson & O. Broberg, 1989. Enclosure experiments with low-dose additions of phosphorus and nitrogen in the acidified lake Njuffatet, Central Sweden. Int. Revue ges. Hydrobiol. 74: 611-631.

Boavida, M. J. & R. T. Heath, 1988. Is alkaline phosphatase always important in phosphate regeneration? Arch. Hydrobiol. 111: 507-518.

Boon, P. I., 1990. Organic matter degradation and nutrient regeneration in Australian freshwaters: II. Spatial and temporal variation, and relation with environmental conditions. Arch. Hydrobiol. 117: 405-436.

Bouldin, D. R., R. J. Johnson, C. Burda & C. W. Kao, 1974. Losses of inorganic nitrogen from aquatic systems. J. envir. Qual. 3: 107-114.

Boström, B., J. M. Andersen, S. Fleischer & M. Jansson, 1988a. Exchange of phosphorus across the sediment-water interface. Hydrobiologia 170: 229-244.

Boström, B., M. Jansson & C. Forsberg, 1982. Phosphorus release from lake sediments. Arch. Hydrobiol. Beih. Ergebn. Limnol. 18: 5-59.

Boström, B., G. Persson & B. Broberg, 1988b. Bioavailability of different phosphorus forms in freshwater systems. Hydrobiologia 170: 133-155.

Boström, B. & K. Pettersson, 1982. Different patterns of phosphorus release from lake sediments in laboratory experiments. Hydrobiologia 91: 415-429.

Bowen, S. H., 1981. Digestion and assimilation of periphytic detrital aggregate by *Tilapia mossambica*. Trans. am. Fish. Soc. 110: 241-247.

Boyd, C. E., 1972. Sources of CO_2 for nuisance blooms of algae. Weed Sci. 20: 492-497.

Boyd, C. E., 1976a. Nitrogen fertilizer effects on production of *Tilapia* in ponds fertilized with phosphorus and potassium. Aquaculture 7: 385-390.

Boyd, C. E., 1976b. Chemical and textural properties of muds from different depths in ponds. Hydrobiologia 48: 141-144.

Boyd, C. E., 1976c. Lime requirements and application in fish ponds. FAO Techn. Conf. on Aquaculture, Fishing News Book, Ltd., Farnham, England: 120-122.

Boyd, C. E., 1979. Water quality in warmwater fish ponds. Auburn University, Craftmaster Printers, Opelika, Alabama, 359 pp.

Boyd, C. E., 1981a. Solubility of granular inorganic fertilizers for fish ponds. Trans. am. Fish. Soc. 110: 451-454.

Boyd, C. E., 1981b. Comparison of five fertilization programs for fish ponds. Trans. am. Fish. Soc. 110: 541-545.

Boyd, C. E., 1981c. Liquid fertilizers for fish ponds. Highlights of Agricultural Research 28 (4): 3.

Boyd, C. E. & M. J. Cuenco, 1980. Refinement of the lime requirement procedure for fish ponds. Aquaculture 21: 293-299.

Boyd, C. E. & Y. Musig, 1981. Orhophosphate uptake by phytoplankton and sediment. Aquaculture 22: 165-173.

Boyd, C. E., Y. Musig & L. Tucker, 1981. Effects of three phosphorus fertilizers on phosphorus concentrations and phytoplankton production. Aquaculture 22: 175-180.

Boyd, C. E., E. E. Prather & R. W. Parks, 1975. Sudden mortality of a massive phytoplankton bloom. Weed Sci. 23: 61-67.

Boyd, C. E., R. P. Romaire & E. Johnston, 1978. Predicting early morning dissolved oxygen concentrations in channel catfish ponds. Trans. am. Fish. Soc. 107: 484-492.

Boyd, C. E. & J. W. Sowles, 1978. Nitrogen fertilisation of ponds. Trans. am. Fish. Soc. 107: 737-741.

Boyd, C. E. & C. S. Tucker, 1979. Emergency aeration of fish ponds. Trans. am. Fish. Soc. 108: 299-306.

Boyd, C. E. & L. Tucker, 1980. Determination of filtrable orthophosphate in water from fish ponds. Trans. am. Fish. Soc. 109: 314-318.

Brady, N. C., 1984. The nature and properties of soil. Ninth Edition. Macmillan Publ. Co., New York, 750 pp.

Brandl, Z., B. Desortova, J. Hrbacek, J. Komárková, V. Vyhnalek, J. Seďa & M. Straškraba, 1989. Seasonal changes of zooplankton and phytoplankton and their mutual relations in some Czechoslovak reservoirs. Arch. Hydrobiol. Beih. Ergebn. Limnol. 33: 597-604.

Brett, J. R., 1979. Environmental factors and growth. In W. S. Hoar, D. J. Randall & J. R. Brett. Fish physiology. Vol. VIII. Bioenergetics and growth. Academic Press, N.Y., San Francisco, London: 599-675.

Broberg O. & G. Persson, 1988. Particulate and dissolved phosphorus forms in freshwater: composition and analysis. Hydrobiologia 170: 61-90.

Broberg, O. & K. Pettersson, 1988. Analytical determination of orthophosphate in water. Hydrobiologia 170: 45-59.

Brooks, J. L. & S. I. Dodson, 1965. Predation, body size and the composition of plankton. Science 150: 26-35.

Brown, E. J. & R. F. Harris, 1978. Kinetics of algal transient phosphate uptake and the cell quota concept. Limnol. Oceanogr. 23: 35-40.

Buck, D. H., R. J. Baur & C. R. Rose, 1978. Utilization of swine manure in a polyculture of Asian and North American fishes. Trans. am. Fish. Soc. 107: 216-222.

Buckman, H. O. & N. C. Brady, 1964. The nature and properties of soil. The MacMillan Company, New York, 516 pp.

Burns, C. W., D. J. Forsyth, J. F. Haney, M. R. James, W. Lampert & R. D. Pridmore, 1989. Coexistence and exclusion of zooplankton by *Anabaena minutissima* var. *attenuata* in Lake Rotongaio, New Zealand. Arch. Hydrobiol. Beih. Ergebn. Limnol. 32: 63-82.

Burns, C. W. & J. J. Gilbert, 1986. Effects of daphnid size and density on interference between *Daphnia* and *Keratella cochlearis*. Limnol. Oceanogr. 31: 848-858.

Burns, R. P. & R. R. Stickney, 1980. Growth of *Tilapia aurea* in ponds receiving poultry wastes. Aquaculture 20: 117-121.

Buttner, J. K., 1986. *Corbicula* as a biological filter and polyculture organism in catfish rearing ponds. Progr. Fish-Cult. 48: 136-139.

Callender E., 1982. Benthic phosphorus regeneration in the Potomac River Estuary. Hydrobiologia 91: 431-446.

Capone, D. G. & R. P. Kiene, 1988. Comparison of microbial dynamics in marine and freshwater sediments: contrasts in anaerobic carbon catabolism. Limnol. Oceanogr. 33: 725-749.

Cappenberg, T. E., 1988. Quantification of aerobic and anaerobic carbon mineralization at the sediment-water interface. Arch. Hydrobiol. Beih. Ergebn. Limnol. 31: 307-317.

Cappenberg, T. E., K. A. Hordijk, G. J. Jonkheer & J. P. M. Lauwen, 1982. Carbon flow across the sediment-water interface in Lake Vechten, The Netherlands. Hydrobiologia 91: 161-168.

Carlton, R. G. & R. G. Wetzel, 1988. Phosphorus flux from lake sediments: effect of epipelic algal oxygen production. Limnol. Oceanogr. 33: 562-570.

Carmouze, J.-P. & P. Caumette, 1985. Les effets de la pollution organique sur les biomassses et activités du phytoplancton et des bactéries hétérotrophes dans la lagune Ebrié (Côte d'Ivoire). Rev. Hydrobiol. trop. 18: 183-211.

Carmouze, J. P., C. Arze & J. Quintamilla, 1984. Le lac Titicaca: stratification physique et métabolisme associé. Rev. Hydrobiol. trop. 7: 3-11.

Carpenter, S. R., J. F. Kitchell, J. R. Hodgson, P. A. Cochran, J. J. Elser, M. M. Elser, D. M. Lodge, D. Kretchmer & X. He, 1987. Regulation of lake primary productivity by food web structure. Ecology 68: 1863-1876.

Carrillo, P., L. Cruz-Pizarro & P. Sanchez-Castillo, 1990. Analysis of phytoplankton-zooplankton relationships in an oligotrophic lake under natural and manipulated conditions. Hydrobiologia 200/201: 49-57.

Cattaneo, A., 1987. Periphyton in lakes of different trophy. Can. J. Fish. aquat. Sci 44: 296-303.

Chakroff, M., 1976. Freshwater fish pond culture and management. VITA Publications, Manuel Series, No. 36 E, Mt. Rainier, USA, 191 pp.

Chang, W. Y. B. & H. Ouyang, 1988. Dynamics of dissolved oxygen and vertical circulation in fish ponds. Aquaculture 74: 263-276.

Chattopadhyay, G. N. & L. N. Mandal, 1980. Effects of different levels of water salinity on the decomposition of organic manures in a brackish water fish pond soil. Hydrobiologia 72: 287-292.

Chen, F. Y. & G. A. Prowse, 1964. The effect of living space on the growth rate of fish. Ichthyologica 3: 11-20. Quoted by Edwards *et al.* (1981)

Chrost, R. J., 1989. Characterization and significance of ß-glucosidase activity in lake water. Limnol. Oceanogr. 34: 660-672.

Chrost, R. J., W. Siuda & G. Z. Halemejko, 1974. Longterm studies on alkaline phosphatase activity (APA) in a lake with fish-aquaculture in relation to lake eutrophication and phosphorus cycle. Arch. Hydrobiol./Suppl. 70: 1-32.

Coche, A., 1986. Pisciculture continentale - Le sol. Méthodes simples pour l'aquaculture. Collection FAO: Formation N° 6. FAO, Rome, 174 pp.

Coche, A. G. & H. Van der Wal, 1981. Water for freshwater fish culture. FAO Training Series No. 4. FAO, Rome, 110 pp.

Cole, B. A. & C. E. Boyd, 1986. Feeding rate, water quality and channel catfish production in ponds. Progr. Fish-Cult. 48: 25-29.

Compton, R. G., K. L. Pritchard & P. R. Umoin, 1989. The dissolution of calcite in acid water: mass transport versus surface control. Freshwat. Biol. 22: 285-288.

Cook, R. B., C. A. Kelly, D. W. Schindler & M. A. Turner, 1986. Mechanisms of hydrogen ion neutralization in an experimentally acidified lake. Limnol. Oceanogr. 31: 134 148.

Cooke, J. G. & R. E. White, 1987a. The effect of nitrate in streamwater on the relationship between denitrification and nitrification in a stream-sediment microcosm. Freshwat. Biol. 18: 213-226.

Cooke, J. G. & R. E. White, 1987b. Spatial distribution of denitrifying activity in a stream draining an agricultural catchment. Freshwat. Biol. 18: 509-519.

Costa-Pierce, B. A., 1990. Letter to the Editor. Aquabyte 3 (2): 18-19.

Costa-Pierce, B. A. & D. B. Craven, 1987. Estimating microbial production and growth rates in aquaculture ponds using rates of RNA and DNA synthesis. Aquaculture 66: 69-78.

Costa-Pierce, B. A., S. P. Malecha & E. A. Laws, 1985. Effects of polyculture and manure fertilization on water quality and heterotrophic productivity in *Macrobrachium rosenbergii* ponds. Trans. am. Fish. Soc. 114: 826-836.

Couillard, D. & Y. Lefebvre, 1985. Analysis of water quality indices. J. environ. Mgt 21: 161-179.

Crisman, T. L. & J. R. Beaver, 1990. Applicability of planktonic biomanipulation for managing eutrophication in the subtropics. Hydrobiologia 200/201: 177-185.

Cuenco, M. L., R. R. Stickney & W. E. Grant, 1985. Fish bioenergetics and growth in aquaculture ponds : III. Effects of intraspecific competition, stocking rate, stocking size and feeding rate on fish productiity. Ecol. Modelling 28: 73-95.

Cuffney, T. F., J. B. Wallace & G. J. Lugthart, 1990. Experimental evidence quantifying the rôle of benthic invertebrates in organic matter dynamics of headwater streams. Freshwat. Biol. 23: 281-299.

Cuker, B. E., 1987. Field experiment on the influence of suspended clay and P on the plankton of a small lake. Limnol. Oceanogr. 32: 840-847.

Currie, D. J., 1984. Microscale nutrient patches: do they matter to the phytoplankton. Limnol. Oceanogr. 29: 211-214.

Currie, D. J., E. Bentzen & J. Kalff, 1986. Does algal-bacterial phosphorus partioning vary among lakes? A comparative study of orthophosphate uptake and alkaline phosphatase activity in freshwater. Can. J. Fish. aquat. Sci 43: 311-318.

Currie, D. J. & J. Kalff, 1984a. A comparison of the abilities of freshwater algae and bacteria to acquire and retain phosphorus. Limnol. Oceanogr. 29: 298-310.

Currie, D. J. & J. Kalff, 1984b. The relative importance of bacterioplankton and phytoplankton in phosphorus uptake in freshwater. Limnol. Oceanogr. 29: 311-321.

Davison, W., 1982. Transport of iron and manganese in relation to the shapes of their concentration depth profiles. Hydrobiologia 91: 463-471.

Dawidowicz, P., 1990a. The effect of *Daphnia* on filament length of blue-green algae. Hydrobiologia 191: 265-268.

Dawidowicz, P., 1990b. Effectiveness of phytoplankton control by large-bodied and small-bodied zooplankton. Hydrobiologia 200/201: 43-47.

Degani, G., C. Dosoretz & D. Levanon, 1984. The influence of cow manure on growth rates of *Oreochromis aureus* and *Clarias lazera* in Israel small outdoor tanks. Bamidgeh 36: 114-120.

De Haan, H. & T. De Boer, 1986. Geochemical aspects of aqueous iron, phosphorus and dissolved organic carbon in the humic Lake Tjeukemeer, The Netherlands. Freshwat. Biol. 16: 661-672.

dela Cruz, C. R., 1983. Fishpond engineering: A technical manual for small- and medium scale coastal fish farms in Southeast Asia. South China Sea Fisheries Development and Coordination Programme, SCS Manual No. 5, Manila, Philippines, 180 pp.

Delmendo, M. N., 1980. A review of integrated live-stock-fowl-fish farming systems. In Pullin, R. S. V. & Z. H. Shehadeh (eds), Integrated agriculture- aquaculture farming system. I.C.L.A.R.M., Manila, Philippines. Conf. Proc. 4: 59-71.

De Pinto, J. V., T. C. Young, J. S. Bonner & P. W. Rodgers, 1986. Microbial recycling of phytoplankton phosphorus. Can. J. Fish. aquat. Sci 43: 336-342.

Dévai, G., 1990. Ecological background and importance of the change of chironomid fauna (Diptera: Chironomidae) in shallow Lake Balaton. Hydrobiologia 191: 189-198.

Diab, S. & M. Shilo, 1986. Transformation of nitrogen in sediments of fish ponds in Israel. Bamidgeh 38: 67-88.

Diab, S. & M. Shilo, 1988. Effects of on the activity and survival of *Nitrosomonas sp.* and *Nitrobacter sp.* isolates from fish ponds. Israel. J. Aquacult. 40: 50-56.

Dokulil, M., A. Herzig & A. Jagsch, 1990. Trophic relationships in the pelagic zone of Mondsee, Austria. Hydrobiologia, 191: 199-212.

Doremus, C. & L. S. Cleresci, 1982. Microbial metabolism in surface sediments and its rôle in the immobilization of phosphorus in oligotrophic lake sediments. Hydrobiologia 91: 261-268.

Drenner, R. W., J. R. Mummert, F. de Noyelles Jr. & D. Kettle, 1984a. Selective particle ingestion by a filter-feeding fish and its impact on phytoplankton community structure. Limnol. Oceanogr. 29: 941-948.

Drenner, R. W., F. de Noyelles Jr. & D. Kettle, 1982a. Selective impact of filter-feeding gizzard shad on zooplankton community structure. Limnol. Oceanogr. 27: 965-968.

Drenner, R. W., W. J. O'Brien & J. R. Mummer, 1982b. Filter-feeding rates of gizzard shad. Trans. am. Fish. Soc. 111: 210-215.

Drenner, R. W., S. B. Taylor, X. Lazzaro & D. Kettle, 1984b. Particle grazing and plankton community impact of an omnivorous cichlid. Trans. am. Fish. Soc. 113: 397-402.

Droop, M. R., 1974. The nutrient status of algal cells in continuous culture. J. Mar. Biol. Assoc. U.K. 54: 825-855.

Dumont, H. J., 1991. The predation-controlled uniqueness of tropical zooplankton: less Cladoceran diversity, but more Limnomedusae, flatworms, mites and numerous prey defenses. Hydrobiologia (in press).

Dumont, H. J. & L. De Meester, 1990. Are contrasting patterns of vertical migration in zooplankton the result of differential natural selection? Rev. Brasil. Biol. 50: 867-874.

Dussart, B. H., 1980. Copépodes. In Durand, J.-R. & C. Lévêque (eds), Flore et faune aquatiques de l'Afrique sahélo-soudanienne, 1. O.R.S.T.O.M., Paris. Documn Techqs 44: 333-356.

Edwards, P., 1980. A review of recycling organic wastes into fish, with emphasis on the tropics. Aquaculture 21: 261-279.

Edwards, P., O.A. Sinchumpasak & M. Tabucanon, 1981. The harvest of microalgae from the effluent of a sewage fed high rate stabilisation pond by *Tilapia nilotica*. Part 2 : Studies of the fish ponds. Aquaculture 23: 107-147.

Eggers, D. M., 1982. Planktivore preference by prey size. Ecology 63: 381-390.

Einsele, W., 1936. Über die Beziehungen des Eisenkreislaufs zum Phosphorkreislauf im eutrophen See. Arch. Hydrobiol. 29: 664-686.

Ejsmont-Karabin, J., M. Bubień & T. Weglеńska, 1989. The effects of zooplankton on phosphorus cycling in man-made Lake Zegrzyńskie. Arch. Hydrobiol. Beih. Ergebn. Limnol. 31: 493-501.

Ejsmont-Karabin, J. & I. Spodniewska, 1990. Influence on phytoplankton biomass in lakes of different trophy by phosphorus in lake water and its regeneration by zooplankton. Hydrobiologia 191: 123-128.

El Samra, M. I. & J. Oláh, 1979. Significance of nitrogen fixation in fish ponds. Aquaculture 18: 367-372.

Elser, J. J., H. J. Carney & C. R. Goldman, 1990. The zooplankton-phytoplankton interface in lakes of contrasting trophic status: an experimental comparison. Hydrobiologia 200/201: 69-82.

Elser, J. J., M. M. Elser, N. A. MacKay & S. R. Carpenter,

1988. Zooplancton- mediated transitions between N- and P-limited algal growth. Limnol. Oceanogr. 33: 1-14.

Elser, J. J. & N. A. MacKay, 1989. Experimental evaluation of effects of zooplankton biomass and size distribution on algal biomass and productivity in three nutrient-limited lakes. Arch. Hydrobiol. 114: 481-496.

Emerson, K., R. C. Russo, R. E. Lund & R. V. Thurston, 1975. Aqueous ammonia equilibrium calculations : effects of pH and temperature. J. Fish. Res. Bd Can. 32: 2379-2383.

Enell, M. & S. Löfgren, 1988. Phosphorus in interstitial water: methods and dynamics. Hydrobiologia 170: 103-132.

Eren, Y., T. Tsur & Y. Avnimelech, 1977. Phosphorus fertilization of fish ponds in Upper Galilee. Bamidgeh 29: 87-92.

Erez, J., M. D. Krom & T. Neuwirth, 1990. Daily oxygen variations in marine fish ponds, Elat, Israel. Aquaculture 84: 289-305.

Evans, T. G., 1988. A framework for discussing seasonal succession and coexistence of phytoplankton species. Limnol. Oceanogr. 33: 1027-1036.

FAO, 1967. Actes du symposium mondial FAO sur la pisciculture en étang à temperature élevée, Rome (Italie), 18-25 mai 1966. Vol. 1 - Rapport. FAO Fish. Rep. (Fr.), 44, Vol.1: 57 pp.

FAO, 1976. Report of the Symposium on Aquaculture in Africa. Accra, Ghana, 30 September - 2 October 1975. CIFA Tech. Pap., 4: 36 pp.

FAO, 1981. Développement de l'aquaculture continentale en Chine. FAO Doc. Tech. Pêches, 215: 152 pp.

Fairchild, G. W. & A. C. Everett, 1988. Effects of nutrient (N, P, C) enrichment upon periphyton standing crop, species composition and primary production in an oligotrophic softwater lake. Freshwat. Biol. 19: 57-70.

Faybenro, O. A. & D. H. J. Sydenham, 1988. Evaluation of *Clarias isheriensis* (Sydenham) under semi-intensive management in ponds. Aquaculture 74: 287-291.

Fernando, C. H., 1980. The freshwater zooplankton of Sri Lanka with a discussion of tropical freshwater zooplankton composition. Int. Revue ges. Hydrobiol. 65: 85-125.

Fernando, C. H., C. Tudorancea & S. Mengestou, 1990. Invertebrate zooplankton predator composition and diversity in tropical lentic waters. Hydrobiologia 198: 13-31.

Feuillade, M. & J. Feuillade, 1990. Heterotrophic capabilities of the blue-green algae *Oscillatoria rubescens*. Arch. Hydrobiol. 117: 61-76.

Fish, G. R., 1955. The food of Tilapia in East Africa. Uganda J. 19: 85-89. Quoted by Trewawas (1983).

Flynn, K. J., 1988. The concept of "primary production" in aquatic ecology. Limnol. Oceanogr. 33: 1215-1216.

Folt, C. L., 1987. An experimental analysis of costs and benefits of zooplankton aggregation. In W. C. Kerfoot & A. Sih (eds), Predation: direct and indirect impacts on aquatic communities. The University Press of New England, Hanover (N.H.); Lond.: 300-332.

Fong, S.-C. & H. P. Ju, 1987. Energy value of biomass within benthic algae of milkfish ponds. Aquaculture 64: 31-38.

Förstner, U., 1982. Accumulative phases for heavy metals in limnic sediments. Hydrobiologia 91: 269-284.

Forsyth, D. J., M. R. James & M. Cryer, 1990. Alteration of seasonal and diel patterns in vertical migration of zooplankton by *Anabaena* and planktivorous fish. Arch. Hydrobiol. 117: 385-404.

Foy, R. H. & A. G. Fitzsimons, 1987. Phosphorus inactivation in a eutrophic lake by the direct addition of ferric aluminium sulphate: changes in phytoplankton populations. Freshwat. Biol. 17: 1-17.

Fitzgerald, G. P., 1970. Aerobic lake muds for the removal of phosphorus from lake water. Limnol. Oceanogr. 15: 550-555.

Fraleigh, P. C. & J. C. Burnham, 1988. Myxococcal predation on cyanobacterial populations: nutrient effects. Limnol. Oceanogr. 33: 476-486.

Francko, D. A., 1985. Phytoplankton metabolism and cyclic nucleotides. II. Nucleotide-induced perturbations of alkaline phosphatase activity. Arch. Hydrobiol. 100: 409-421.

Franko, D. A., 1986. Epilimnetic phosphorus cycling : influence of humic materials and iron on coexisting major mechanism. Can. J. Fish. aquat. Sci 43: 302-310.

Friday, L. E., 1987. The diversity of macroinvertebrate and macrophyte community in ponds. Freshwat. Biol. 18: 87-104.

Froelich, P. N., 1988. Kinetic control of dissolved phosphate in natural rivers and estuaries: a primer on the phosphate buffer mechanism. Limnol. Oceanogr. 33: 649-668.

Fryer, G., 1987. Quantitative and qualitative: numbers and reality in the study of living organisms. Freshwat. Biol. 17: 177-189.

Fukuhara, H. & M. Sakamoto, 1988. Ecological significance of bioturbation of zoobenthos community in nitrogen release from bottom sediments in a shallow eutrophic lake. Arch. Hydrobiol. 113: 425-445.

Fulton, R. S. III., 1988. Grazing on filamentous algae by herbivorous zooplankton. Freshwat. Biol. 20: 263-271.

Fulton, R. S. III & H. W. Paerl, 1987. Effect of colonial morphology on zooplankton utilization of algal resources during blue-green algal (*Microcystis aeruginosa*) blooms. Limnol. Oceanogr. 32: 634-644.

Gächter, R., J. S. Meyer & A. Mares, 1988. Contribution of bacteria to release and fixation of phosphorus in lake sediments. Limnol. Oceanogr. 33: 1542-1558.

Gage, M. A. & E. Gorham, 1985. Alkaline phosphatase activity and cellular phosphorus as an index of the phosphorus status of phytoplankton in Minnesota lakes. Freshwat. Biol. 15: 227-233.

Ganf, G. G. & A. J. Horne, 1975. Diurnal stratification, photosynthesis and nitrogen-fixation in a shallow equatorial Lake (Lake George, Uganda). Freshwat. Biol. 5: 13-59.

Gaardner, T. & H. H. Gran, 1927. Investigations of the production of planktion in the Oslo fjord. Rapp. Proc.-Verb. Cons. Int. Expl. Mer 42: 3-48.

Gardner, W. S., T. F. Nalepa & J. M. Malczyk, 1987. Nitrogen mineralization and denitrification in Lake Michigan sediments. Limnol. Oceanogr. 32: 1226-1238.

Gawler, M., G. Balvay, P. Blanc, J. C. Druart & J. C. Pelletier, 1988. Plankton ecology of Lake Geneva: a test of the PEG-model. Arch. Hydrobiol. 114: 161-174.

Gerhart, D. Z. & G. E. Likens, 1975. Enrichement experiments for determining nutrient limitation : four methods compared. Limnol. Oceanogr. 20: 649-653.

Gerritsen J. & S. W. Bradley, 1987. Electrophoretic mobility of natural particles and cultural organisms in freshwater. Limnol. Oceanogr. 32: 1049-1058.

Gilbert, J. J., 1989. The effect of *Daphnia* interference on a natural rotifer and ciliate community: short-term bottle experiment. Limnol. Oceanogr. 34: 606-617.

Gilbert, J. J., 1990. Differential effects of *Anabaena affinis* on cladocerans and rotifers: mechanisms and implications. Ecology 71: 1727-1740.

Gliwicz, Z. M., 1985. Predation or food limitation: an ultimate reason for extinction of planktonic cladoceran species. Arch. Hydrobiol. Beih. Ergebn. Limnol. 21: 419-430.

Gliwicz, Z. M., 1990. Why do cladocerans fail to control algal blooms? Hydrobiologia 200/201: 83-97.

Gliwicz, Z. M., A. Ghilarov & J. Pijanowska, 1981. Food and predation as major factors limiting two natural populations of *Daphnia cucullata* Sars. Hydrobiologia 80: 155-177.

Golachowska, J., 1986. Diurnal fluctuations of phosphorus forms in lake water and seston in Pluß-See. Pol. Arch. Hydrobiol. 33: 165-175.

Goldman, J. C., D. A. Caron & M. R. Dennett, 1987. Regulation of gross efficiency and ammonium regeneration in bacteria by substrate C:N ratio. Limnol. Oceanogr. 32: 1239-1252.

Golterman, H. L., 1982. Loading concentration models for phosphate in shallow lakes. Hydrobiologia 91: 169-174.

Golterman, H. L., 1988. Reflections on fractionation and bioavailability of sediment bound phosphate. Arch. Hydrobiol. Beih. Ergebn. Limnol. 30: 1-4.

Golterman, H. L. & F. A. Kouwe, 1980. Chemical budgets and nutrient pathways. In E. D. Le Cren & R. H. Lowe-McConnell (eds), The functioning of freshwater ecosystems. Cambridge University Press, Cambridge. International Biological Programme 22: 85-140.

Gophen, M., 1990. Biomanipulation: retrospective and future development. Hydrobiologia 200/201: 1-11.

Gophen, M., R. Drenner, G. Vinyard & P. Spaturu, 1983. The cichlid fishery of Lake Kinneret: history and management recommendations. In L. Fishelson & Z. Yaron (eds), Tel Aviv University, Tel Aviv, Israel. Proc. Int. Symp. on Tilapia in Aquaculture: 1-7.

Green, B. W., R. P. Phelps & H. R. Alvarenga, 1989. The effect of manures and chemical fertilizers on the production of *Oreochromis niloticus* in earthen ponds. Aquaculture 76: 37-42.

Grover, J. P., 1989a. Phosphorus-dependent growth kinetics of eleven species of freshwater algae. Limnol. Oceanogr. 34: 341-348.

Grover, J. P., 1989b. Effects of Si:P supply ratios, supply variability, and selective grazing in the plankton: an experiment with a natural algal and protistan assemblage. Limnol. Oceanogr. 34: 349-367.

Güde, H., 1988. Direct and indirect influence of crustacean zooplankton on bacterioplankton of Lake Constance. Hydrobiologia 159: 63-73.

Guiral, D., 1984. Devenir de la matière organique particulaire dans un milieu eutrophe tropical. Rev. Hydrobiol. trop. 17: 191-206.

Guiral, D., R. Arfi & J. P. Torréton, 1989. Mécanismes et incidences écologiques de l'homogénéisation annuelle de densité dans un milieu eutrophe stratifié. Hydrobiologia 183: 195-210.

Gulati, R. D., 1990. Structural and grazing responses of zooplankton community to biomanipulation of some Dutch water bodies. Hydrobiologia 200/201: 99-118.

Gunatilaka, A., 1982. Phosphate adsorption kinetics of resuspended sediments in a shallow lake, Neusiedlersee, Austria. Hydrobiologia 91: 293-298.

Hall, P. O. J., L. G. Anderson, M. M. Rutgers van der Loeff, B. Sundby & S. F. G. Westerlund, 1989. Oxygen uptake kinetics in the benthic boundary layer. Limnol. Oceanogr. 34: 734-746.

Hansson, L. A., 1988. Effects of competitive interactions on the biomass development of planktonic and periphytic algae in lakes. Limnol. Oceanogr. 33: 121-128.

Hansson, L. A., 1989. The influence of a periphytic biolayer on phosphorus exchange between substrate and water. Arch. Hydrobiol. 115: 21-26.

Hansson, L. A., L. Johansson & L. Persson, 1987. Effects of fish grazing on nutrient release and succession of primary producers. Limnol. Oceanogr. 32: 723-729.

Hamilton, D. T. & W. D. Taylor, 1987. Short-term effects of zooplankton manipulations on phosphate uptake. Can. J. Fish. aquat. Sci 45: 1038-1044.

Hardy, F. J., K. S. Shortreed & J. G. Stockner, 1986. Bacterioplankton, phytoplankton and zooplankton communities in a British Columbia coastal lake before and after nutrient reduction. Can. J. Fish. aquat. Sci 43: 1504-1514.

Harris, G. H., 1980. The measurement of photosynthesis in natural populations of phytoplankton. In I. Morris (ed.), The physiological ecology of phytoplankton. Blackwell Scientific Publication, Oxford. Studies in Ecology, Volume 7: 129-187.

Harrison, W. G., 1978. Experimental measurements of nitrogen remineralization in coastal waters. Limnol. Oceanogr. 23: 684-694.

Hart, B. T., 1982. Uptake of trace metals by sediments and suspended particulates: a review. Hydrobiologia 91: 299-313.

Hasselrot, B. & H. Hultberg, 1984. Liming of acidified Swedish lakes and streams and its consequences for aquatic ecosystems. Fisheries 9: 4-9.

Havel, J. E., 1987. Predator-induced defenses: a review. In W. C. Kerfoot & A. Sih (eds), Predation: direct and indirect impacts on aquatic communities. The University Press of New England, Hanover (N.H.); Lond.: 263-278.

Hayes, F. R. & J. E. Philips, 1958. Lake water and sediment : IV. Radiophosphorus equilibrium with mud, plants and bacteria under oxidized and reduced conditions. Limnol. Oceanogr. 3: 459-475.

Healey, F. D., 1978. Physiological indicators of nutrient deficiency in algae. Mitt. Int. Ver. Theor. Angew. Limnol. 21: 34-41.

Healey, F. P., 1985. Interacting effects of light and nutrient limitation on the growth rate of *Synechococcus linearis* (Cyanophyceae). J. Phycol. 21: 134-146. Quoted by Jansson *et al.* (1988)

Heath, R. T., 1986. Dissolved organic phosphorus compounds: do they satisfy planktonic phosphate demand in summer? Can. J. Fish. aquat. Sci 43: 343-350.

Hecky, R. E. & P. Kilham, 1988. Nutrient limitation of phytoplankton in freshwater and marine environments: a review of recent evidence on the effects of enrichment. Limnol. Oceanogr. 33: 796-822.

Hecky, R. E., R. H. Spigel & G. W. Coulter, 1991. The nutrient regime. In G. W. Coulter (ed.), Lake Tanganika and its life. Oxford Univ. Press, London, 76-89.

Heinen, J. M., G. C. G. Fornshell & P. J. Foley, 1988. Removal of floating filamentous algae from ponds with a shallow seine. Progr. Fish-Cult. 50: 187.

Hendrey, G. R. & F. R. Wright, 1976. Acid precipitation in Norway: Effects on aquatic fauna. J. Great Lakes Res. 2 (Suppl. 1): 192-207. Quoted by Mulholland *et al.*, 1986.

Hepher, B., 1958. On the dynamics of phosphorus added to fishponds in Israel. Limnol. Oceanogr. 3: 84-100.

Hepher, B., 1962. Primary production in fishponds and its application to fertilization experiments. Limnol. Oceanogr. 6: 131-136.

Hepher, B., 1966. Some aspects of the phosphorus cycle in fishponds. Verh. int. Ver. Limnol. 16: 1293-1297.

Hessen, D. O., T. Andersen & A. Lyche, 1989. Differential grazing and resource utilization of zooplankton in a humic lake. Arch. Hydrobiol. 114: 321-347.

Hessen, D. O., T. Andersen & A. Lyche, 1990. Carbon metabolism in a humic lake: pool sizes and cycling through zooplankton. Limnol. Oceanogr. 35: 84-99.

Heyman, U. & A. Lundgren, 1988. Phytoplankton biomass and production in relation to phosphorus: some conclusions from field studies. Hydrobiologia 170: 211-227.

Hickling, C. F., 1962. Fish Culture. Faber & Faber, London, 295 pp.

Hieltjes, A. H. M. & L. Lijklema, 1980. Fractionation of inorganic phosphates in calcareous sediments. J. environ. Qual. 9: 405-407.

Hillbricht-Ilkowska, A. & T. Weglenska, 1973. Experimentally increased fish stock in the pond type Lake Warniak. VII. Numbers, biomass and production of zooplankton. Ekol. pol. 21: 533-552.

Hino, S., 1988. Extracellular release of organic matter associated with the physiological state of freshwater blue-green alga. Arch. Hydrobiol. 113: 307-317.

Hogendoorn, H., 1983. Growth and production of the African catfish, *Clarias lazera* (C. & V.) III. Bioenergetic relations of body weight and feeding level. Aquaculture 35: 1-17.

Hollerman, W. D. & C. E. Boyd, 1985. Effects of draining on water quality and production of channel catfish in ponds. Aquaculture 46: 45-54.

Hollerman, W. D. & C. E. Boyd, 1986. Effects of annual draining on the limnology of sunfish ponds. Progress. Fish-Cult. 48: 120-124.

Holtan, H., L. Kamp-Nielsen & A. O. Stuanes, 1988. Phosphorus in soil, water and sediment: an overview. Hydrobiologia 170: 19-34.

Horne, A. J. & M. L. Commins, 1989a. Phytoplanktonic monotony. I. Low spatial heterogeneity for nitrogen fixation, algae, nutrients, and temperature, but not for zooplankton, in Lake Rotongaio, New Zealand. Arch. Hydrobiol. Beih. Ergebn. Limnol. 32: 83-97.

Horne, A. J. & M. L. Commins, 1989b. Phytoplanktonic monotony. II. Patterns of temporal stability in phytoplankton patches in Lake Rotongaio, New Zealand. Arch. Hydrobiol. Beih. Ergebn. Limnol. 32: 99-112.

Horne, A. J. & A. B. Viner, 1971. Nitrogen fixation and its significance in tropical lake George, Uganda. Nature 232: 417-418.

Houk, V., 1989. Seasonal changes of phytoplankton development near the dam of the Zelivka-Reservoir (Czechoslovakia). Arch. Hydrobiol. Beih. Ergebn. Limnol. 33: 397-407.

Howard-Williams, C., 1985. Cycling and retention of nitrogen and phosphorus in wetlands: a theoretical and applied perspective. Freshwat. Biol. 15: 391-431.

Howarth, R. W., R. Marino, J. Lane & J. J. Cole, 1988a. Nitrogen fixation in freshwater, estuaries, and marine ecosystems. 1. Rates and importance. Limnol. Oceanogr. 33: 669-687.

Howarth, R. W., R. Marino & J. J. Cole, 1988b. Nitrogen fixation in freshwater, estuaries, and marine ecosystems. 2. Biogeochemical controls. Limnol. Oceanogr. 33: 688-701.

Hrbáček, J., 1962. Species composition and the amount of zooplankton in relation to the fish stock. Rozpr. ČSAV, Ser. mat. nat. sci. 72: 1-117.

Huet, M. & J. A. Timmermans, 1970. Traité de pisciculture. Ed. Ch. de Wijngaert, Bruxelles, 718 pp.

Hul, M., 1987. The effect of domestic sewage on the structure of the microbenthic ciliate communities in the Łyna River. Pol. Arch. Hydrobiol. 34: 567-578.

Humphries, S. E. & V. D. Lyne, 1988. Cyanophyte blooms: the role of cell buoyancy. Limnol. Oceanogr. 33: 79-91.

Hunt, D. & C. E. Boyd, 1981. Alkalinity losses from ammonium fertilizers used in fish ponds. Trans. am. Fish. Soc. 110: 81-85.

Huntsman, S. A. & W. G. Sunda, 1980. The role of trace metals in regulating phytoplankton growth. In I. Morris (ed.), The physiological ecology of phytoplankton. Blackwell Scientific Publication, Oxford. Studies in Ecology Volume 7: 285-328.

Hutchinson, G. E., 1961. The paradox of the plankton. Am. Nat. 95: 137-145.

Iltis, A., 1980. Les algues. In J.-R. Durand & C. Lévêque (eds), Flore et faune aquatiques de l'Afrique sahélo-soudanienne.O.R.S.T.O.M., Paris. DocumnTechqs 44: 9-61.

Irvine, K., B. Moss & J. Stansfield, 1990. The potential of artificial refugia for maintaining a community of large-bodied cladocera against fish predation in a shallow eutrophic lake. Hydrobiologia 200/201: 379-389.

Jackson, G. A., 1989. Simulation of bacterial attraction and adhesion to falling particles in an aquatic environment. Limnol. Oceanogr. 34: 514-530.

Jacquet, J.-M., E. Davaud, F. Rapin & J.-P. Vernet, 1982. Basic concepts and associated statistical methodology in the geochemical study of lake sediments. Hydrobiologia 91: 139-146.

Jana, B. B., S. K. Roy & G. N. Patel, 1983. Relationship between the nitrification and denitrification and the fish growth in the fish ponds under polyculture system. Pol.

Arch. Hydrobiol. 30: 65-68.
Jana, B. B. & U. K. De, 1983. Primary productivity of phytoplankton, environmental covariates and fish growth in West Bengal ponds under a polyculture system. Int. Revue ges. Hydrobiol. 68: 45-58.
Jansson, M., 1976. Phosphatases in lake water: Characterization of enzymes from phytoplankton and zooplankton by gel filtration. Science 194: 320-321. Quoted by Jansson *et al.*, 1988.
Jansson, M., 1988. Phosphate uptake and utilization by bacteria and algae. Hydrobiologia 170: 177-189.
Jansson, M., H. Olsson & K. Pettersson, 1988. Phosphatases: origin, characteristics and function in lakes. Hydrobiologia 170: 157-175.
Jeffries, H. J. & J. H. Lawton, 1985. Predator-prey ratios in communities of freshwater invertebrates: the role of enemy free space. Freshwat. Biol. 15: 105-112.
Jeppesen, E., 1982. Diurnal variation in the oxygen uptake of river sediments *in* by use of continuous flow-through systems. Hydrobiologia 91: 189-195.
Jeppesen, E., J. P. Jensen, P. Kristensen, M. Søndergaard, E. Mortensen, O. Sortkjær & K. Obrik, 1990a. Fish manipulation as a lake restoration tool in shallow, eutrophic temperate lakes. 2: Threshold levels, long-term stability and conclusions. Hydrobiologia 200/201: 219-227.
Jeppesen, E., M. Søndergaard, E. Mortensen, P. Kristensen, B. Riemann, H. J. Jensen, J. P. Müller, O. Sortkjær, J. P. Jensen, K. Christoffersen, S. Bosselmann & E. Dall, 1990b. Fish manipulation as a lake restoration tool in shallow, eutrophic temperate lakes. 1: Cross-analysis of three Danish case-studies. Hydrobiologia 200/201: 205-218.
Jeppesen, E., M. Søndergaard, O. Sortkjær, E. Mortensen & P. Kristensen, 1990c. Interactions between phytoplankton, zooplankton and fish in a shallow, hypertrophic lake: a study on phytoplankton collapses in Lake Sobygård, Denmark. Hydrobiologia 191: 139-148.
John, D. M., 1986. The inland waters of tropical West Africa. Arch. Hydrobiol. Beih. Ergebn. Limnol. 23: 1-244.
Johnsen, G. H. & P. J. Jakobsen, 1987. The effect of food limitation on vertical migration in *Daphnia longispina*. Limnol. Oceanogr. 32: 873-880.
Johnson, R. K., B. Boström & W. van de Bund, 1989. Interactions between *Chironomus plumosus* (L.) and the microbial community in the surficial sediments of a shallow, eutrophic lake. Limnol. Oceanogr. 34: 992-1003.
Jones, R. I., 1990. Phosphorus transformations in the epilimnion of humic lakes: biological uptake of phosphate. Freshwat. Biol. 23: 323-337.
Jones, R. I., K. Salonen & H. De Haan, 1988. Phosphorus transformations in the epilimnion of humic lakes: abiotic interactions between dissolved humic materials and phosphate. Freshwat. Biol. 19: 357-369.
Jørgensen, B. B. & J. Sørensen, 1985. Seasonal cycles of O_2, NO_3 and SO_4^{2-} reduction maximum in spring. Mar. Ecol. Prog. Ser. 24: 65-74.
Jurkowska, J., 1987. Estimation of phytoplankton biomass and production, and organic matter decomposition in three small field-ponds. Pol. Arch. Hydrobiol. 34: 35-49.
Kaiser, G. E. & F. W. Wheaton, 1983. Nitrification filters for aquatic culture systems : State of art. J. World Maricul. Soc. 14: 302-324.
Kajak, Z., 1988. Considerations on benthos abundance in freshwaters, its factors and mechanism. Int. Revue ges. Hydrobiol. 73: 5-19.
Kajak, Z., J. I. Rybak, I. Spodniewska & W. A. Godlewska-Lipowa, 1975. Influence of the planktivorous fish, *Hypothalmichthys molitrix* (Val.) on the plankton and benthos of the eutrophic lake. Pol. Arch. Hydrobiol. 22: 301-310.
Kamp-Nielsen, L., H. Meyer & S. E. Jørgensen, 1982. Modelling the influence of bioturbation on the vertical distribution of sedimentary phosphorus in L. Esrom. Hydrobiologia 91: 197-206.
Kaspar, H. F., 1985. The denitrification capacity of sediment from a hypereutrophic lake. Freshwat. Biol. 15: 449-453.
Kelderman, P. & A. M. van de Repe, 1982. Temperature dependence of sediment- water exchange in Lake Grevelingen, S.W. Netherlands. Hydrobiologia 91: 489-490.
Kelly, C. A. & J. W. M. Rudd, 1989. Reply to comment by Urban and Baker. Limnol. Oceanogr. 34: 1146-1148.
Kelly, C. A., J. W. M. Rudd & D. W. Schindler, 1988. Carbon and electron flow via methanogenesis, SO_4^{2-}, NO_3^-, Fe^{3+}, and Mn^{4+} reduction in the anoxic hypolimnia of three lakes. Arch. Hydrobiol. Beih. Ergebn. Limnol. 31: 333-344.
Kerfoot, W. C., 1987. Cascading effects and indirect pathways. In W. C. Kerfoot & A. Sih (eds), Predation: direct and indirect impacts on aquatic communities. The University Press of New England, Hanover (N.H.); Lond.: 57-70.
Kerfoot, W. C., D. L. Kellogg, Jr & J. R. Strickler, 1980. Visual observations of live zooplankton: evasion, escape, and chemical defenses. In W. C. Kerfoot (ed.), Evolution and ecology of zooplankton communities. Special Symp. Vol. 3, Am. Soc. Limnol. Oceanogr., University Press of New England, Hanover, New Hampshire, London: 10-27.
Kerekes, J. J., A. C. Blouin & S. T. Beauchamp, 1990. Trophic response to phosphorus in acidic and non-acidic lakes in Nova Scotia, Canada. Hydrobiologia 191: 105-110.
Ketchum, B. & A. C. Redfield, 1949. Some physical and chemical characteristics of algal growth in mass culture. J. cell. comp. Physiol. 33: 281-299.
Kilham, P., 1990. Mechanisms controlling the chemical composition of lakes and rivers: data from Africa. Limnol. Oceanogr. 35: 80-83.
Kilham, P. & R. E. Hecky, 1988. Comparative ecology of marine and freshwater phytoplankton. Limnol. Oceanogr. 33: 776-795.
Kilham, P. & S. S. Kilham, 1980. The evolutionary ecology of phytoplankton. In I. Morris (ed.), The physiological ecology of phytoplankton. Blackwell Scientific Publication, Oxford. Studies in Ecology Volume 7: 571-597.
Kilham, P. & S. S. Kilham, 1990. Endless summer: internal loading processes dominate nutrient cycling in tropical lakes. Freshwat. Biol. 23: 379-389.
Kilham, P., S. S. Kilham & R. E. Hecky, 1986. Hypothesized resource relationships among African planktonic diatoms. Limnol. Oceanogr. 31: 1169-1181.
Kilham, S. S., 1986. Dynamics of lake Michigan natural

phytoplankton communities in continuous cultures along a Si:P loading gradient. Can. J. Fish. aquat. Sci. 43: 351-360.

Kirk, J. T. O., 1986. Light and photosynthesis in aquatic ecosystems. Cambridge University Press, Cambridge, London, N.Y., 401 pp.

Kirk, K. L. & J. J. Gilbert, 1990. Suspended clay and the population dynamics of planktonic rotifers and cladocerans. Ecology 71: 1741-1755.

Kitchell, J. F. & S. R. Carpenter, 1987. Piscivores, planktivores, fossils, and phorbins. In W. C. Kerfoot & A. Sih (eds), Predation: direct and indirect impacts on aquatic communities. The University Press of New England, Hanover (N.H.); Lond.: 132-146.

Klapwijk, A. & W. J. Snodgrass, 1982. Experimental measurement of sediment nitrification and denitrification in Hamilton Harbour, Canada. Hydrobiologia 91: 207-216.

Klapwijk, S. P., J. W. Kroon & M.-L. Meijer, 1982. Available phosphorus in lake sediments in the Netherlands. Hydrobiologia 91: 491-500.

Kling, G. W., 1988. Comparative transparency, depth of mixing, and stability of stratification in lakes of Cameroon, West Africa. Limnol. Oceanogr. 33: 27-40.

Komákorvá, J., 1989. Changes in phytoplankton assemblage during the spring period in the moderately eutrophic Římov reservoir (Czechoslovakia). Arch. Hydrobiol. Beih. Ergebn. Limnol. 33 : 419-433.

Konopka, A., 1983. The effect of nutrient limitations and its interaction with light upon the products of photosynthesis in *Merismopedia tenuissima* (Cyanophyceaea). J. Phycol. 19: 403-409. Quoted by Jansson *et al.* (1988)

Konopka, A., 1989. Metalimnetic cyanobacteria in hard-water lakes: buoyancy regulation and physiological state. Limnol. Oceanogr. 34: 1174-1184.

Konopka, A. & M. Schnur, 1980. Effect of light intensity upon macromolecular synthesis in cyanobacteria. Microb. Ecol. 6: 291-301. Quoted by Jansson *et al.* (1988)

Konopka, A. & M. Schnur, 1981. Biochemical composition and photosynthetic carbon metabolism of nutrient limited cultures of *Merismopedia tenuissima* (Cyanophyceae). J. Phycol. 17: 118-122. Quoted by Jansson *et al.* (1988)

Korzeniewski, K. & A. Moczulska, 1985. Sorption of cations by bottom sediments of Szczytno Małe and Łętowo Lakes, with cage trout culture. Pol. Arch. Hydrobiol. 32: 131-144.

Krom, M. D., J. Erez, C. B. Porter & S. Ellner, 1989a. Phytoplancton nutrient uptake dynamics in earthen marine fishponds under winter and summer conditions. Aquaculture 76: 237-253.

Krom, M. D., A. Neori & J. van Rijn, 1989b. Importance of water flow rate in controlling water quality processes in marine and freshwater fish ponds. Israel. J. Aquacult. 41: 23-33.

Kronert, U., G. Hörstgen-Schwark & H.-J. Langholz, 1989. Prospect of selecting for late maturity in tilapia (*Oreochromis niloticus*). I. Family studies under laboratory conditions. Aquaculture 77: 113-121.

Kuenzler, E., 1970. Dissolved organic phosphorus excretion by marine phytoplankton. J. Phycol. 6: 7-13. Quoted by Jansson *et al.* (1988).

Kurihara, Y., 1989. Ecology of some ricefields in Japan as exemplified by some benthic fauna, with notes on management. Int. Revue ges. Hydrobiol. 74: 507-548.

Kwiatkiwski, R. E. & J. C. Roff, 1976. Effects of acidity on the phytoplankton and primary productivity of selected northern Ontario lakes. Can. J. Bot. 54: 2546-2561. Quoted by Mulholland *et al.*, 1986.

Lam, D. C. L., W. M. Schertzer & A. S. Fraser, 1982. Mass balance models of phosphorus in sediments and water. Hydrobiologia 91: 217-225.

Lamarra, V. A., 1975. Digestive activities of carp as a major contributor to the nutrient loading of lakes. Verh. int. Ver. Limnol. 19: 2461-2468.

Lammens, E. H. R. R., R. D. Gulati, M.-L. Meijer & E. van Donk, 1990. The first biomanipulation conference: a synthesis. Hydrobiologia 200/201: 619-627.

Lampert, W., 1987. Vertical migration of freshwater zooplankton: indirect effects of vertebrate predators on algal communities. In W. C. Kerfoot & A. Sih (eds), Predation: direct and indirect impacts on aquatic communities. The University Press of New England, Hanover (N.H.); Lond.: 291-299.

Lampert, W., W. Fleckner, H. Rai & B. E. Taylor, 1986. Phytoplankton control by grazing zooplankton: a study on the spring clear-water phase. Limnol. Oceanogr. 31: 478-490.

Landau, R., 1983. Factors influencing tilapia biomass in Lake Kinneret. In L. Fishelson & Z. Yaron (eds), Tel Aviv University, Tel Aviv, Israel. Proc. Int. Symp. on Tilapia in Aquaculture: 8-17.

Lazard, J., 1980. Le développement de la pisciculture intensive en Côte d'Ivoire : Exemple de la ferme piscicole pilote de Natio-Kobadara (Korhogo). C.T.F.T., Notes et Documents, Nouvelle Série 21: 1-44.

Lazzaro, X., 1987. A review of planktivorous fishes: their evolution, feeding behaviour, selectivities, and impacts. Hydrobiologia 146: 97-167.

Lazzaro, X., R. W. Drenner, R. A. Stein & J. D. Smith, 1991a. Planktivores and plankton dynamics: effects of fish biomass and planktivore type. Can. J. Fish. aquat. Sci. (in press).

Lazzaro, X., R. W. Drenner, R. A. Stein & J. D. Smith, 1991b. Interactions between gizzard shad and bluegill as regulators of aquatic communities: a mesocosm experiment. Trans. am. Fish. Soc. (in press).

Lean, D. R. S., F. R. Pick, S. F. Mitchell, M. T. Downes, P. H. Woods & E. White, 1989. Lake Ontaro enclosure experiments: test ecosystems to evaluate plankton phosphorus and nitrogen deficiency. Arch. Hydrobiol. Beih. Ergebn. Limnol. 32: 195-211.

Leavitt, P. R., S. R. Carpenter & J. F. Kitchell, 1989. Whole-lake experiments: the annual record of fossil pigments and zooplankton. Limnol. Oceanogr. 34: 700-717.

Legendre, M., M. Pagano & L. Saint-Jean, 1987. Peuplements et biomasse zooplanctonique dans les étangs de pisciculture lagunaire (Layo, Côte d'Ivoire). Etude de la recolonisation après la mise en eau. Aquaculture 67: 321-341.

Lehman, J. T., 1988. Ecological principles affecting community structure and secondary production by zooplankton in marine and freshwater environments.

Limnol. Oceanogr. 33: 931-945.
Lehman, J. T. & D. Scavia, 1984. Measuring the ecological significance of microscale nutrient patches. Limnol. Oceanogr. 29: 214-216.
Lehninger, A. L., 1965. Bioenergetics. The molecular basis of biological energy transformations. W. A. Benjamin, Inc., N.Y., Amsterdam, 258 pp.
Lennox, L. J., 1984. Lough Ennell : laboratory studies on sediment phosphorus release under varying mixing, aerobic and anaerobic conditions. Freshwat. Biol. 14: 183-187.
Leventer, H. & B. Teltsch, 1990. The contribution of silver carp (*Hypophthalmichthys molitrix*) to the biological control of Netofa reservoirs. Hydrobiologia 191 : 47-55.
Levitan, C., 1987. Formal stability analysis of a planktonic freshwater community. In W. C. Kerfoot & A. Sih (eds), Predation: direct and indirect impacts on aquatic communities. The University Press of New England, Hanover (N.H.); Lond.: 71-100.
Li, S., 1987. Energy structure and efficiency of a typical Chinese integrated fish farm. Aquaculture 65: 105-118.
Liu, Q. & H. Seki, 1988. The effect of a "Bio-filter" aquatic treatment plan on Doh-Hoh-Numa Bog, Tsukuba City, Japan: rapid oligotrophication without damage to the phytoplancton community. Arch. Hydrobiol. 114: 63-82.
Lijklema, L. & A. H. M. Hieltjes, 1982. A dynamic phosphate budget model for a eutrophic lake. Hydrobiologia 91: 227-233.
Lindmark, G. K., 1982. Acidified lakes: sediment treatment with sodium carbonate - a remedy? Hydrobiologia 91: 537-547.
Lipschultz, F., S. C. Wofsy & L. E. Fox, 1985. The effects of light and nutrients on rates of ammonium transformation in an eutrophic river. Mar. Chem. 16 : 329-335.
Little, D. & J. Muir, 1987. A guide to integrated warmwater aquaculture. Publ.Inst. Aquaculture, Univ. of Sterling, Sterling, UK, 224 pp.
Logan, T. J., 1982. Mechanisms for release of sediment-bound phosphate to water and the effect of agricultural land management on fluvial transport of particulate and dissolved phosphate. Hydrobiologia 91: 519-530.
Lund, J. W. G., 1965. The ecology of the freshwater phytoplankton. Biol. Rev. 40 : 231-293.
Lyche, A., B. A. Faafeng & A. Braband, 1990. Predictability and possible mechanisms of plankton response to reduction of planktivorous fish. Hydrobiologia 200/201 : 251-261.
Mann, K. H., 1988. Production and use of detritus in various freshwater, estuarine, and coastal marine ecosystems. Limnol. Oceanogr. 33: 910-930.
Marsden, M. W., 1989. Lake restoration by reducing external phosphorus loading: the influence of sediment phosphorus release. Freshwat. Biol. 21: 139-162.
Marvalin, O., L. Aluya & C. Amblard, 1989. Importance relative des fractions bactériennes libres et fixes en milieu lacustre eutrophe. Arch. Hydrobiol. 115: 371-390.
Matveev, V. E. & E. G. Balseiro, 1990. Contrasting responses of two Cladocerans to changes in the nutritional value of nannoplankton. Freshwat. Biol. 23: 197-204.
Mazumder, A. & M. D. Dickman, 1989. Factors affecting the spatial and temporal distribution of phototrophic sulfur bacteria. Arch. Hydrobiol. 116: 209-226.
Mazumder, A., D. J. McQueen, W. D. Taylor & D. R. S. Lean, 1986. Effects of fertilization and planktivorous fish (yellow perch) predation on size distribution of particulate phosphorus and assimilated phosphate: large enclosure experiments. Limnol. Oceanogr. 33: 421-430.
Mazumder, A., D. J. McQueen, W. D. Taylor, D. R. S. Lean & M. D. Dickman, 1990. Micro- and mesoplankton grazing on natural pico- and nanoplankton in contrasting plankton communities produced by planktivore manipulation and fertilization. Arch. Hydrobiol. 118: 257-282.
McAllister, C. D., N. Shah & J. D. H. Strickland, 1964. Some interactions of temperature, light intensity and nutrient concentration during the continuous culture of *Nitzschia closterum* and *Tetraselmis sp.* Limnol. Oceanogr. 9: 79-86. Quoted by Jansson *et al.* (1988)
McAndrew, B. J., 1988. Tilapia hatchery systems. In H. R. King & K. H. Ibrahim (eds), Village level aquaculture development in Africa. Proc. Comm. Cons. Workshop on Village Level Aquaculture Development in Africa. Freetown, Sierra Leone, 14-20 February 1985. Commonwealth Secretariat, London: 137-146.
McCarthy, J. J., 1980. Nitrogen. In I. Morris (ed.), The physiological ecology of phytoplankton. Blackwell Scientific Publication, Oxford. Studies in Ecology Volume 7: 191-233.
McFarland, M. A. & D. W. Toetz, 1988. Nitrogen fixation (acetylene reduction) in Lake Hefner, Oklahoma. Arch. Hydrobiol. 114: 213-230.
McGeachin R. & R. R. Stickney, 1982. Manuring rates for production of blue tilapia in simulated sewage lagoons receiving laying hen waste. Prog. Fish-Cult. 44: 25-28.
McKnight, D. M. & F. M. M. Morel, 1980. Copper complexation by siderophores from filamentous blue-green algae. Limnol. Oceanogr. 25: 62-71.
McQueen, D. J. & J. R. Post, 1988. Cascading trophic interactions: uncoupling at the zooplankton-phytoplankton link. Hydrobiologia 159: 277-296.
McQueen, D. G., J. R. Post & E. L. Mills, 1986. Trophic relations in freshwater pelagic ecosystems. Can. J. Fish. aquat. Sci. 43: 1571-1581.
Meade, J. W., 1985. Allowable ammonia for fishculture. Prog. Fish-Cult. 47: 135-145.
Meffert, M.-E., 1989. Planktic unsheathed filaments (Cyanophyceae) with polar and central gas vacuoles. II. Biology, population dynamics and biotopes of *Limnothrix redekei* (Van Goor) Meffert. Arch. Hydrobiol. 116: 257-282.
Meijer, M.-L., E. H. R. R. Lammens, A. J. P. Raat, M. P. Grimm & S. H. Hosper, 1990a. Impact of cyprinids on zooplankton and algae in ten drainable ponds. Hydrobiologia 191 : 275-284.
Meijer, M.-L., M. W. de Haan, A. W. Breukelaar & H. Buiteveld, 1990b. Is reduction of the benthivorous fish an important cause of high transparency following biomanipulation in shallow lakes? Hydrobiologia 200/201: 303-315.
Mélard, Ch., 1986. Les bases biologiques de l'élevage intensif du tilapia du Nil. Cahiers Ethol. appl. 6 (3): 1-224.
Mesfin, M., C. Tudorancea & R. M. Baxter, 1988. Some

limnological observations on two Ethiopian hydroelectric reservoirs: Koka (Shewa administrative district) and Finchaa (Walega administrative district). Hydrobiologia 157: 47-55.

Miller, J. C., 1987. Evidence for the use of non-detrital dissolved organic matter by microheterotrophs on plant detritus in a woodland stream. Freshwat. Biol. 18: 483-494.

Miller, J. W., 1976. Fertilization and feeding practices in warmwater pond fishculture in Africa. FAO/CIFA symposium on aquaculture in Africa, 1975. Accra, Ghana. CIFA Techn. Pap. 4, suppl.1: 512-541.

Miller, T. E. & W. C. Kerfoot, 1987. Redefining indirect effects. In W. C. Kerfoot & A. Sih (eds), Predation: direct and indirect impacts on aquatic communities. The University Press of New England, Hanover (N.H.); Lond.: 33-37.

Milstein, A., M. Feldlite, N. Mozes & Y. Avnimelech, 1989. Limnology of reservoirs used for fish farming and crop irrigation, with integrated free and cage fish culture. Israel. J. Aquacult. 41: 12-22.

Miura, T., 1990. The effects of planktivorous fishes on the plankton community in a eutrophic lake. Hydrobiologia 200/201: 567-579.

Moav, R., G. Wohlfarth, G. L. Schroeder, G. Hulata & H. Barash, 1977. Intensive polyculture of fish in freshwater ponds. I. Substitution of expensive feeds by liquid cow manure. Aquaculture 10: 25-43.

Monod, J., 1942. Recherches sur la croissance des cultures bactériennes. Hermann, Paris, 210 pp.

Moore, W. J., 1956. Physical chemistry. Longmans, Greens & Co., London. 633 p.

Moriarty, D. J. W., 1973. The physiology of digestion of blue-green algae in the cichlid fish *Tilapia nilotica*. J. Zool. 171: 25-39.

Moriarty, D. J. W., 1990. Letter to the Editor. Aquabyte 3 (2): 18.

Moriarty, D. J. W., J. P. E. C. Darlington, I. G. Dunn, C. M. Moriarty & M. P. Tevlin, 1973. Feeding and grazing in Lake George, Uganda. Proc. R. Soc., Lond. B 184: 299-319.

Moriarty, D. J. W. & C. M. Moriarty, 1973. The assimilation of carbon from phytoplankton by two herbivorous fishes: *Tilapia nilotica* and *Haplochromis nigripinnis*. J. Zool. 171: 41-56.

Morris, D. P. & W. M. Lewis Jr., 1988. Phytoplankton nutrient limitation in Colorado mountain lakes. Freshwat. Biol. 20: 315-327.

Mortimer, C. H., 1941. The exchange of dissolved substances between mud and water in lakes. Part I & II. J. Ecol. 29: 280-329.

Mortimer, C. H., 1942. The exchange of dissolved substances between mud and water in lakes. Part III & IV. J. Ecol. 30: 147-201.

Moss, B., 1990. Engineering and biological approaches to the restoration from eutrophication of shallow-lakes in which aquatic plant communities are important components. Hydrobiologia 200/201: 367-377.

Msiska, O. V., 1981. Quality and nutrient release of poultry manure immersed in water. J. Limnol. Soc. Sth. Afr. 7: 59-62.

Msiska, O. V., 1983. Yields of *Sarotherodon mossambicus* using inorganic fertilizers and rice bran in Southern Malawi. Aquaculture 32: 201-205.

Msiska, O. V. & M. A. Cantrell, 1985. Influence of poultry manure on growth of *Oreochromis shiranus chilwae*. Aquaculture 44: 67-73.

Mulholland, P. J., Elwood, J. W., A. V. Palumbo & R. J. Stevenson, 1986. Effect of stream acidification on periphyton composition, chlorophyll, and productivity. Can. J. Fish. aquat. Sci. 43: 1846-1858.

Murad, A. & C. E. Boyd, 1987. Experiment on fertilization of sport-fish ponds. Progr. Fish-Cult. 49: 100-107.

Murphy, T. P., Lean, D. R. S. & C. Nalewajko, 1976. Blue-green algae: their excretion of iron-selective chelators enables them to dominate other algae. Science 198: 900-902.

Nalewajko, C. & D. R. S. Lean, 1980. Phosphorus. In I. Morris (ed.), The physiological ecology of phytoplankton. Blackwell Scientific Publication, Oxford. Studies in Ecology Volume 7: 235-258.

Nalewajko, C., B. Paul, K. Lee & H. Shear, 1986. Light history, phosphorus status and the occurence of light stimulation or inhibition of phosphate uptake in Lake Superior phytoplankton and bacteria. Can. J. Fish. aquat. Sci. 43: 329-335.

Nembrini, G., J. A. Capobianco, J. Garcia & J. M. Jaquet, 1982. Interaction between interstitial water and sediment in two cores of Lake Léman, Switzerland. Hydrobiologia 91: 363-375.

Ng, W. J., T. S. Sim, S. L. Ong, K. Kho, L. M. Ho, S. H. Tay & C. C. Goh, 1990. The effect of *Elodea densa* on aquaculture water quality. Aquaculture 84 (3/4): 267-276.

Nicklisch, A. & J.-G. Kohl, 1989. The influence of light on the primary production of two planktic blue-green algae. Arch. Hydrobiol. Beih. Ergebn. Limnol. 33: 451-455.

Nilssen, J.P., 1984. Tropical lakes - functional ecology and future development: the need for a process-oriented approach. Hydrobiologia 113: 231-242.

Noriega-Curtis, P., 1979. Primary productivity and related fish yield in intensely manured fishponds. Aquaculture 17: 335-344.

Nriagu, J. O., 1978. Dissolved silica in pore water of Lake Ontario, Erie, and Superior sediments. Limnol. Oceanogr. 23: 53-67.

O'Brien, W.J., 1987. Planktivory by freshwater fish: thrust and parry in the pelagia. In W. C. Kerfoot & A. Sih (eds), Predation: direct and indirect impacts on aquatic communities. The University Press of New England, Hanover (N.H.); Lond.: 3-16.

Oláh, J., V. R. P. Sinha, S. Ayyappan, C. S. Purushothaman & S. Radhayshyam, 1987. Sediment oxygen consumption in tropical undrainable fishponds. Int. Revue ges. Hydrobiol. 72: 297-305.

Oldorf, W., W. Kronert, J. Balarin, R. Haller, G. Hörstgen-Schwark & H. J. Langholz, 1989. Prospect of selecting for late maturity in tilapia (*Oreochromis niloticus*). II. Strain comparisons under laboratory and field conditions. Aquaculture 77: 123-133.

Olsen, Y., A. Jensen, H. Reinertsen, K. Y. Borsheim, M. Heldal & A. Langeland, 1986. Dependence of the rate of release of phosphorus on the P:C ratio in the food supply, as calculated by a recycling model. Limnol. Oceanogr. 31: 36-44.

Olsen, Y., G. Knutsen & T. Lien, 1983. Characteristics of phosphorus limitation in *Chlamydomonas reinhardtii* (Chlorophyceae) and its palmelloids. J. Phycol. 19: 313-319. Quoted by Jansson *et al.*, 1988.

Olsson, H., 1990. Phosphatase activity in relation to phytoplankton composition and pH in Swedish lakes. Freshwat. Biol. 23: 353-362.

Ongley, E. D., M. C. Bynoe & J. B. Percival, 1982. Physical and geochemical characteristics of suspended solids, Wilton Creek, Ontario. Hydrobiologia 91: 41-57.

Opuszyński, K., L. Myskowski, G. Okoniewska, W. Opuszyńska, M. Szlamińska, J. Wolnicki & M. Woźniewski, 1989. Rearing of common carp, grass carp, silver carp and bighead carp larvae using zooplancton and/or different dry feeds. Pol. Arch. Hydrobiol. 36: 217-230.

Ostrofsky, M. L., 1987. Phosphorus species in the surficial sediments of lakes of Eastern North America. Can. J. Fish. aquat. Sci. 44: 960-966.

Paasche, E., 1980. Silicon. In I. Morris (ed.), The physiological ecology of phytoplankton. Blackwell Scientific Publication, Oxford. Studies in Ecology Volume 7: 259-284.

Pace, M. L., 1986. An empirical analysis of zooplankton community size structure across lake trophic gradients. Limnol. Oceanogr. 31: 45-55.

Paerl, H. W., 1985. Microzone formation: its role in the enhancement of aquatic N_2 fixation. Limnol. Oceanogr. 30: 1246-1252.

Paerl, H. W., 1988. Nuisance phytoplankton blooms in coastal, estuarine, and inland waters. Limnol. Oceanogr. 33: 823-847.

Paine, R. T., 1988. Food webs: road maps of interactions or grist for theoretical development. Ecology 69: 1641-1654.

Pańczakowa, J. & T. Szyszka, 1986. Concentrations of nitrogen and phosphorus in waters of two hypertrophic lakes and their effects on the quantity of phytoplankton biomass. Pol. Arch. Hydrobiol. 33: 151-164.

Paterson, D. M., 1989. Short-term changes in the erodibility of intertidal cohesive sediments related to the migratory behavior of epipelic diatoms. Limnol. Oceanogr. 34: 223-296.

Pearshall, W. H., 1924. Phytoplankton and environment in the English Lake District. Revue Algologique 1: 54-67. Quoted by Shoesmith & Brook, 1983.

Pearsall, W. H. & C. H. Mortimer, 1939. Oxidation and reduction potentials in water logged soils, natural waters and muds. J. Ecol. 27: 483-501.

Peters G. T., J. R. Webster & E. F. Benfield, 1987. Microbial activity associated with seston in headwater streams: effects of nitrogen, phosphorus and temperature. Freshwat. Biol. 18: 405-413.

Peterson, C. G., 1987. Gut passage and insect grazer selectivity of lotic diatoms. Freshwat. Biol. 18: 455-460.

Pettersson K., B. Boström & O.-S. Jacobsen, 1988. Phosphorus in sediments - speciation and analysis. Hydrobiologia 170: 91-101.

Philippart, J.-C. & J.-C. Ruwet, 1982. Ecology and distribution of tilapias. In R. S. V. Pullin & R. H. Lowe-McConnell (eds), The biology and culture of tilapias. ICLARM Conf. Proc. 7, International Centre for Living Aquatic Resources Management, Manila, Philippines: 15-59.

Pick, F. R., 1989. Species-specific phytoplankton response to nutrient enrichment in limnetic enclosures. Arch. Hydrobiol. Beih. Ergebn. Limnol. 32: 177-187.

Piedrahita, R. H., 1988. Introduction to computer modelling of aquaculture pond ecosystems. Aquac. Fish. Manage. (Aquacultural and Fisheries Management) 19:1-12.

Piedrahita, R. H., 1990. Calibration and validation of TAP, an aquaculture pond water quality model. Aquac. Engin. (Aquacultural Engineering) 9:75-96.

Piedrahita, R. H., D. E. Brune, G. Tchobanoglous & G. T. Orlob, 1984. A general model of the aquaculture pond ecosystem. J. World. Maricul. Soc. 15: 355-366.

Pomeroy, L. W., E. E. Smith & C. M. Grant, 1965. The exchange of phosphate between estuarine water and sediments. Limnol. Oceanogr. 10: 167-172.

Pomeroy, L. R. & W. J. Wiebe, 1988. Energetics of microbial food webs. Hydrobiologia 159: 7-18.

Porter, K. G., 1988. Phagotrophic phytoflagellates in microbial food webs. Hydrobiologia 159: 89-97.

Post, J. R. & D. J. McQueen, 1987. The impact of planktivorous fish on the structure of a plankton community. Freshwat. Biol. 17: 79-89.

Preisendorfer, R. W., 1986. Secchi disk science: visual optics of natural waters. Limnol. Oceanogr. 31: 909-926.

Prowse, G. A., 1964. Some limnological problems in tropical fish ponds. Verh. int. Ver. Limnol. 15: 480-484.

Psenner, R., R. Pucsko & M. Sager, 1985. Fraktionierung organischer und anorganischer Phosphorverbindungen von Sedimenten. Versuch einer Definition ökologisch wichtiger Fraktionen. Arch. Hydrobiol. Suppl. 70: 11-115.

Qin, J. & S. T. Threskeld, 1990. Experimental comparison of the effects of benthivorous fish and planktivorous fish on plankton community structure. Arch. Hydrobiol. 119: 121-141.

Rappaport, U. & S. Sarig, 1978. The results of manuring on intensive growth fish farming at the Ginosar Station ponds in 1977. Bamidgeh 30: 27-36.

Rai, H. & T. R. Jakobsen, 1990. Phosphate (^{32}P)-uptake capabilities of natural picoplankton and ultraplankton communities in lakes of differing degrees of eutrophication. Arch. Hydrobiol. 118: 421-435.

Rhee, G.-Y., 1978. Effects of N:P atomic ratios and nitrate limitation on algal growth, cell composition and nitrate uptake. Limnol. Oceanogr. 23: 10-25.

Richardson, W. B., S. A. Wickham & S. T. Threlkeld, 1990. Foodweb response to the experimental manipulation of a benthivore (*Cyprinus carpio*), zooplanktivore (*Menidia beryllina*) and benthic insects. Arch. Hydrobiol. 119: 143-165.

Riessen, H. P. & W. G. Sprules, 1990. Demographic costs of antipredator defenses in *Daphnia pulex*. Ecology 71: 1536-1546.

Rippey, B. & D. H. Jewson, 1982. The rates of sediment-water exchange of oxygen and sediment bioturbation in Lough Neagh, Northern Irland. Hydrobiologia 91: 377-382.

Romaire, R. P. & C. E. Boyd, 1979. Effects of solar

radiation on the dynamics of dissolved oxygen in channel catfish ponds. Trans. am. Fish. Soc. 108: 473-478.

Romaire, R. P., C. E. Boyd & W. J. Collins, 1978. Predicting nighttime dissolved oxygen decline in ponds used for *Tilapia* culture. Trans. am. Fish. Soc. 107: 804-808.

Rosseland, B. O. & O. K. Skogheim, 1984. Attempts to reduce effects of acidification on fishes in Norway by different mitigation techniques. Fisheries 9: 10-16.

Rothhaupt, K. O., 1990. Ressource competition of herbivorous zooplankton: a review of approaches and perspectives. Arch. Hydrobiol. 118: 1-29.

Sanchez, P. A., 1976. Properties and management of soils in the tropics. John Wiley & Sons, N.Y., 618 pp.

Sanders, J. G., S. J. Cibik, C. F. D'Elia & W. R. Boynton, 1987. Nutrient enrichment studies in a coastal plain estuary: changes in phytoplankton species composition. Can. J. Fish. aquat. Sci. 43: 83-90.

Sanders, R. W. & R. G. Porter, 1990. Bacterivorous flagellates as food resource for the freshwater crustacean zooplankter *Daphnia ambigua*. Limnol. Oceanogr. 35: 188-191.

Sanders, R. W., K. G. Porter, S. J. Bennett & A. E. DeBiase, 1989. Seasonal pattern of bacterivory by flagellates, ciliates, rotifers, and cladocerans in a freshwater planktonic community. Limnol. Oceanogr. 34: 673-687.

Sanni, S. & S. B. Waervagen, 1990. Oligotrophication as a result of planktivorous fish removal with rotenone in the small, eutrophic, Lake Mosvatn, Norway. Hydrobiologia 200/201: 263-274.

Sarig, S. & Y. Arieli, 1980. Growth capacity of tilapia in intensive culture. Bamidgeh 32: 57-65.

Saunders, G. W., 1972. Potential heterotrophy in a natural population of *Oscillatoria agardhii* var. *isothrix* Skuja. Limnol. Oceanogr. 17: 704-711.

Saunders, G. W. (Coord.), 1980. Organic matter and decomposers. In E. D. Le Cren & R. H. Lowe-McConnell (eds), The functioning of freshwater ecosystems. Cambridge University Press, Cambridge. International Biological Programme 22: 341-392.

Scheda, S. M. & B. C. Cowell, 1988. Rotifer grazers and phytoplankton: seasonal experiments on natural communities. Arch. Hydrobiol. 114: 31-44.

Scheffer, M., 1990. Multiplicity of stable states in freshwater systems. Hydrobiologia 200/201: 475-486.

Schindler, D. W., 1978. Factors regulating phytoplankton production and standing crop in the world's freshwaters. Limnol. Oceanogr. 23: 478-486.

Schindler, D. W., 1987. Detecting ecosystem responses to anthropogenic stress. Can. J. Fish. aquat. Sci. 44 (Suppl. 1): 6-25.

Schindler, D. W., R. H. Hesslein & M. A. Turner, 1987. Exchange of nutrients between sediments and water after 15 years of experimental eutrophisation. Can. J. Fish. aquat. Sci. 44: 26-33.

Schindler, D. W., K. H. Mills, D. F. Malley, D. L. Findlay, J. A. Shearer, I. J. Davies, M. A. Turner, G. A. Linsey & D. R. Cruikshank, 1985. Long-term ecosystem stress: the effects of years of experimental acidification on a small lake. Science 228: 1395-1401.

Schroeder, G. L., 1980. Fish farming in manure-loaded ponds. In R. S. V. Pullin & Z. H. Shehadeh (eds.), Integrated agriculture-aquaculture farming systems. I.C.L.A.R.M., Manila, Philippines. I.C.L.A.R.M. Conference Proceedings 4: 73-86.

Schroeder, G. L., 1987. Carbon and nitrogen budgets in manured fish ponds on Israel's coastal plain. Aquaculture 62: 259-279.

Schuurkes, J. A. A. R. & C. J. Kok, 1988. In vitro studies on sulphate reduction and acidification in sediments of shallow soft water lakes. Freshwat. Biol. 19: 417-426.

Seďa, J., J. Kubečka & Z. Brandl, 1989. Zooplankton structure and fish population development in the Římov Reservoir, Czechoslovakia. Arch. Hydrobiol. Beih. Ergebn. Limnol. 33: 605-609.

Ségalen, P., 1964. Le fer dans le sol. O.R.S.T.O.M., Paris. Initiations - Documn Techqs 4: 150 pp.

Ségalen, P., 1973. L'aluminium dans les sols. O.R.S.T.O.M., Paris. Initiations - Documn Techqs 22: 281 pp.

Segupta, S. & B. B. Jana, 1987. Effect of aeration on the primary productivity of the phytoplancton in experimental tanks. Aquaculture 62: 131-142.

Seitzinger, S. P., 1988. Denitrification in freshwater and coastal marine ecosystems: ecological and geochemical significance. Limnol. Oceanogr. 33: 702-724.

Senft, W. H., 1978. Dependance of light-saturated rates of algal photosynthesis on intracellular concentrations of phosphorus. Limnol. Oceanogr. 23: 709-718.

Seymour, E. A., 1980. The effects and control of algal blooms in fish ponds. Aquaculture 19: 55-74.

Shapiro, J., 1988. Introductory lecture at the international symposium "Phosphorus in Freshwater Ecosystems" Uppsala, Sweden in October 1985. Hydrobiologia 170: 9-17.

Shapiro, J., 1990. Biomanipulation: the next phase - making it stable. Hydrobiologia 200/201: 13-27.

Shapiro, J., V. Lamarra & M. Lynch, 1975. Biomanipulation: an ecosystem approach to lake restoration. In P. L. Brezonik & J. L. Fox (eds), Proc. Symp. on Water Quality Management through Biological Control. University of Florida, Gainesville, 85-96.

Shearer, J. A., E. J. Fee, E. R. DeBruyn & D. R. DeClercq, 1987. Phytoplancton productivity changes in a small, double-basin lake in response to termination of experimental fertilization. Can. J. Fish. aquat. Sci. 44 (Suppl. 1): 47-54.

Sheldon, R. W., P. Nival & F. Rassoulzadegan, 1986. An experimental investigation of a flagellate-ciliate-copepod food chain with some observations relevant to the linear biomass hypothesis. Limnol. Oceanogr. 31: 184-188.

Sherr, B. F., E. B. Sherr & C. S. Hopkinson, 1988. Trophic interactions within pelagic microbial communities: indications of feedback regulation of carbon flow. Hydrobiologia 159: 19-26.

Sherr, E. & B. Sherr, 1988. Role of microbes in pelagic food webs: a revised concept. Limnol. Oceanogr. 33: 1225-1227.

Shoesmith, E. A. & A. J. Brook, 1983. Monovalent-divalent cation ratios and the occurence of phytoplankton, with special reference to the desmids. Freshwat. Biol. 13: 151-155.

Sieburth, J. McN., V. Smetacek & J. Lenz, 1978. Pelagic ecosystem structure: heterotrophic compartments of the plankton and their relationship to plankton size fractions. Limnol. Oceanogr. 23: 1256-1263.

Sigler, J. W., T. C. Bjornn & F. H. Everest, 1984. Effects of chronic turbidity on density and growth of steelheads and coho salmon. Trans. am. Fish. Soc. 113: 142-150.

Sih, A., 1987. Predators and prey lifestyles: an evolutionary and ecological overview. In W. C. Kerfoot & A. Sih (eds), Predation: direct and indirect impacts on aquatic communities. The University Press of New England, Hanover (N.H.); Lond.: 203-224.

Simon, M., 1987. Biomass and production of small and large free-living and attached bacteria in Lake Constance. Limnol. Oceanogr. 32: 591-607.

Sin, A. W. & M. T. Chiu, 1983. The intensive monoculture of the tilapia hybrid *Sarotherodon nilotica* (male) x *S. mossambica* (female) in Hong-Kong. In L. Fishelson & Z. Yaron (eds), Tel Aviv University, Tel Aviv, Israel. Proc. Int. Symp. on Tilapia in Aquaculture: 506-516.

Sinke, A. J. C. & T. E. Cappenberg, 1988. Influence of bacterial processes on the phosphorus release from sediments in the eutrophic Loosdrecht Lakes, The Netherlands. Arch. Hydrobiol. Beih. Ergebn. Limnol. 30: 5-13.

Sinke, A. J., A. A. Cornelese, P. Keizer, O. F. R. Van Tongeren & T. E. Cappenberg, 1990. Mineralization, pore water chemistry and phosphorus release from peaty sediments in the eutrophic Loosdrecht lakes, The Netherlands. Freshwat. Biol. 23: 587-599.

Smayda, T. J., 1980. Phytoplankton species succession. In: I. Morris (ed.), The physiological ecology of phytoplankton. Blackwell Scientific Publication, Oxford. Studies in Ecology Volume 7: 493-570.

Smayda, T., 1990. The influence of lime and biological activity on sediment pH, redox and phosphorus dynamics. Hydrobiologia 192: 191-203.

Smith, D. F., 1986. Small-scale spatial heterogeneity in dissolved nutrient concentrations. Limnol. Oceanogr. 31: 167-171.

Smith, D. W., 1988. Phytoplankton and catfish culture: a review. Aquaculture 74: 167-189.

Smith, D. W. & R. H. Piedrahita, 1988. The relation between phytoplankton and dissolved oxygen in fish ponds. Aquaculture 68: 249-265.

Smith, R. E. & J. Kalff, 1985. Phosphorus competition among phytoplankton: a reply. Limnol. Oceanogr. 30: 440-444.

Smith, S. V., J. T. Hollibaugh, S. T. Dollar & S. Vink, 1989. Tomales Bay, California: a case for carbon-controlled nitrogen cycling. Limnol. Oceanogr. 34: 37-52.

Smith, S. V., S. Serruya, Y. Geifman & T. Berman, 1989. Internal sources and sinks of water, P, N, Ca, and Cl in Lake Kinneret, Israel. Limnol. Oceanogr. 34: 1202-1213.

Sommer, U., 1984. The paradox of the plankton: fluctuations of phosphorus availability maintain diversity of phytoplankton in flow-through culture. Limnol. Oceanogr. 29: 633-636.

Sommer, U., 1988. Phytoplankton succession in microcosm experiments under simultaneous grazing pressure and resource limitation. Limnol. Oceanogr. 33: 1037-1054.

Sommer, U., 1989. Nutrient status and nutrient competition of phytoplankton in a shallow, hypertrophic lake. Limnol. Oceanogr. 34: 1162-1173.

Sommer, U., Z. M. Gliwicz, W. Lampert & A. Duncan, 1986. The PEG-model of seasonal succession of planktonic events in fresh waters. Arch. Hydrobiol. 106: 433-471.

Søndergaard, M., 1989. Phosphorus release from a hypertrophic lake sediment: experiments with intact sediment cores in a continuous flow system. Arch. Hydrobiol. 116: 45-59.

Søndergaard, M., E. Jeppesen, P. Kristensen & O. Sortkjær, 1990. Interactions between sediment and water in shallow and hypertrophic lake: a study on phytoplankton collapses in Lake Søbygård, Denmark. Hydrobiologia 191: 129-138.

Spotte, S., 1979. Seawater aquarium : the captive environment. John Wiley & Sons, N.Y., 413 pp.

Stabel, H.-H., 1986. Calcite precipitation in Lake Constance: chemical equilibrium, sedimentation, and nucleation by algae. Limnol. Oceanogr. 31: 1081-1093.

Stack, V. T., Jr., 1973. Stabilization oxygen demand. Biological methods for the assessment of water quality. ASTM STP 528, American Society for Testing and Materials: 209-220.

Starling, F. L. R. M. & A. J. A. Rocha, 1990. Experimental study of the impacts of planktivorous fishes on plankton community and eutrophication of a tropical Brazilian reservoir. Hydrobiologia 200/201: 581-591.

Steemann Nielsen, E., 1952. The use of radioactive carbon (^{14}C) for measuring organic production in the sea. J. Cons. perm. Int. Expl. Mer 18: 117-140.

Steemann Nielsen, E. & V. K. Hansen, 1959. Light adaptation in marine phytoplankton populations and its interrelation with temperature. Physiol. Pl. 12: 353-370. Quoted by Jansson *et al.* (1988)

Stemberger, R. S. & J. J. Gilbert, 1987. Defenses of planktonic rotifers against predators. In W. C. Kerfoot & A. Sih (eds), Predation: direct and indirect impacts on aquatic communities. The University Press of New England, Hanover (N.H.); Lond.: 227-239.

Stenson, J. A. E., 1987. Variation in capsule size of *Holopedium gibberum* (Zaddach): a response to invertebrate predation. Ecology 68 (4): 928-934.

Sterner, R. W., 1989. Resource competition during seasonal succession toward dominance by cyanobacteria. Ecology 70: 229-245.

Sternik, K. H., 1983. Untersuchungen zur Phosphor-Abgabe und insbesondere zur Orthophosphat-Exkretion junger Karpfen (*Cyprinus carpio*). Ein Beitrag zur Phosphat-Remobilisierung in Gewässern. Arch. Hydrobiol./Suppl. 66: 1-82.

Stickney, R. R., J. H. Hesby, R. B. McGeachin & W. A. Isbell, 1979. Growth of *Tilapia nilotica* in ponds with different histories of organic fertilization. Aquaculture 17: 189-194.

Stockner, J. G., 1988. Phototrophic picoplankton: an overview from marine and freshwater ecosystems. Limnol. Oceanogr. 33: 765-775.

Stockner, J. G. & N. J. Antia, 1986. Algal picoplankton from marine and freshwater ecosystems : a multidisciplinary perspective. Can. J. Fish. aquat. Sci. 43: 2472-2503.

Stockner, J. G. & K. S. Shortreed, 1988. Responses of *Anabaena* and *Synechococcus* to manipulation of nitrogen: phosphorus ratios in a lake fertilization experiment.

Limnol. Oceanogr. 33: 1348-1361.

Storch, T. A. & G. W. Saunders, 1978. Phytoplankton extracellular release and its relation to the seasonal cycle of dissolved organic carbon in a eutrophic lake. Limnol. Oceanogr. 23: 112-119.

Stumm, W. & J. J. Morgan, 1981. Aquatic chemistry. An introduction emphasizing chemical equilibria in natural waters. 2nd Ed., John Wiley & Sons, N.Y., 780 pp.

Straškraba, M., 1980. The effects of physical variables on freshwater production : analyses based on models. In E. D. Le Cren & R. H. Lowe-McConnell (eds), The functioning of freshwater ecosystems. Cambridge University Press, Cambridge. International Biological Programme 22: 13-84.

Strickler, J. R., 1970. Über das Schwimmverhalten von Cyclopoiden bei Verminderungen der Bestrahlungsstärke. Schweiz. Z. Hydrol. 32: 150-180.

Strickler, J. R., 1975. Swimming of planktonic *Cyclops* species (Copepoda, Crustacea): pattern, movements and their controls. In T. Y.-T. Wu *et al.* (eds), Swimming and flying in nature, Vol. 2, Plenum: 599-613.

Sugita, H., K. Oshima, T. Fushino & Y. Deguchi, 1987. Substrate specificity of heterotrophic bacteria in the water and sediment of a carp culture pond. Aquaculture 64: 39-46.

Suttle, C. A., J. A. Fuhrman & D. G. Capone, 1990. Rapid ammonium cycling and concentration-dependent partitioning of ammonium and phosphate: implications for carbon transfer in planktonic communities. Limnol. Oceanogr. 35: 424-433.

Suttle, C. A. & P. J. Harrison, 1988. Ammonium and phosphate uptake rates, N:P supply ratios, and evidence for N and P limitation in some oligotrophic lakes. Limnol. Oceanogr. 33: 186-202.

Sweerts, J. P., J. W. M. Rudd & C. A. Kelly, 1986. Metabolic activities in flocculent surface sediments and underlying sandy littoral sediments. Limnol. Oceanogr. 31: 330-338.

Szumiec, M. A., 1989. The rôle of carp, temperature and morphometry in solar radiation transmission in ponds. Arch. Hydrobiol. Beih. Ergebn. Limnol. 31: 571-577.

Talling, J. F., 1969. The incidence of vertical mixing, and some biological and chemical consequences, in tropical African lakes. Verh. Internat. Verein. Limnol. 17: 998-1012.

Talling, J. F. & I. B. Talling, 1965. The chemical composition of African lake waters. Int. Revue ges. Hydrobiol. 50: 421-463.

Tapiador, D. D., H. F. Henderson, M. N. Delmendo & H. Tsutsui, 1978. Pêche et aquaculture continentales en Chine. FAO Doc. Tech. Pêches, 168: 86 pp.

Tarapchak, S. J. & L. R. Herche, 1986. Phosphate uptake by microorganisms in lake water : deviations from simple Michaelis-Menten kinetics. Can. J. Fish. aquat. Sci. 43: 319-328.

Tarapchak, S. J. & C. Nalewajko, 1986a. Introduction : Phosphorus - plankton dynamics symposium. Can. J. Fish. aquat. Sci. 43: 293-301.

Tarapchak, S. J. & C. Nalewajko, 1986b. Synopsis : Phosphorus - plankton symposium. Can. J. Fish. aquat. Sci. 43: 416-419.

Taylor, F. J. R., 1980. Basic biological features of phytoplankton cells. In In I. Morris (ed.), The physiological ecology of phytoplankton. Blackwell Scientific Publication, Oxford. Studies in Ecology, Volume 7: 3-55.

Theiss, J., K. Zielinski & H. Lang, 1990. Biomanipulation by introduction of herbivorous zooplankton. A helpful shock for eutrophic lakes? Hydrobiologia 200/201: 58-68.

Thompson, P. A., M. E. Levasseur & P. J. Harrison, 1989. Light-limited growth on ammonium vs. nitrate: what is the avantage for marine phytoplankton. Limnol. Oceanogr. 34: 1048-1061.

Threlkeld, S. T., 1988. Planktivory and planktivore biomass effects on zooplankton, phytoplankton and the trophic cascade. Limnol. Oceanogr. 33: 1362-1375.

Tilman, D., R. Kiesling, R. Sterner, S. S. Kilham & F. A. Johnson, 1986. Green, blue-green and diatom algae: taxonomic differences in competitive ability for phosphorus, silicon and nitrogen. Arch. Hydrobiol. 106: 473-485.

Tilzer, M. M., 1989. The productivity of phytoplankton and its control by resource availability. A review. In P. D. Kumar (ed.), Phycotalk, Varansi Hindi University, Varansi: 1-40.

Tilzer, M. M., 1990. Environmental and physiological control of phytoplankton productivity in large lakes. In M. M. Tilzer & C. Serruya (eds), Large lakes: ecological structure and function. Springer.

Tipping, E., C. Woof & M. Ohnstad, 1982. Forms of iron in the oxygenated waters of Eastwaite Water, U. K. Hydrobiologia 91: 383-393.

Tranvik, L. J. & J. McN. Sieburth, 1989. Effects of flocculated humic matter on free and attached pelagic microorganisms. Limnol. Oceanogr. 34: 688-699.

Trewawas, E., 1983. Tilappine fishes of the genera *Sarotherodon*, *Oreochromis*, and *Danakilia*. British Museum, London.

Trifonova, I. S., 1989. Changes in community structure and productivity of phytoplankton as indicator of lake reservoir eutrophication. Arch. Hydrobiol. Beih. Ergebn. Limnol. 33: 363-371.

Tucker, C. S., 1984. Potassium permanganate demand of pond waters. Prog. Fish- Cult. 46: 24-28.

Tucker, C. S., 1989. Method for estimating potassium permanganate disease treatment rate for channel catfish in ponds. Progr. Fish-Cult. 51: 24-26.

Tucker, C. S. & C. E. Boyd, 1977. Relationships between potassium permanganate treatment and water quality. Trans. am. Fish. Soc. 106: 481-488.

Tucker, C. S. & C. E. Boyd, 1978. Consequences of periodic applications of copper sulfate and simazine for phytoplankton control in catfish ponds. Trans. am. Fish. Soc. 107: 316-320.

Tucker, L., C. E. Boyd & E. W. McCoy, 1979. Effects on feeding rate on water quality, production of channel catfish, and economic returns. Trans. am. Fish. Soc. 108: 389-396.

Tudorancea, C., C. H. Fernando & J. C. Paggi, 1988. Food and feeding ecology of *Oreochromis niloticus* (Linneaus, 1758) juveniles in Lake Awassa (Ethiopia). Arch. Hydrobiol. (Monogr. Beitr.), Suppl. 79: 267-289.

Tudorancea, C. & A. D. Harrison, 1988. The benthic community of the saline lakes Abijata and Shala (Ethiopia).

Hydrobiologia 158: 117-123.
Tudorancea, C. & A. Zullini, 1989. Associations and distribution of benthic nematodes in the Ethiopian Rift Valley lakes. Hydrobiologia 179: 81-96.
Twinch, A. L. & C. M. Breen, 1982. Vertical stratification in sediments from a young oligotrophic South African impoundment: implication in phosphorus cycling. Hydrobiologia 91: 395-404.
Uehlinger, U. & J. Bloesch, 1987. The influence of crustacean zooplankton on the size structure of the algal biomass and suspended and settling seston (biomanipulation in limnocorral II). Int. Revue ges. Hydrobiol. 72: 473-486.
Uehlinger, U. & J. Bloesch, 1989. Primary production of different phytoplankton size classes in an oligo-mesotrophic Swiss lake. Arch. Hydrobiol. 116: 1-21.
Urban, N. R. & L. A. Baker, 1989. Alkalinity generation associated with the formation of ester sulfates. Limnol. Oceanogr. 34: 1144-1146.
Vadstein, O., B. O. Harkjerr, A. Jensen, Y. Olsen & H. Reinertsen, 1989. Cycling of organic carbon in the photic zone of a eutrophic lake with special reference to the heterotrophic bacteria. Limnol. Oceanogr. 34: 840-855.
Vadstein, O., A. Jensen, Y. Olsen & H. Reinertsen, 1988. Growth and phosphorus status of limnetic phytoplankton and bacteria. Limnol. Oceanogr. 33: 489-503.
van Eck, G. T. M., 1982. Forms of phosphorus in particulate matter from the Hollands Diep/Haringvliet, The Netherlands. Hydrobiologia 91: 665-681.
van Liere, L. & L. R. Mur, 1982. The influence of groundwater-movement on the phosphorus release from sediments, as measured in a continuous flow system. Hydrobiologia 91: 511-518.
Vanni, M. J., 1986. Competition in zooplankton communities: suppression of small species by *Daphnia pulex*. Limnol. Oceanogr. 31: 1039-1056.
Vanni, M. J., 1987a. Effects of nutrients and zooplankton size on the structure of a phytoplankton community. Ecology 68: 624-635.
Vanni, M. J., 1987b. Indirect effect of predators on age-structured prey populations: planktivorous fish and zooplankton. In W. C. Kerfoot & A. Sih (eds), Predation: direct and indirect impacts on aquatic communities. The University Press of New England, Hanover (N.H.); Lond.: 149-160.
Vanni, M. J. & D. L. Findlay, 1990. Trophic cascades and phytoplankton community structure. Ecology 71: 921-937.
Vanni, M. J. & J. Temte, 1990. Seasonal patterns of grazing and nutrient limitation of phytoplankton in a eutrophic lake. Limnol. Oceanogr. 35: 697-709.
Van Rijn, J., S. Diab & M. Shilo, 1987. Phytoplankton succession in relation to the nitrogen régime in shallow, brackish-water fishponds. Arch. Hydrobiol. 111: 183-196.
Van Rijn, J., S. R. Stutz, S. Diab & M. Shilo, 1986. Chemical, physical and biological parameters of superintensive concrete fish ponds. Bamidgeh 38: 35-43.
Verani, J. R., C. S. R. M. Pinto, P. Paiva & Y. A. Tabata, 1983. Experimental studies on intensive fish culture of the all male hybrid of *Sarotherodon niloticus* (female) x *S. hornorum* (male) stocked various levels. In L. Fishelson & Z. Yaron (eds), Tel Aviv University, Tel Aviv, Israel. Proc. Int. Symph. on Tilapia in Aquaculture: 517-523.
Verhulst, P. F., 1838. Notice sur la loi que la population suit dans son accroissement. Corresp. Math. Phys. 10: 113-121.
Verreth, J., 1990. The accuracy of population density estimates of a horizontally distributed zooplankton community in Dutch fish ponds. Hydrobiologia 203: 53-61.
Vieillefon, J., 1977. Les sols des mangroves et des tannes de Basse Casamance (Sénégal). Importance du comportement géochimique du soufre dans leur pédogenèse. O.R.S.T.O.M., Paris, Mémoires 83: 291 pp.
Vincent, W. F., 1989. Cyanobacterial growth and dominance in two eutrophic lakes: review and synthesis. Arch. Hydrobiol. Beih. Ergebn. Limnol. 32: 239-254.
Vincent, W. F. & S. J. Dryden, 1989. Phytoplankton succession and cyanobacterial dominance in a eutrophic lake of the mid-temperate zone (Lake Okaro, New Zealand). Arch. Hydrobiol. Beih. Ergebn. Limnol. 32: 137-163.
Viner, A. B., 1989. Fluorescence and carbon fixation by phosphorus deficient phytoplankton in a Western Australian reservoir. Arch. Hydrobiol. 115: 1-19.
Viner, A. B. & I. R. Smith, 1973. Geographical, historical and physical aspects of Lake George. Proc. R. Soc. Lond. B 184: 235-270.
Vincke, M. M. J., 1988. Developing productive systems under village conditions. In H. R. King & K. H. Ibrahim (eds), Village level aquaculture development in Africa. Proc. Comm. Cons. Workshop on Village Level Aquaculture Development in Africa. Freetown, Sierra Leone, 14-20 February 1985. Commonwealth Secretariat, London: 39-57.
Vostradovský, J., J. Křížek, O. Albertová, L. Růžička & M. Vostradovská, 1989. The changes in fish communities and biomanipulation in water supply reservoirs. Arch. Hydrobiol. Beih. Ergebn. Limnol. 33: 587-594.
Walsby, A. E., R. Kinsman, B. W. Ibelings & C. S. Reynolds, 1991. Highly buoyant colonies of the cyanobacterium *Anabaena lemmermannii* from persistent surface waterblooms. Arch. Hydrobiol. 121: 261-280.
Walsby, A. E. & C. S. Reynolds, 1980. Sinking and floating. In I. Morris (ed.), The physiological ecology of phytoplankton. Blackwell Scientific Publication, Oxford. Studies in Ecology Volume 7: 371-412.
Wei, S. L. & E. A. Laws, 1989. Spatial and temporal variation of water column measurements in aquaculture ponds. Aquaculture 78: 253-266.
Weilenmann, U., C. R. O'Melia & W. Stumm, 1989. Particle transport in lakes: models and measurements. Limnol. Oceanogr. 34: 1-18.
Weisse, T., 1990. Trophic interactions among heterotrophic microplankton, nanoplankton, and bacteria in Lake Constance. Hydrobiologia 191: 111-122.
Welcomme, R. L., 1972. The inland waters of Africa. Les eaux intérieures d'Afrique. FAO CIFA Techn. Pap. 1, 117 pp.
Welcomme, R. L., 1983. River basins. FAO Fish. Techn. Pap. 202, 60 pp.
Weninger, G., 1985. Principal freshwater-types and comparative hydrochemistry of tropical running water systems. Rev. Hydrobiol. trop. 18: 79-110.
Westlake, D. F. (Coordinator), 1980. Primary production. In

E. D. Le Cren & R. H. Lowe-McConnell (eds), The functioning of freshwater ecosystems. Cambridge University Press, Cambridge. International Biological Programme 22: 141-246.

Wetzel, R. G., 1983. Limnology. 2nd Ed., Saunders, Philadelphia.

White, E., M. T. Downes, G. W. Payne, S. Pickmere, P. Woods & J. Rueter, 1989. Community responses of the late summer phytoplankton of Lake Okaro (New Zealand) to experimental changes in nutrient abundance: evidence of nitrogen limitation in the presence of phosphorus deficiency. Arch. Hydrobiol. Beih. Ergebn. Limnol. 32: 165-175.

Williamson, C. E., 1987. Predator-prey interactions between omnivorous diaptomid copepods and rotifers: the rôle of prey morphology and behavior. Limnol. Oceanogr. 33: 167-177.

Williamson, C. E. & N. M. Butler, 1986. Predation on rotifers by the suspension-feeding calanoid copepod *Diaptomus pallidus*. Limnol. Oceanogr. 31: 393-402.

Wurtsbaugh, W. A., W. F. Vincent, R. Alfaro Tapia, C. L. Vincent & P. J. Richerson, 1985. Nutrient limitation of algal growth and nitrogen fixation in a tropical alpine lake, Lake Titicaca. Freshwat. Biol. 15: 185-195.

Wynne, D. & T. Berman, 1990. The influence of environmental factors on nitrate reductase activity in freshwater phytoplankton. Hydrobiologia 194: 235-245.

Yan, N. D., 1979. Phytoplankton community of an acidified, heavy metal-contaminated lake near Sudbury, Ontario: 1973-1977. Water Air Soil Pollut. 11: 43-55. Quoted by Mulholland *et al.*, 1986.

Yentsch, C. E., 1980. Light attenuation and phytoplankton photosynthesis. In I. Morris (ed.), The physiological ecology of phytoplankton. Blackwell Scientific Publication, Oxford. Studies in Ecology Volume 7: 95-127.

Young, T. C. & J. V. DePinto, 1982. Algal-availability of particulate phosphorus from diffuse and point sources in the lower Great Lakes basin. Hydrobiologia 91: 111-119.

Young, T. C. & D. L. King, 1980. Interacting limits to algal growth: light, phosphorus, and carbon dioxide availability. Water Res. 14: 409-412. Quoted by Jansson *et al.* (1988)

Yusoff, F. M. & C. D. McNabb, 1987. Effects of nutrient availability on primary productivity and fish production in fertilized tropical ponds. Aquaculture 78: 303-319.

Zaret, T. M., 1980. Predation and freshwater communities. Yale Univ. Press, New Haven, 187 pp.

Zhang, F. L., Y. Zhu & X. Y. Zhou, 1987. Studies on the ecological effects of varying the size of fish ponds loaded with manures and feeds. Aquaculture 60: 107-116.

Zlotnik, I. & Z. Dubinsky, 1989. The effect of light and temperature on DOC excretion by phytoplankton. Limnol. Oceanogr. 34: 831-839.

Zohar, G., U. Rappaport, Y. Avnimelech & S. Sarig, 1984. Results of the experiments carried out in the Genosar experimental station in 1983. Cultivation of tilapia in high densities and with periodic flushing of the pond water. Bamidgeh 36: 63-69.

Zurbuch, P. E., 1984. Neutralization of acidified streams in West Virginia. Fisheries 9: 42-47.

Zweig, R. D., 1989. Evolving water quality in a common carp and blue tilapia high production pond. Hydrobiologia 171: 11-21.

Index

GPSR Compliance

The European Union's (EU) General Product Safety Regulation (GPSR) is a set of rules that requires consumer products to be safe and our obligations to ensure this.

If you have any concerns about our products, you can contact us on ProductSafety@springernature.com

In case Publisher is established outside the EU, the EU authorized representative is:

Springer Nature Customer Service Center GmbH
Europaplatz 3
69115 Heidelberg, Germany

Batch number: 10762209

Printed by Printforce, the Netherlands